18 Unconventional Essays on the Nature of Mathematics

18 Unconventional Essays on the Nature of Mathematics

Reuben Hersh
Editor

Reuben Hersh
Department of Mathematics and Statistics
University of New Mexico
Albuquerque, NM
USA

Cover illustration: Photographs of three contributing authors: Alfréd Rényi, Gian-Carlo Rota, and Leslie A. White.

Mathematics Subject Classification MSC 2000: 01A05 51–03

Library of Congress Control Number: 2005925514

ISBN-10: 0-387-25717-9
ISBN-13: 978-0387-25717-4

Printed on acid-free paper.

Printed in the United States of America. (SPI/EB)

9 8 7 6 5 4 3 2 1

springeronline.com

Contents

Introduction

This book comes from the Internet. Browsing the Web, I stumbled on philosophers, cognitive scientists, sociologists, computer scientists, even mathematicians!—saying original, provocative things about mathematics. And many of these people had probably never heard of each other! So I have collected them here. This way, they can read each other's work. I also bring back a few provocative oldies that deserve publicity.

The authors are philosophers, mathematicians, a cognitive scientist, an anthropologist, a computer scientist, and a couple of sociologists. (Among the mathematicians are two Fields Prize winners and two Steele Prize winners.) None are historians, I regret to say, but there are two historically oriented articles. These essays don't share any common program or ideology. The standard for admission was: Nothing boring! Nothing trite, nothing trivial! Every essay is challenging, thought-provoking, and original.

Back in the 1970s when I started writing *about* mathematics (instead of just doing mathematics), I had to complain about the literature. Philosophy of science was already well into its modern revival (largely stimulated by the book of Thomas Kuhn). But philosophy of mathematics still seemed to be mostly foundationist ping-pong, in the ancient style of Rudolf Carnap or Willard Van Ormond Quine. The great exception was *Proofs and Refutations* by Imre Lakatos. But that exciting book was still virtually unknown and unread, by either mathematicians or philosophers. (I wrote an article entitled "Introducing Imre Lakatos" in the *Mathematical Intelligencer* in 1978.)

Since then, what a change! In the last few years newcomers—linguists, neuroscientists, cognitive scientists, computer scientists, sociologists–are bringing new ideas, studying mathematics with new tools. (George Lakoff-Rafael Núñez, Stanislas Dehaene, Brian Butterworth, Keith Devlin).

In previous centuries, old questions–"What Is Man?" "What Is Mind?" "What Is Language?"–were transformed from philosophical questions, free for speculation, into scientific problems. The subjects of linguistics, psychology and anthropology detached from philosophy to become autonomous disciplines. Maybe the question, "What is mathematics?" is coming into recognition as a scientific problem.

In 1981, in *The Mathematical Experience,* speaking about the prevailing alternative views of the nature of mathematics, Phil Davis and I asked, "Do we really have to choose between a formalism that is falsified by our everyday experience, and a Platonism that postulates a mythical fairyland where the uncountable and the inaccessible lie waiting to be observed by the mathematician whom God blessed with a good enough intuition? It is reasonable to propose a different task for mathematical philosophy, not to seek indubitable truth, but to give an account of mathematical knowledge as it really is–fallible, corrigible, tentative, and evolving, as is every other kind of human knowledge. Instead of continuing to look in vain for foundations, or feeling disoriented and illegitimate for lack of foundations, we have tried to look at what mathematics really is, and account for it as a part of human knowledge in general. We have tried to reflect honestly on what we do when we use, teach, invent, or discover mathematics." (p. 406)

Before long, the historian Michael Crowe said these words were "a programme that I find extremely attractive." In 1986-1987 Crowe visited Donald Gillies at King's College in London. Gillies had been a student of Imre Lakatos. In 1992 Gillies published an anthology, *Revolutions in Mathematics,* where historians of mathematics like Crowe collaborated with philosophers of mathematics like Gillies.

Such collaboration developed further in Emily Grosholz and Herbert Breger's anthology, *The Growth of Mathematical Knowledge* (Kluwer, 2000). Emily Grosholz wrote, "during the last decade, a growing number of younger philosophers of mathematics have turned their attention to the history of mathematics and tried to make use of it in their investigations. The most exciting of these concern how mathematical discovery takes place, how new discoveries are structured and integrated into existing knowledge, and what light these processes shed on the existence and applicability of mathematical objects." She mentions books edited by Philip Kitcher and William Aspray, by Krueger, and by Javier Echeverria. She finds the Gillies volume "perhaps the most satisfactory synthesis."

History of mathematics is today a lively and thriving enterprise. It is tempting to start listing my favorite historians, but I will limit myself to singling out the monumental work by Sanford L. Segal, *Mathematicians Under the Nazis* (Princeton, 2003).

Already in my 1979 article "Some Proposals for Reviving the Philosophy of Mathematics," (reprinted in Thomas Tymoczko's anthology, *New Directions in the Philosophy of Mathematics,* Birkhauser, 1986),and at greater length in my two subsequent books, I explained that, contrary to fictionalism, mathematical objects do exist—*really!* But, contrary to Platonism, their existence is not transcendental, or independent of humanity. It is created by human activity, and is part of human culture. I cited the 1947 essay by the famous anthropologist Leslie White, which is reprinted here. And at last, in 2003 and 2004, a few philosophers are also recognizing that mathematical objects are real and are our creations. (Jessica Carter, "Ontology and Mathematical Practice," *Philosophia Mathematica* 12 (3), October 2004; M. Panza, "Mathematical Proofs," *Synthese,* 134. 2003; M. Muntersbjorn, "Representational innovation and mathematical ontology," *Synthese,* 134, 2003). I know of recent conferences in Mexico, Belgium, Denmark, Italy, Spain, Switzerland, and Hungary on philosophical issues of mathematical practice.

While others are starting to pay more attention to our ways, we mathematicians ourselves are having to look more deeply at what we are doing. A Special Interest Group on philosophy is now active in the Mathematical Association of America. A famous proposal by Arthur Jaffe and Frank Quinn in 1993 in the *Bulletin of the American Mathematical Society,* to accept not-so-rigorous mathematics by labeling it as such, provoked a flood of controversy. (The essay by William Thurston in this volume was his contribution to that controversy.)

After the rest of this book had gone to the editor at Springer, I found an article on the Web by Jonathan M. Borwein, the leader of the Centre for Experimental and Constructive Mathematics at Simon Fraser University in Vancouver. He quoted approvingly this five-point manifesto of mine:

"*1. Mathematics is human.* It is part of and fits into human culture. It does not match Frege's concept of an abstract, timeless, tenseless, objective reality.

2. *Mathematical knowledge is fallible.* As in science, mathematics can advance by making mistakes and then correcting or even re-correcting them. The "fallibilism" of mathematics is brilliantly argued in Lakatos' *Proofs and Refutations.*
3. *There are different versions of proof or rigor.* Standards of rigor can vary depending on time, place, and other things. The use of computers in formal proofs, exemplified by the computer-assisted proof of the four color theorem in 1977, is just one example of an emerging nontraditional standard of rigor.
4. *Empirical evidence, numerical experimentation and probabilistic proof all can help us decide what to believe in mathematics.* Aristotelian logic isn't necessarily always the best way of deciding.
5. *Mathematical objects are a special variety of a social-cultural–historical object.* Contrary to the assertions of certain post-modern detractors, mathematics cannot be dismissed as merely a new form of literature or religion. Nevertheless, many mathematical objects can be seen as shared ideas, like Moby Dick in literature, or the Immaculate Conception in religion."

(R. Hersh, "Fresh Breezes in the Philosophy of Mathematics," *American Mathematical Monthly,* August-September 1995, 589-594; quoted by Jonathan Borwein in "The Experimental Mathematician: The Pleasure of Discovery and the Role of Proof," prepared for the *International Journal of Computers for Mathematical Learning*, July 2004.)

As more and more important proofs approach and go beyond the limits of conventional verification, mathematicians are having to face honestly the embarrassing ambiguity and temporal dependence of our central sacred icon—rigorous proof.

In his 1986 anthology Thomas Tymoczko called attention to the troublesome philosophical issues raised by the recent proof of the famous four-color theorem. This was the first time the solution of a major mathematical problem had relied essentially on machine computation.

Today, the status of several other famous problems raises even more prominent and severe difficulties. The story of Thomas Hales' "99% accepted" proof of the Kepler conjecture makes clear that something new and strange is happening in the very center of the mathematical enterprise (see George Szpiro, *Kepler's Conjecture,* Wiley, 2003). (Then there is also the on-going decades-long ups-and-downs, the many thousands of pages proof, of the classification of simple finite groups. See Ron Solomon, "On Finite Simple Groups and their Classification," *Notices of the AMS* 42 (2), February 1995, 231-239.)

Johannes Kepler in 1611 considered how spherical balls can be packed to fill space as densely as possible. There are three natural ways to pack spheres, and it's clear which of the three is best. Kepler guessed that this way is in fact the best possible. It turns out that this is fiendishly hard to prove. Wu-Yi Hsiang of the University of California, Berkeley, claimed to have a proof in 1993, but he failed to convince his colleagues and competitors. He has not relinquished his claim and continues to hold to it. Thomas Hales of the University of Michigan announced a proof by a different method in 1997. His proof follows suggestions made earlier by Laszlo Fejes-Toth, and it involves, like the famous computer proof of the four-color theorem, computer checking of thousands of separate cases, many of them individually very laborious. The *Annals of Mathematics* invited Hales to submit his manuscript. It is 250 pages long. A committee of 12 experts was appointed to referee the paper, coordinated by Gabor Fejes-Toth, Laszlo's son. After four years, the committee announced that they had found no errors, but still could not certify the correctness. They simply ran out of energy and gave up. Robert Macpherson, the editor of the *Annals,* wrote, "The news from the referees is bad, from my perspective. They have not been able to certify the correctness of the proof, and will not be able to certify it in the future, because they have run out of energy to devote to the problem. This is not what I had hoped for." He reluctantly acceded to their decision, and accepted the theoretical part of Hales' paper, leaving the computer part for publication elsewhere. Hales then announced that he was affiliating with a group of computer scientists known as the QED Project. This dormant project had as its original stated goal: to computerize all of mathematics! Hales' new project, the Flyspeck Project, proposes to do a computer coding and verification of his proof of the Kepler conjecture—a proof which, in ordinary mathematical form, was already too long and complicated to be completely checked, in four years, by a committee of 12 leading human experts. Project Flyspeck is expected to involve the work of hundreds of people and take 20 years. I do not know anyone who thinks either that this project can be completed, or that even if claimed to be complete it would be universally accepted as a convincing proof of Kepler's conjecture. Donald Mackenzie's article in this volume sheds some light on these issues.

Such a story undermines our faith that mathematical proof will remain as we have always thought of it–that after reasonable time and effort, its correctness must be definitely decidable by unanimous consensus of competent specialists.

In fact, even without regard to Hales' theorem, it is easy to see that in principle there must be an upper bound on the length and complexity of the longest proof that at any time can be completely checked and verified by the mathematical community. In principle it is possible for a recognized, established mathematician to submit a proof longer than this upper bound. What should be the status of such a proof? Should it be accepted for publication? What degree of conviction or credibility should we attribute to its conclusion? Should it depend on our estimation of the reliability of its creator? May we use it as a building block in our own research? What if it has "applications" in physics? Such judgments are made every day in the "real world" of ordinary life. But in mathematics???!!!

As I explained in my book, *What Is Mathematics, Really?* (1997, Oxford) the words "mathematical proof" have two different meanings—and the difference is not usually acknowledged. One meaning, found in logic texts and philosophy journals, is "a sequence of formalized statements, starting with unproved statements about undefined terms, and proceeding by steps permitted in first-order predicate calculus." The other meaning, not found in a precise or formal statement anywhere, is "an argument accepted as conclusive by the present-day mathematical community." The problem is to clarify and understand—not justify!!—the second meaning. A first stab at clarification might be, "an argument accepted as conclusive by the highest levels of authority in the present-day mathematical community." Such a clarification rests on several implicit hypotheses:

(1) that there *is* a "mathematical community."
(2) that this "community" has accepted "high levels of authority."
(3) that these "high authorities" have a legitimacy based on some generally accepted rationale.
(4) that the highest level of authority can agree on what to accept.
(5) that arguments accepted as proofs by the recognized highest levels of authority in the mathematical community will remain accepted, at least for a very long time, at least with very high probability.

What seems to be threatening is that increasing length and complexity of proposed proofs, whether involving heavy use of computers or not, may go beyond the capacity of recognized authorities to reach a convincingly informed consensus.

The first of our eighteen articles is Alfréd Rényi's "A Socratic Dialogue on Mathematics" (*Dialogues on Mathematics,* Holden-Day, 1967.) Rényi was a famous probabilist and number theorist, co-creator with Pal Erdos of the subject random graphs, and for many years director of the Institute of Mathematics in Budapest. This is a most inviting, charming and thought-provoking tour de force and jeu d'esprit. It poses the basic problem, and answers it in a way that invites further questioning and deeper development. I have heard that Prof. Rényi used to give live performances of this work, assisted by his daughter Zsuzsanna, to whom he dedicated the book from which this excerpt is taken.

The next article, by the logician-philosopher Carlo Cellucci of Rome, is the introductory chapter to his book, *Filosofia e matematica.* He simply lists 13 standard assumptions about mathematics (what he calls "the dominant view") and demolishes all of them. A most impressive and stimulating performance. I eagerly await the translation of his whole book into English.

William Thurston's friendly, down-to-earth article, "On Proof and Progress in Mathematics," provides a rare, invaluable glimpse for outsiders at some aspects of mathematical creation at the highest level. Its frank, unpretentious look at what really is done, what really happens at that level is told in a style and language accessible to anyone. It was published in the *Bulletin of the AMS*, one of the responses to the Jaffe-Quinn proposal mentioned above.

The U.S.-based English philosopher Andrew Aberdein's article, "The Informal Logic of Mathematical Proof," draws on "informal logic," a subject that was revived by Stephen Toulmin. "Informal logic" is closely allied to "rhetoric". An old article by Phil Davis and myself called "Rhetoric and Mathematics" may be relevant to Aberdein's article. (It appeared in *The Rhetoric of the Human Sciences,* edited by John S. Nelson, Allan Megill and Donald N. McCloskey, University of Wisconsin Press, 1987 and also as a chapter in our book, *Descartes' Dream*).

The article by the Israeli-French mathematician Yehuda Rav is "Philosophical Problems of Mathematics in the Light of Evolutionary Epistemology." He shows that the human ability and inclination to mathematize can be understood as the result of natural selection. It is necessary and advantageous for our survival as a species. It was first published in the journal *Philosophica,* and reprinted in the anthology *Math Worlds: Philosophical and Social Studies of Mathematics and Mathematics Education,* edited by Sal Restivo, Jean Paul van Bendegem, and Roland Fischer (Albany; State University of New York Press, 1993.)

The English-American mathematician-turned cognitive scientist, Brian Rotman, provides a surprising insight into mathematics in the language of semiotics. His article, "Toward a Semiotics of Mathematics," clarifies what you do when you write mathematics. Three different personae participate: first of all, there is the disembodied pure thinker, the impersonal voice who calls himself "we." Secondly, there is also an imaginary automaton, who in imagination ("in principle") carries out any calculations or algorithms that "we" mention. And yes, there is also an actual live flesh-and-blood human being, who is sitting in your chair. This article first appeared in *Semiotica* 72-1/2 (1988).

Donald Mackenzie's article, "Computers and the Sociology of Mathematical Proof," gives a detailed history of the computer scientist's search for program correctness, and thereby shines a searchlight on the notion of mathematical certainty. It was presented at a conference at the University of Roskilde, Denmark, in 1998, whose proceedings were published as *New Trends in the History and Philosophy of Mathematics,* University Press of Southern Denmark, 2004.

"From G.H.H. and Littlewood to XML and Maple: Changing Needs and Expectations in Mathematical Knowledge Management," by computer scientist Terry Stanway of Vancouver, BC, looks at mathematics from a new vantage point, as a problem of information storage and retrieval. Stanway is connected to the Centre for Experimental and Constructive Mathematics at Simon Fraser University, led by Jonathan and Peter Borwein. Their remarkable work integrates theory and computation in surprising and fruitful ways.

"Do Real Numbers Really Move? Language, Thought, and Gesture: The Embodied Cognitive Foundations of Mathematics" is by the Chilean-Swiss cognitive scientist Rafael Núñez, now at the University of California, San Diego. Building on his previous work with Berkeley linguist George Lakoff (*Where Mathematics Comes from,* Basic Books, 2000), he uses rigorous study of how we unconsciously produce millisecond-precise hand gestures as we talk mathematics—*literal* "hand-waving"!—to prove that mathematics is indeed built from embodied metaphor. It appears in a collection, *Embodied Artificial Intelligence,* edited by F. Iida et al., Springer, 2004.

Timothy Gowers of Cambridge university contributes "Does Mathematics Need a Philosophy?" He writes as a dedicated teacher who is an outstanding creator of mathematics. Does a mathematician and a teacher of mathematics need a philosophy of mathematics? The answer, of course, turns out to be: "Yes and No."

The philosopher Jody Azzouni of Boston and Brooklyn analyzes "How and Why Mathematics Is Unique as a Social Practice." Addressing "mavericks" such as myself, he finds some agreement, and some unanswered questions.

Gian-Carlo Rota held the unheard of title, "Professor of Applied Mathematics and Philosophy" at M.I.T. Gianco was at once a leading combinatorialist and a deep phenomenologist. "The Pernicious Influence of Mathematics upon Philosophy" sticks some pins into academic analytic philosophy. This essay was originally a talk at a session on philosophy of mathematics, organized by me in 1990 at the national meeting of the American Association for Advancement of Science in New Orleans. It was previously published in *Synthese*, 88 (2), August 1991; and in Rota's book, *Indiscrete Thoughts,* Birkhauser, 1997.

Jack Schwartz was Rota's mentor at Yale, as recounted in Gian-Carlo's book. Jack's mischievous shocker, "The Pernicious Influence of Mathematics on Science" must have stimulated Gian-Carlo's parallel demolition piece. (It appeared in *Discrete Thoughts,* Birkhauser, 1986, which was coedited by

Rota, Schwartz, and Mark Kac.) Jack is co-author, with Nelson Dunford, of the prize-winning three-volume "bible" of functional analysis, *Linear Operators*. Having completed that gigantic task, he later transformed himself into a leading authority in computer science.

Alfonso C. Avila del Palacio, of the University of Durango, Mexico, asks, "What Is Philosophy of Mathematics Looking for?" He finds that mathematicians, philosophers and historians can't agree because they are asking different questions. His article is updated and translated from one that he published in *Syntesis* No. 3 (1997), *Revista de la Universidad Autonoma de Chihuahua,* Mexico.

Andrew Pickering of Urbana, a physicist who went to Edinburgh to become a sociologist, tells the story of Hamilton's invention of the quaternion as an exemplary tale. His article, "Concepts and the Mangle of Practice Constructing Quaternions," uses this historical material to bring mathematical practice in line with scientific practice, making the almost unprecedented move of aligning philosophy of mathematics with philosophy of science! It appeared in the collection, *Mathematics, Science, and Postclassical Theory,* edited by Barbara Herrnstein Smith and Arkady Plotnitsky, Duke University Press, 1997.

The philosopher Eduard Glas of Delft, in the Netherlands, in his article "Mathematics as Objective Knowledge and as Human Practice," shows that the famous philosopher of science Karl Popper should also be considered a philosopher of mathematics. Indeed, Popper's concept of science and mathematics as problem-solving avoids the traps of Platonism and formalism. Glas backs up his case with the illuminating history of descriptive and projective geometry in revolutionary and post-revolutionary France.

"The Locus of Mathematical Reality: An Anthropological Footnote" is by the famous anthropologist Leslie White, and is the oldest piece in our collection. White gave a clear, simple answer to the basic question about the nature of mathematical existence. He was a close friend to the leading topologist Raymond Wilder, and inspired Wilder's writings on a cultural approach to the nature of mathematics. His article first appeared in the journal, *Philosophy of Science,* in October, 1947, and was reprinted in James R. Newman's huge four-volume anthology, *The World of Mathematics* (Simon and Schuster, 1956).

Perhaps the most recalcitrant issue in philosophy of mathematics is what Eugene Wigner famously called the "unreasonable effectiveness" of mathematics. Mark Steiner has written a challenging book, *The Applicability of Mathematics as a Philosophical Problem,* that claims to refute naturalism on the basis of that effectiveness. My attempt to study this question, "Inner Vision, Outer Truth," is my contribution to this collection. It is reprinted from *Mathematics and Science,* edited by Ronald Mickens, World Scientific, 1990.

The three articles by Aberdein, Azzouni and Glas all come from a conference I attended in Brussels in 2002, organized by Jean Paul von Bendegem

(*Theories of Mind, Social Science and Mathematical Practice,* Kluwer, 2005). Mackenzie's is from a conference in Denmark. The journal *Philosophia Mathematica* publishes a wide variety of perspectives. Recent books, from several different directions, have explored the nature of live mathematics. In English, I know of: *Where Mathematics Comes from* by George Lakoff and Rafael E. Núñez, *The Number Sense* by Stanislas Dehaene, *The Math Gene* by Keith Devlin, *What Counts* by Brian Butterworth, *Towards a Philosophy of Real Mathematics* by David Corfield, *Doing Mathematics* by Martin H. Krieger, *Social Constructivism as a Philosophy of Mathematics* by Paul Ernest, *Indiscrete Thoughts* by Gian-Carlo Rota, *Converging Realities* by Roland Omnès, and a forthcoming book by Alexandre Borovik. There is also my own *What Is Mathematics, Really?* (and, with Phil Davis, *Descartes' Dream* and *The Mathematical Experience.*) All in different ways go beyond traditional philosophizing, try to grapple with mathematical knowledge and activity as actual phenomena, as part of the real world. (We must mention forerunners–George Polya, Imre Lakatos, Karl Popper, Ludwig Wittgenstein, Raymond Wilder, Hans Freudenthal, Hao Wang, Philip Kitcher.) I know that there are also recent relevant books in several other languages. But I must leave it to others to compile those lists.

The articles here are largely limited to the cognitive aspect of mathematics. Of course the emotional, social and political aspects are also vitally important. The forthcoming book, *Loving and Hating Mathematics,* coauthored with Vera John-Steiner, turns to those aspects of mathematical life.

Although these articles were written independently, without contact between the authors, it's not surprising that there are several mutual references. Rav quotes Rényi and White; Aberdein quotes Thurston, Rav, and Rota; Azzouni, Núñez, Avila and Cellucci quote Hersh. But I have made no effort to weave these articles together, or to summarize what they add up to, or to announce what needs to be done next. The goal is to show the possibilities of thinking fresh, sticking close to actual practice, fearlessly letting go of standard shibboleths.

Reuben Hersh
Santa Fe, NM

About the Authors

Andrew Aberdein is Assistant Professor of Logic and Humanities at Florida Institute of Technology. Before coming to Florida he studied at the University of St. Andrews, from where he graduated with a Ph.D. in Logic & Metaphysics in 2001. He then taught at the Universities of Edinburgh and Dundee. In spite of these persistent Scottish connections, and his own name, he is actually English. His research is mostly concerned with the interplay of logic and the philosophy of science. He is particularly interested in the foundations of formal reasoning and the applications of informal reasoning.

Alfonso C. Avila del Palacio was a Professor at the National University of Mexico for 15 years and is presently a Researcher in the Institute of Social Sciences, University Juarez of the State of Durango, Mexico. He is the author of the book, *Estructura Matematica de la Teoria Kenesiana,* [2000], Mexico: Fondo de Cultura Economica.

Carlo Cellucci is Professor of Logic at the University of Rome "La Sapienza." He is author of *Teoria della dimostrazione* (1978), *Le ragioni della logica* (1998), and *Filosofia e matematica* (2002) and editor of *La filosofia della matematica* (1967) and *Il paradiso di Cantor* (1978).

Jody Azzouni writes philosophy, among other things. He has published three books, *Metaphysical Myths, Mathematical Practice: The Ontology and Epistemology of the Exact Sciences* (Cambridge, 1994), *Knowledge and Reference in Empirical Science* (Routledge, 2000), and *Deflating Existential Consequence: A Case for Nominalism* (Oxford, 2004).

Eduard Glas is Associate Professor at the Department of Applied Mathematics of Delft University of Technology in charge of teaching mathematics and a program on "Mathematics and Society." He published books and articles on the history and philosophy of mathematics and of science.

William Timothy Gowers is Rouse Ball Professor of Mathematics at Cambridge University, and Fellow of Trinity College. In 1996 he received the Prize of the European Mathematical Union, and in 1998 he won the Fields Prize of the International Congress of Mathematicians. He makes extensive use of methods from combinatorial theory to solve many long-standing open problems in functional analysis.

Reuben Hersh is a student of Peter Lax, a protege of Hao Wang and Gian-Carlo Rota, and a coauthor with Larry Bobisud, Y.W.Chen, Robert Cogburn, Paul Cohen, Martin Davis, Philip J. Davis, James A. Donaldson, Archie Gibson, Priscilla Greenwood, Richard J. Griego, Vera John-Steiner, Tosio Kato, Elena Anne Marchisotto, George Papanicolaou, and Mark Pinsky. *The Mathematical Experience,* coauthored with Phil Davis, won a National Book Award.

Donald MacKenzie studied applied mathematics and social studies of science at the University of Edinburgh and in 1975 was appointed to a lectureship in sociology there; he now holds a personal chair. His first book was *Statistics in Britain, 1865-1930: The Social Construction of Scientific Knowledge* (Edinburgh: Edinburgh University Press, 1981); his most recent is *Mechanizing Proof: Computing, Risk, and Trust* (Cambridge, Mass.: MIT Press, 2001). His current research is on the development of the modern theory of finance and its interactions with the financial markets.

"Do Real Numbers really move? Language, Thought and Gesture: The Embodied Cognitive Foundations of Mathematics" is by the Chilean-Swiss cognitive scientist *Rafael Núñez,* now at the University of California, San Diego. Building on his previous work with Berkeley linguist George Lakoff (*Where Mathematics Comes from*. Basic Books, 2000), he uses rigorous study of how we unconsciously produce millisecond-precise hand gestures as we talk mathematics (literal "hand-waving"!) to prove that mathematics is indeed built from embodied metaphor.

Andrew Pickering is Professor of Sociology at the University of Illinois at Urbana-Champaign. He is the author of *Constructing Quarks: A Sociological History of Particle Physics* and *The Mangle of Practice: Time, Agency, and Science,* and the editor of *Science as Practice and Culture*. He is currently writing a book on the history of cybernetics in Britain.

Yehuda Rav was born in 1930, in Vienna, Austria and grew up in Israel. He came in 1951 to the US for his undergraduate and graduate studies at Columbia University, majoring in mathematics and wrote a thesis in biological automata theory. He taught at Columbia University and Hofstra University before emigrating to France in 1967. From that date until his retirement in 1995, he was on the faculty of the University of Paris, Orsay Center. He has published numerous articles in logic, set theory, philosophy of mathematics, and history of cybernetics, as well as more than 200 reviews in *Mathematical Reviews*. He is still active in scholarly work, with a special interest in the neurocognitive sciences and evolutionary epistemology.

Alfréd Rényi earned his Ph.D. at Szeged under Frigyes Riesz. He learned from Lipot Fejer in Budapest, and worked on number theory with Linnik in Russia. With Pal Erdos he created the theory of random graphs. He was a founder and for 20 years the director of the Hungarian Academy of Sciences. In his youth, during the Nazi occupation of Hungary, he escaped from a Labor Camp, hid for six months to avoid capture, and rescued his parents from the Budapest ghetto by stealing a German soldier's uniform in a bath house, and walking into the ghetto to march his parents out.

Gian-Carlo Rota was Professor of Applied Mathematics and Philosophy at MIT. He received the Steele Prize from the American Mathematical Society in 1988; the citation singled out his 1964 paper "On the Foundations of Combinatorial Theory" as "the single paper most responsible for the revolution that has incorporated combinatorics into the mainstream of modern mathematics." He was a member of the National Academy of Sciences, and in 1992 was awarded the Medal for Distinguished Service from the National Security Agency. He was a member of the Heidegger Circle. In his Introduction to *Indiscrete Thoughts,* where the present essay appeared, he wrote, "The truth offends...The paper (reprinted five times in four languages) was taken as a personal insult by several living philosophers."

Brian Rotman is a writer and academic. His writing includes numerous articles and reviews in the *Guardian, Times Literary Supplement, London Review of Books,* a series of stage plays, and published books, among which are *Signifying Nothing: the Semiotics of Zero* and *Ad Infinitum ... the Ghost in Turing's Machine.* He is currently Professor in the Department of Comparative Studies, Ohio State University.

Jack Schwartz (Jacob T.) is Professor of Computer Science and Mathematics at the Courant Institute and Director of the Center for Multimedia Technologies at New York University. With Nelson Dunford, he won the Steele Prize of the American Mathematical Society in 1981 for their three-volume "bible" of functional analysis. Since then he has made major contributions to parallel computation, robotics, and multimedia. His present major interest is the human brain He is a member of the National Academy of Sciences.

Terry Stanway is a teacher of mathematics at Vancouver Technical Secondary School in Vancouver, British Columbia. He is associated with the CoLab at Simon Fraser University's Centre for Experimental and Constructive Mathematics. His research interests include the effect of new media on mathematical research and community formation.

Bill Thurston (aka William Paul Thurston) is Professor of Mathematics at Cornell University. He won the Fields Prize of the International Congress of Mathematicians in 1982. In the lecture on Thurston's work at the Congress in the following year, Professor Wall said: "Thurston has fantastic geometric insight and vision: his ideas have completely revolutionised the study of topology in 2 and 3 dimensions." While Thurston was director of the Mathematical Sciences Research Institute in Berkeley in 1993, he organized meetings between mathematics researchers and high school teachers.

Leslie A. White was professor of anthropology at the University of Michigan, starting in 1943. He devoted years of research to the Pueblo Indians of Arizona and New Mexico, and published a series of monographs about them. Later he wrote on cultural evolution. His major works are *The Science of Culture* (1949) and *The Evolution of Culture: The Development of Civilization to the Fall of Rome* (1959). One biographer writes, "His participation in the Socialist Labor Party brought him to the attention of the FBI during the height of the Cold War, and near-legendary scholarly and political conflicts surrounded him at the University of Michigan."

1

A Socratic Dialogue on Mathematics

ALFRÉD RÉNYI

SOCRATES Are you looking for somebody, my dear Hippocrates?

HIPPOCRATES No, Socrates, because I have already found him, namely you. I have been looking for you everywhere. Somebody told me at the agora that he saw you walking here along the River Ilissos; so I came after you.

SOCRATES Well then, tell me why you came, and then I want to ask you something about our discussion with Protagoras. Do you still remember it?

HIPPOCRATES How can you ask? Since that time not a single day has passed without my thinking about it. I came today to ask your advice because that discussion was on my mind.

SOCRATES It seems, my dear Hippocrates, that you want to talk to me about the very question I wish to discuss with you; thus the two subjects are one and the same. It seems that the mathematicians are mistaken in saying that two is never equal to one.

HIPPOCRATES As a matter of fact, Socrates, mathematics is just the topic I want to talk to you about.

SOCRATES Hippocrates, you certainly know that I am not a mathematician. Why did you not take your questions to the celebrated Theodoros?

HIPPOCRATES You are amazing, Socrates, you answer my questions even before I tell you what they are. I came to ask your opinion about my becoming a pupil of Theodoros. When I came to you the last time, with the intention of becoming a pupil of Protagoras, we went to him together and you directed the discussion so that it became quite clear that he did not know the subject he taught. Thus I changed my mind and did not follow him. This discussion helped me to see what I should not do, but did not show me what I should do. I am still wondering about this. I visit banquets and the palaestra with young men of my age, I dare say I have a pleasant time, but this does not satisfy me. It disturbs me to feel myself ignorant. More precisely, I feel that the knowledge I have is rather uncertain. During the discussion with Protagoras, I realized that my knowledge about familiar notions like virtue, justice and courage was far from satisfactory. Nevertheless, I think it is great progress that I now see clearly my own ignorance.

SOCRATES I am glad, my dear Hippocrates, that you understand me so well. I always tell myself quite frankly that I know nothing. *The difference between me and most other people is that I do not imagine I know what in reality I do not know.*

HIPPOCRATES This clearly shows your wisdom, Socrates. But such knowledge is not enough for me. I have a strong desire to obtain some certain and solid knowledge, and I shall not be happy until I do. I am constantly pondering what kind of knowledge I should try to acquire. Recently, Theaitetos told me that certainty exists only in mathematics and suggested that I learn mathematics from his master, Theodoros, who is the leading expert on numbers and geometry in Athens. Now, I should not want to make the same mistake I made when I wanted to be a pupil of Protagoras. Therefore tell me, Socrates, shall I find the kind of sound knowledge I seek if I learn mathematics from Theodoros?

SOCRATES If you want to study mathematics, O son of Apollodoros, then you certainly cannot do better than go to my highly esteemed friend Theodoros. But you must decide for yourself whether or not you really do want to study mathematics. Nobody can know your needs better than you yourself.

HIPPOCRATES Why do you refuse to help me, Socrates? Perhaps I offended you without knowing it?

SOCRATES You misunderstand me, my young friend. I am not angry; but you ask the impossible of me. Everybody must decide for himself what he wants to do. *I can do no more than assist as a midwife at the birth of your decision.*

HIPPOCRATES Please, my dear Socrates, do not refuse to help me, and if you are free now, let us start immediately.

SOCRATES Well, if you want to. Let us lie down in the shadow of that plane-tree and begin. *But first tell me, are you ready to conduct the discussion in the manner I prefer? I shall ask the questions and you shall answer them. By this method you will come to see more clearly what you already know, for it brings into blossom the seeds of knowledge already in your soul.* I hope you will not behave like King Darius who killed the master of his mines because he brought only copper out of a mine the king thought contained gold. I hope you do not forget that a miner can find in a mine only what it contains.

HIPPOCRATES I swear that I shall make no reproaches, but, by Zeus, let us begin mining at once.

SOCRATES All right. Then tell me, do you know what mathematics is? I suppose you can define it since you want to study it.

HIPPOCRATES I think every child could do so. Mathematics is one of the sciences, and one of the finest.

SOCRATES I did not ask you to praise mathematics, but to describe its nature. For instance, if I asked you about the art of physicians, you would answer that this art deals with health and illness, and has the aim of healing the sick and preserving health. Am I right?

HIPPOCRATES Certainly.

SOCRATES Then answer me this: does the art of the physicians deal with something that exists or with something that does not exist? If there were no physicians, would illness still exist?

HIPPOCRATES Certainly, and even more than now.

SOCRATES Let us have a look at another art, say that of astronomy. Do you agree with me that astronomers study the motion of the stars?

HIPPOCRATES To be sure.

SOCRATES And if I ask you whether astronomy deals with something that exists, what is your answer?

HIPPOCRATES My answer is yes.

SOCRATES Would stars exist if there were no astronomers in the world?

HIPPOCRATES Of course. And if Zeus in his anger extinguished all mankind, the stars would still shine in the sky at night. But why do we discuss astronomy instead of mathematics?

SOCRATES Do not be impatient, my good friend. Let us consider a few other arts in order to compare them with mathematics. How would you describe the man who knows about all the creatures living in the woods or in the depths of the sea?

HIPPOCRATES He is a scientist studying living nature.

SOCRATES And do you agree that such a man studies things which exist?

HIPPOCRATES I agree.

SOCRATES And if I say that every art deals with something that exists, would you agree?

HIPPOCRATES Completely.

SOCRATES Now tell me, my young friend, what is the object of mathematics? What things does a mathematician study?

HIPPOCRATES I have asked Theaitetos the same question. He answered that a mathematician studies numbers and geometrical forms.

SOCRATES Well, the answer is right, but would you say that these things exist?

HIPPOCRATES Of course. How can we speak of them if they do not exist?

SOCRATES Then tell me, if there were no mathematicians, would there be prime numbers, and if so, where would they be?

HIPPOCRATES I really do not know what to answer. Clearly, if mathematicians think about prime numbers, then they exist in their consciousness; but if there were no mathematicians, the prime numbers would not be anywhere.

SOCRATES Do you mean that we have to say mathematicians study non-existing things?

HIPPOCRATES Yes, I think we have to admit that.

SOCRATES Let us look at the question from another point of view. Here, I wrote on this wax tablet the number 37. Do you see it?

HIPPOCRATES Yes, I do.

SOCRATES And can you touch it with your hand?

HIPPOCRATES Certainly.

SOCRATES Then perhaps numbers do exist?

HIPPOCRATES O Socrates, you are mocking me. Look here, I have drawn on the same tablet a dragon with seven heads. Does it follow that such a dragon exists? I have never met anybody who has seen a dragon, and I am convinced that dragons do not exist at all except in fairy tales. But suppose I am mistaken, suppose somewhere beyond the pillars of Heracles dragons really do exist, that still has nothing to do with my drawing.

SOCRATES You speak the truth, Hippocrates, and I agree with you completely. But does this mean that even though we can speak about them, and write them down, numbers nevertheless do not exist in reality?

HIPPOCRATES Certainly.

SOCRATES Do not draw hasty conclusions. Let us make another trial. Am I right in saying that we can count the sheep here in the meadow or the ships in the harbor of Pireus?

HIPPOCRATES Yes, we can.

SOCRATES And the sheep and the ships exist?

HIPPOCRATES Clearly.

SOCRATES But if the sheep exist, their number must be something that exists, too?

HIPPOCRATES You are making fun of me, Socrates. Mathematicians do not count sheep; that is the business of shepherds.

SOCRATES Do you mean, what mathematicians study is not the number of sheep or ships, or of other existing things, but the number itself? And thus they are concerned with something that exists only in their minds?

HIPPOCRATES Yes, this is what I mean.

SOCRATES You told me that according to Theaitetos mathematicians study numbers and geometrical forms. How about forms? If I ask you whether they exist, what is your answer?

HIPPOCRATES Certainly they exist. We can see the form of a beautiful vessel, for example, and feel it with our hands, too.

SOCRATES Yet I still have one difficulty. If you look at a vessel what do you see, the vessel or its form?

HIPPOCRATES I see both.

SOCRATES Is that the same thing as looking at a lamb? Do you see the lamb and also its hair?

HIPPOCRATES I find the simile very well chosen.

SOCRATES Well, I think it limps like Hephaestus. You can cut the hair off the lamb and then you see the lamb without its hair, and the hair without the lamb. Can you separate in a similar way the form of a vessel from the vessel itself?

HIPPOCRATES Certainly not, and I dare say nobody can.

SOCRATES And nevertheless you still believe that you can see a geometric form?

HIPPOCRATES I am beginning to doubt it.

SOCRATES Besides this, if mathematicians study the forms of vessels, shouldn't we call them potters?

HIPPOCRATES Certainly.

SOCRATES Then if Theodoros is the best mathematician would he not be the best potter, too? I have heard many people praising him, but nobody has told me that he understands anything about pottery. I doubt whether he could make even the simplest pot. Or perhaps mathematicians deal with the form of statues or buildings?

HIPPOCRATES If they did, they would be sculptors and architects.

SOCRATES Well, my friend, we have come to the conclusion that mathematicians when studying geometry are not concerned with the forms of existing objects such as vessels, but with forms which exist only in their thoughts. Do you agree?

HIPPOCRATES I have to agree.

SOCRATES Having established that mathematicians are concerned with things that do not exist in reality, but only in their thoughts, let us examine the statement of Theaitetos, which you mentioned, that mathematics gives us more reliable and more trustworthy knowledge than does any other branch of science. Tell me, did Theaitetos give you some examples?

HIPPOCRATES Yes, he said for instance that one cannot know exactly how far Athens is from Sparta. Of course, the people who travel that way agree on the number of days one has to walk, but it is impossible to know exactly how many feet the distance is. On the other hand, one can tell, by means of the theorem of Pythagoras, what the length of the diagonal of a square is. Theaitetos also said that it is impossible to give the exact number of people living in Hellas. If somebody tried to count all of them, he would never get the exact figure, because during the counting some old people would die and children would be born; thus the total number could be only approximately correct. But if you ask a mathematician how many edges a regular dodecahedron has, he will tell you that the dodecahedron is bounded by 12 faces, each having 5 edges. This makes 60, but as each edge belongs to two faces and thus has been counted twice, the number of edges of the dodecahedron is equal to 30, and this figure is beyond every doubt.

SOCRATES Did he mention any other examples?

HIPPOCRATES Quite a few, but I do not remember all of them. He said that in reality you never find two things which are exactly the same. No two eggs are exactly the same, even the pillars of Poseidon's temple are slightly different from each other; but one may be sure that the two diagonals of a rectangle are exactly equal. He quoted Heraclitus who said that everything which exists is constantly changing, and that sure knowledge is only possible about things which never change, for instance, the odd and the even, the straight line and the circle.

SOCRATES That will do. These examples convince me that in mathematics we can get knowledge which is beyond doubt, while in other sciences or in everyday life it is impossible. Let us try to summarize the results of our inquiry into the nature of mathematics. Am I right in saying we came to the conclusion that mathematics studies non-existing things and is able to find out the full truth about them?

HIPPOCRATES Yes, that is what we established.

SOCRATES But tell me, for Zeus's sake, my dear Hippocrates, is it not mysterious that one can know more about things which do not exist than about things which do exist?

HIPPOCRATES If you put it like that, it certainly is a mystery. I am sure there is some mistake in our arguments.

SOCRATES No, we proceeded with the utmost care and we controlled every step of the argument. There cannot be any mistake in our reasoning. But listen, I remember something which may help us to solve the riddle.

HIPPOCRATES Tell me quickly, because I am quite bewildered.

SOCRATES This morning I was in the hall of the second archon, where the wife of a carpenter from the village Pitthos was accused of betraying and, with the aid of her lover, murdering her husband. The woman protested and swore to Artemis and Aphrodite that she was innocent, that she never loved anyone but her husband, and that her husband was killed by pirates. Many people were called as witnesses. Some said that the woman was guilty, others said that she was innocent. It was impossible to find out what really happened.

HIPPOCRATES Are you mocking me again? First you confused me completely, and now instead of helping me to find the truth you tell me such stories.

SOCRATES Do not be angry, my friend, I have serious reásons for speaking about this woman whose guilt it was impossible to ascertain. But one thing is sure. The woman exists. I saw her with my own eyes, and of anyone who was there, many of whom have never lied in their lives, you can ask the same question and you will receive the same answer.

HIPPOCRATES Your testimony is sufficient for me, my dear Socrates. Let it be granted that the woman exists. But what has this fact to do with mathematics?

SOCRATES More than you imagine. But tell me first, do you know the story about Agamemnon and Clytemnestra?

HIPPOCRATES Everybody knows the story. I saw the trilogy of Aeschylus at the theatre last year.

SOCRATES Then tell me the story in a few words.

HIPPOCRATES While Agamemnon, the king of Mycenae, fought under the walls of Troy, his wife, Clytemnestra, committed adultery with Aegisthus, the cousin of her husband. After the fall of Troy, when Agamemnon returned home, his wife and her lover murdered him.

SOCRATES Tell, me Hippocrates, is it quite sure that Clytemnestra was guilty?

HIPPOCRATES I do not understand why you ask me such questions. There can be no doubt about the story. According to Homer, when Odysseus visited the underworld he met Agamemnon, who told Odysseus his sad fate.

SOCRATES But are you sure that Clytemnestra and Agamemnon and all the other characters of the story really existed?

HIPPOCRATES Perhaps I would be ostracized if I said this in public, but my opinion is that it is impossible either to prove or disprove today, after so many centuries, whether the stories of Homer are true or not. But this is quite irrelevant. When I told you that Clytemnestra was guilty, I did not speak about the real Clytemnestra–if such a person ever lived–but about the Clytemnestra of our Homeric tradition, about the Clytemnestra in the trilogy of Aeschylus.

SOCRATES May I say that we know nothing about the real Clytemnestra? Even her existence is uncertain, but as regards the Clytemnestra who is a character in the triology of Aeschylus, we are sure that she was guilty and murdered Agamemnon because that is what Aeschylus tells us.

HIPPOCRATES Yes, of course. But why do you insist on all this?

SOCRATES You will see in a moment. Let me summarize what we found out. It is impossible in the case of the flesh and blood woman who was tried today in Athens to establish whether she is guilty, while there can be no doubt about the guilt of Clytemnestra who is a character in a play and who probably never existed. Do you agree?

HIPPOCRATES Now I am beginning to understand what you want to say. But it would be better if you drew the conclusions yourself.

SOCRATES The conclusion is this: we have much more certain knowledge about persons who exist only in our imagination, for example about characters in a play, than about living persons. If we say that Clytemnestra was guilty, it means only that this is how Aeschylus imagined her and presented her in his play. The situation is exactly the same in mathematics. We may be sure that the diagonals of a rectangle are equal because this follows from the definition of a rectangle given by mathematicians.

HIPPOCRATES Do you mean, Socrates, that our paradoxical result is really true and one can have a much more certain knowledge about non-existent things–for instance about the objects of mathematics–than about the real objects of nature? I think that now I also see the reason for this. The notions which we ourselves have created are by their very nature completely known to us, and we can find out the full truth about them because they have no other reality outside our imagination. However, the objects which exist in the real world are not identical with our picture of them, which is always incomplete and approximate; therefore our knowledge about these real things can never be complete or quite certain.

SOCRATES That is the truth, my young friend, and you stated it better than I could have.

HIPPOCRATES This is to your credit, Socrates, because you led me to understand these things. I see now not only that Theaitetos was quite right in telling me I must study mathematics if I want to obtain unfailing knowledge, but also why he was right. However, if you have guided me with patience up to now, please do not abandon me yet because one of my questions, in fact the most important one, is still unanswered.

SOCRATES What is this question?

HIPPOCRATES Please remember, Socrates, that I came to ask your advice as to whether I should study mathematics. You helped me to realize that mathematics and only mathematics can give me the sort of sound knowledge I want. But what is the use of this knowledge? It is clear that if one obtains some knowledge about the existing world, even if this knowledge is incomplete and is not quite certain, it is nevertheless of value to the individual as well as to the state. Even if one gets some knowledge about things such as the stars, it may be useful, for instance in navigation at night. But what is the use of knowledge of non-existing things such as that which mathematics offers? Even if it is complete and beyond any doubt, what is the use of knowledge concerning things which do not exist in reality?

SOCRATES My dear friend, I am quite sure you know the answer, only you want to examine me.

HIPPOCRATES By Heracles, I do not know the answer. Please help me.

SOCRATES Well, let us try to find it. We have established that the notions of mathematics are created by the mathematician himself. Tell me, does this mean that the mathematician chooses his notions quite arbitrarily as it pleases him?

HIPPOCRATES As I told you, I do not yet know much about mathematics. But it seems to me that the mathematician is as free to choose the objects of his study as the poet is free to choose the characters of his play, and as the poet invests his characters with whatever traits please him, so can the mathematician endow his notions with such properties as he likes.

SOCRATES If this were so, there would be as many mathematical truths as there are mathematicians. How do you explain, then, that all mathematicians study the same notions and problems? How do you explain that, as often happens, mathematicians living far from each other and having no contact independently discover the same truths? I never heard of two poets writing the same poem.

HIPPOCRATES Nor have I heard of such a thing. But I remember Theaitetos telling me about a very interesting theorem he discovered on incommensurable distances. He showed his results to his master, Theodoros, who produced a letter by Archytas in which the same theorem was contained almost word for word.

SOCRATES In poetry that would be impossible. Now you see that there is a problem. But let us continue. How do you explain that the mathematicians of different countries can usually agree about the truth, while about questions concerning the state, for example, the Persians and the Spartans have quite opposite views from ours in Athens, and, moreover, we here do not often agree with each other?

HIPPOCRATES I can answer that last question. In matters concerning the state everybody is personally interested, and these personal interests are often in contradiction. This is why it is difficult to come to an agreement. However, the mathematician is led purely by his desire to find the truth.

SOCRATES Do you mean to say that the mathematicians are trying to find a truth which is completely independent of their own person?

HIPPOCRATES Yes, I do.

SOCRATES But then we were mistaken in thinking that mathematicians choose the objects of their study at their own will. It seems that the object of their study has some sort of existence which is independent of their person. We have to solve this new riddle.

HIPPOCRATES I do not see how to start.

SOCRATES If you still have patience, let us try it together. Tell me, what is the difference between the sailor who finds an uninhabited island and the painter who finds a new color, one which no other painter has used before him?

HIPPOCRATES I think that the sailor may be called a discoverer, and the painter an inventor. The sailor discovers an island which existed before him, only it was unknown, while the painter invents a new color which before that did not exist at all.

SOCRATES Nobody could answer the question better. But tell me, the mathematician who finds a new truth, does he discover it or invent it? Is he a discoverer as the sailor or an inventor as the painter?

HIPPOCRATES It seems to me that the mathematician is more like a discoverer. He is a bold sailor who sails on the unknown sea of thought and explores its coasts, islands and whirlpools.

SOCRATES Well said, and I agree with you completely. I would add only that to a lesser extent the mathematician is an inventor too, especially when he invents new concepts. But every discoverer has to be, to a certain extent, an inventor too. For instance, if a sailor wants to get to places which other sailors before him were unable to reach, he has to build a ship that is better than the ships other sailors used. The new concepts invented by the mathematicians are like new ships which carry the discoverer farther on the great sea of thought.

HIPPOCRATES My dear Socrates, you helped me to find the answer to the question which seemed so difficult to me. The main aim of the mathematician is to explore the secrets and riddles of the sea of human thought. *These exist independently of the person of the mathematician, though not from humanity as a whole.* The mathematician has a certain freedom to invent new concepts as tools, and it seems that he could do this at his discretion. However, he is not quite free in doing this because the new concepts have to be useful for his work. The sailor also can build any sort of ship at his discretion, but, of course, he would be mad to build a ship which would be crushed to pieces by the first storm. Now I think that everything is clear.

SOCRATES If you see everything clearly, try again to answer the question: what is the object of mathematics?

HIPPOCRATES We came to the conclusion that *besides the world in which we live, there exists another world, the world of human thought*, and the mathematician is the fearless sailor who explores this world, not shrinking back from the troubles, dangers and adventures which await him.

SOCRATES My friend, your youthful vigor almost sweeps me off my feet, but I am afraid that in the ardor of your enthusiasm you overlook certain questions.

HIPPOCRATES What are these questions?

SOCRATES I do not want to disappoint you, but I feel that your main question has not yet been answered. We have not yet answered the question: what is the use of exploring the wonderful sea of human thought?

HIPPOCRATES You are right, my dear Socrates, as always. But won't you put aside your method this time and tell me the answer immediately?

SOCRATES No, my friend, even if I could, I would not do this, and it is for your sake. The knowledge somebody gets without work is almost worthless to him. We understand thoroughly only that which-perhaps with some outside help-we find out ourselves, just as a plant can use only the water which it sucks up from the soil through its own roots.

HIPPOCRATES All right, let us continue our search by the same method, but at least help me by a question.

SOCRATES Let us go back to the point where we established that the mathematician is not dealing with the number of sheep, ships or other existing things, but with the numbers themselves. Don't you think, however, that what the mathematicians discover to be true for pure numbers is true for the number of existing things too? For instance, the mathematician finds that 17 is a prime number. Therefore, is it not true that you cannot distribute 17 living sheep to a group of people, giving each the same number, unless there are *17* people?

HIPPOCRATES Of course, *it* is true.

SOCRATES Well, how about geometry? Can it not be applied in building houses, in making pots or in computing the amount of grain a ship can hold?

HIPPOCRATES Of course, it can be applied, though it seems to me that for these practical purposes of the craftsman not too much mathematics is needed. The simple rules known already by the clerks of the pharaohs in Egypt are sufficient for most such purposes, and the new discoveries about which Theaitetos spoke to me with such overflowing fervor are neither used nor needed in practice.

SOCRATES Perhaps not at the moment, but they may be used in the future.

HIPPOCRATES I am interested in the present.

SOCRATES If you want to be a mathematician, you must realize you will be working mostly for the future. Now, let us return to the main question. We saw that knowledge about another world of thought, about things which do not exist in the usual sense of the word, can be used in everyday life to answer questions about the real world. Is this not surprising?

HIPPOCRATES More than that, it is incomprehensible. It is really a miracle.

SOCRATES Perhaps it is not so mysterious at all, and if we open the shell of this question, we may find a real pearl.

HIPPOCRATES Please, my dear Socrates, do not speak in puzzles like the Pythia.

SOCRATES Tell me then, are you surprised when somebody who has travelled in distant countries, who has seen and experienced many things, returns to his city and uses his experience to give good advice to his fellow citizens?

HIPPOCRATES Not at all.

SOCRATES Even if the countries which the traveller has visited are very far away and are inhabited by quite a different sort of people, speaking another language, worshipping other gods?

HIPPOCRATES Not even in that case, because there is much that is common between different people.

SOCRATES Now tell me, if it turned out that the world of mathematics is, in spite of its peculiarities, in some sense similar to our real world, would you still find it miraculous that mathematics can be applied to the study of the real world?

HIPPOCRATES In that case no, but I do not see any similarity between the real world and the imaginary world of mathematics.

SOCRATES Do you see that rock on the other side of the river, there where the river broadens out and forms a lake?

HIPPOCRATES I see it.

SOCRATES And do you see the image of the rock reflected in the water?

HIPPOCRATES Certainly I do.

SOCRATES Then tell me, what is the difference between the rock and its reflection?

HIPPOCRATES The rock is a solid piece of hard matter. It is made warm by the sun. If you touched it, you would feel that it is rough. The reflected image cannot be touched; if I put my hand on it, I would touch only the cool water. As a matter of fact, the reflected image does not really exist; it is illusion, nothing else.

SOCRATES Is there nothing in common between the rock and its reflected image?

HIPPOCRATES Well, in a certain sense, the reflected image is a faithful picture of the rock. The contour of the rock, even its small abutments, are clearly visible in the reflected image. But what of it? Do you want to say that the world of mathematics is a reflected image of the real world in the mirror of our thinking?

SOCRATES You said it, and very well.

HIPPOCRATES But how is that possible?

SOCRATES Let us recall how the abstract concepts of mathematics developed. We said that the mathematician deals with pure numbers, and not with the numbers of real objects. But do you think that somebody who has never counted real objects can understand the abstract notion of number? When a child learns counting, he first counts pebbles and small sticks. Only if he knows that two pebbles and three pebbles make five pebbles, and the same about sticks or coins, is he able to understand that two and three make five. The situation is essentially the same with geometry. The child arrives at the notion of a sphere through experiences with round objects like balls. Mankind developed all fundamental notions of mathematics in a similar way. These notions are crystallized from a knowledge of the real world, and thus it is not surprising but quite natural that they bear the marks of their origin, as children do of their parents. And exactly as children when they grow up become the supporters of their parents, so any branch of mathematics, if it is sufficiently developed, becomes a useful tool in exploring the real world.

HIPPOCRATES Now it is quite clear to me how a knowledge of the non-existent things of the world of mathematics can be used in everyday life. You rendered me a great service in helping me to understand this.

SOCRATES I envy you, my dear Hippocrates, because I still wonder about one thing which I should like to have settled. Perhaps you can help me.

HIPPOCRATES I would do so with pleasure, but I am afraid you are mocking me again. Do not make me ashamed by asking my help, but tell me frankly the question which I overlooked.

SOCRATES You will see it yourself if you try to summarize the results of our discussion.

HIPPOCRATES Well, when it became clear why mathematics is able to give certain knowledge about a world different from the world in which we live, about the world of human thought, the question remained as to the use of this knowledge. Now we have found that the world of mathematics is nothing else but a reflection in our mind of the real world. This makes it clear that every discovery about the world of mathematics gives us some information about the real world. I am completely satisfied with this answer.

SOCRATES If I tell you the answer is not yet complete, I do so not because I want to confuse you, but because I am sure that sooner or later you will raise the question yourself and will reproach me for not having called your attention to it. You would say: "Tell me, Socrates, what is the sense of studying the reflected image if we can study the object itself?"

HIPPOCRATES You are perfectly right; it is an obvious question. You are a wizard, Socrates. You can totally confuse mc by a few words, and you can knock down by an innocent-looking question the whole edifice which we have built with so much trouble. I should, of course, answer that if we are able to have a look at the original thing, it makes no sense to look at the reflected image. But I am sure this shows only that our simile fails at this point. Certainly there is an answer, only I do not know how to find it.

SOCRATES Your guess is correct that the paradox arose because we kept too close to the simile of the reflected image. A simile is like a bow-if you stretch it too far, it snaps. Let us drop it and choose another one. You certainly know that travellers and sailors make good use of maps.

HIPPOCRATES I have experienced that myself. Do you mean that mathematics furnishes a map of the real world?

SOCRATES Yes. Can you now answer the question: what advantage would it be to look at the map instead of looking at the landscape?

HIPPOCRATES This is clear: using the map we can scan vast distances which could be covered only by travelling many weeks or months. The map shows us not every detail, but only the most important things. Therefore it is useful if we want to plan a long voyage.

SOCRATES Very well. But there is something else which occurred to me.

HIPPOCRATES What is it?

SOCRATES There is another reason why the study of the mathematical image of the world may be of use. If mathematicians discover some property of the circle, this at once gives us some information about any object of circular shape. Thus, the method of mathematics enables us to deal with different things at the same time.

HIPPOCRATES What about the following similes: If somebody looks at a city from the top of a nearby mountain, he gets a more comprehensive view than if he walks through its crooked streets; or if a general watches the movements of an enemy army from a hill, he gets a clearer picture of the situation than does the soldier in the front line who sees only those directly opposite him.

SOCRATES Well, you surpass me in inventing new similes, but as I do not want to fall behind, let me also add one parable. Recently I looked at a painting by Aristophon, the son of Aglaophon, and the painter warned me, "If you go too near the picture, Socrates, you will see only colored spots, but you will not see the whole picture."

HIPPOCRATES Of course, he was right, and so were you, when you did not let us finish our discussion before we got to the heart of the question. But I think it is time for us to return to the city because the shadows of night are falling and I am hungry and thirsty. If you still have some patience, I would like to ask something while we walk.

SOCRATES All right, let us start and you may ask your

HIPPOCRATES Our discourse convinced me fully that I should start studying mathematics and I am very grateful to you for this. But tell me, why are you yourself not doing mathematics? Judging from your deep understanding of the real nature and importance of mathematics, it is my guess that you would surpass all other mathematicians of Hellas, were you to concentrate on it. I would be glad to follow you as your pupil in mathematics if you accepted me.

SOCRATES No, my dear Hippocrates, this is not my. business. Theodoros knows much more about mathematics than I do and you cannot find a better master than him. As to your question why I myself am not a mathematician, I shall give you the reasons. I do not conceal my high opinion about mathematics. I think that we Hellenes have in no other art made such important progress as in mathematics, and this is only the beginning. If do not extinguish each other in mad wars, we shall obtain wonderful results as discoverers as well as inventors. You asked me why I do not join the ranks of those who develop this great science. As a matter of fact, I am some sort of a mathematician, only of a different kind. An inner voice, you may call it an oracle, to which I always listen carefully, asked me many years ago, "What is the source of the great advances which the mathematicians have made in their noble science?" I answered, "I think the source of the success of mathematicians lies in their methods, the high standards of their logic, their striving without the least compromise to the full truth, their habit of starting always from first principles, of defining every notion used exactly and avoiding self-contradictions." My inner voice answered, "Very well, but why do you think, Socrates, that this method of thinking and arguing can be used only for the study of numbers and geometric forms? Why do you not try to convince your fellow citizens to apply the same high logical standards in every other field, for instance in philosophy and politics, in discussing the problems of everyday private and public life?" From that time on, this has been my goal. I *have demonstrated* (you remember, for instance, our discussion with Protagoras) *that those who are thought to be wise men are mostly ignorant fools. All their arguing lacks solid foundation, since they* use-contrary to mathematicians-*undefined and only half-understood notions*. By this activity I have succeeded in making almost everybody my enemy. This is not surprising because for all people who are sluggish in thinking and idly content to use obscure terms, I am a living reproach. People do not like those who constantly remind them of the faults which they are unable or unwilling to correct. The day will come when these people will fall upon me and exterminate me. But until that day comes, I shall continue to follow my calling. You, however, go to Theodoros.

2

"Introduction" to *Filosofia e matematica**

CARLO CELLUCCI

Mathematics has long been a preferential subject of reflection for philosophers, inspiring them since antiquity in developing their theories of knowledge and their metaphysical doctrines. Given the close connection between philosophy and mathematics, it is hardly surprising that some major philosophers, such as Descartes, Leibniz, Pascal and Lambert, have also been major mathematicians.

In the history of philosophy the reflection on mathematics has taken several forms. Since it is impossible to deal with all of them in a single volume, in this book I will present what seems to me the most satisfactory form today. My own view, however, differs considerably from the dominant view, and on a number of accounts.

1. According to the dominant view, the reflection on mathematics is the task of a specialized discipline, the philosophy of mathematics, starting with Frege, characterized by its own problems and methods, and in a sense "the easiest part of philosophy"[1]. In this view, the philosophy of mathematics "is a specialized area of philosophy, but not merely a specialized area. Many of the questions that arise within it, though by no means all, are particular cases of more general questions that arise elsewhere in philosophy, and occur within the philosophy of mathematics in an especially pure, or especially simplified, form"[2]. Thus, "if you cannot solve these problems, what philosophical problems can you hope to solve?"[3].

   The view expressed in this book is instead that entrusting reflection on mathematics to a specialized discipline poses serious limitations, because one cannot assume that philosophical problems occur in mathematics in an

* Translation of a revised version of the "Introduction" to Cellucci 2002a.

[1] Dummett 1998, p. 190.

[2] *Ibid.*, p. 123.

[3] *Ibid.*, p. 191.

especially pure, or especially simplified, form. The reflection on mathematics entails dealing with such problems in all their impurity and complexity, and cannot be carried out adequately without dealing with them.

The idea that philosophical problems occur in mathematics in an especially pure or especially simplified form depends on the assumption that, whereas applied mathematics draws its "concepts from experience, observation, scientific theories, and even economics", pure mathematics does not; on the contrary, "it is its purity that gives rise to many of the questions" on mathematics "we have been puzzling over"[4]. Pure mathematics "requires no input from experience: it is exclusively the product of thought"[5].

This view is unjustified, however, since, like applied mathematics, pure mathematics draws its concepts from experience, observation, scientific theories and even economics. The questions considered by the reflection on mathematics have, therefore, all the impurity and complexity of which philosophical problems are capable.

2. According to the dominant view, the main problem in the philosophy of mathematics is the justification of mathematics. This problem arises because "our much-valued mathematics rests on two supports: inexorable deductive logic, the stuff of proof, and the set theoretic axioms", which raises "the question of what grounds our faith in logical inference" and "what justifies the axioms of set theory"[6]. To answer such questions one must clarify the foundations of mathematics, providing a justification for them. On the other hand, the philosophy of mathematics does not concern itself with the problem of mathematical discovery, since it is only "concerned with the product of mathematical thought; the study of the process of production is the concern of psychology, not of philosophy"[7].

   The view expressed in this book is instead that the main problem in the reflection on mathematics is discovery. This includes the problem of justification, since discovery is not merely a part of mathematical activity but encompasses the whole of it, and therefore includes justification. Indeed, discovery requires making hypotheses capable of solving given problems, and in order to choose the hypotheses one must carefully evaluate the reasons for and against them. The evaluation process is intertwined with the process of hypothesis-formation, since one must compare alternative hypotheses in order to select one of them. This blurs the distinction between discovery and justification. In fact no such distinction is possible, since there are normally so many possible hypotheses to be formed and evaluated for any given problem that an exhaustive

[4] *Ibid.*, p. 190.
[5] Dummett 2001, p. 10.
[6] Maddy 1997, pp. 1-2.
[7] Dummett 1991, p. 305.

search cannot take place. Given that one cannot first make all possible hypotheses and then evaluate them, making hypotheses and evaluating them must be concurrent processes.

The idea that the main problem in the philosophy of mathematics is the justification of mathematics has made the philosophy of mathematics an increasingly less attractive subject, devoted to the study of questions – such as Frege's: 'What is the number one?' – which seem irrelevant to mathematicians, neglecting those which are more important for understanding mathematics. No wonder, then, that there is widespread disregard and misunderstanding, and often outright antagonism, between philosophers of mathematics and mathematicians. The problem of the justification of mathematics seems unpalatable to the vast majority of mathematicians, who consider it irrelevant to their work.

Moreover, the idea that solving the problem of the justification of mathematics consists of clarifying the foundations of mathematics contradicts mathematical experience, which shows that mathematics is by no means a static structure, based on foundations given once and for all, but is a dynamic process, multifarious and articulated, whose ways of justification are also multifarious and articulated.

3. According to the dominant view, another important problem in the philosophy of mathematics is the existence of mathematical objects. This problem arises because "the point of view of common sense is perhaps that, if a proposition is true, it is because there are entities existing independently of the proposition which have the properties or stand in the relations which the proposition asserts of them"[8]. This "suggests that since mathematical propositions are true, that there are entities in virtue of which the propositions are true. The ontological issue is whether there are such entitities and if so what their nature is"[9]. This problem supplements "the epistemological question" of the justification of mathematics, namely, "how mathematical beliefs come to be completely justified"[10].

   The view expressed in this book is instead that the problem of the existence of mathematical objects is irrelevant to mathematics because, as Locke pointed out, "all the discourses of the mathematicians about the squaring of a circle" – or any other geometrical figure – "concern not the existence of any of those figures", and their proofs "are the same whether there be any square or circle existing the world, or no"[11]. Indeed, it is compatible with mathematical practice that there are no mathematical objects of which its theorems are true. Mathematical objects are simply hypotheses introduced to solve specific problems. To speak of

[8] Lehman 1979, p. 1.
[9] *Ibid.*
[10] *Ibid.*
[11] Locke 1975, p. 566.

mathematical objects is misleading because the term 'object' seems to designate the things the investigation is about, whereas hypotheses are the tools for the investigation. The latter is intended not to study properties of mathematical objecs but to solve problems.

4. According to the dominant view, the philosophy of mathematics does not add to mathematics. Since its main problem is the justification of mathematics, it aims at clarifying the foundations of mathematics, not at expanding mathematics. Thus, "as the philosophy of law does not legislate, or the philosophy of science devise or test scientific hypotheses, so – we must realize from the outset – the philosophy of mathematics does not add to the number of mathematical theorems and theories"[12]. Its "arguments should have no doctrinal or practical impact on mathematics at all"[13]. For "mathematics comes first, then philosophizing about it, not the other way around"[14]. This is simply a special case of the fact that "philosophy does not contribute to the progress of knowledge: it merely clarifies what we already know"[15].

   The view expressed in this book is instead that the reflection on mathematics is relevant to the progress of mathematics. Since its main problem is mathematical discovery, it aims at improving existing methods of discovery and at inventing new ones. In this way the reflectiom on mathematics may contribute to the progress of mathematics, because the improvement in existing methods of discovery and the invention of new ones are of the utmost importance to that aim. Even Frege acknowledges that "a development of method, too, furthers science. Bacon, after all, thought it better to invent a means by which everything could easily be discovered than to discover particular truths, and all steps of scientific progress in recent times have had their origin in an improvement of method"[16].

   That the reflection on mathematics can contribute to the progress of mathematics entails that mathematics does not come first, with philosophizing about it following. On the contrary, they proceed together, both contributing to the advancement of learning.

5. According to the dominant view, the philosophy of mathematics does not require any detailed knowledge of mathematics, because its main aim – the justification of mathematics through a clarification of its foundations – does not require any detailed knowledge of the edifice built up on such foundations. Thus, even "if you have little knowledge of mathematics, you do not need to remedy that defect before interesting

[12] Körner 1960, p. 9.
[13] Wagner 1982, p. 267.
[14] Hersh 1997, p. *xi*.
[15] Dummett 2001, p. 24.
[16] Frege 1967, p. 6.

yourself in the philosophy of mathematics"[17]. You can "very well understand a good deal of the debates on the subject and a good deal of the theories advanced concerning it without an extensive knowledge of its subject-matter"[18]. Similarly, the philosophy of mathematics does not require any detailed knowledge of the history of mathematics, because "the etiology of mathematical ideas, however interesting, is not something whose study promises to reveal much about the structure of thought: for the most part, the origin and development of mathematical ideas are simply far too determined by extraneous influences"[19]. On the other hand, the philosophy of mathematics requires detailed knowledge of mathematical logic, "not so much as part of the object of study as serving as a tool of inquiry"[20].

The view expressed in this book is instead that the reflection on mathematics does in fact require detailed knowledge of mathematics. Neglecting this has led the philosophy of mathematics to deal with marginal issues, deliberately excluding the broader ones. The philosophy of mathematics has done so on the assumption that, although the broader questions are "more interesting, more pressing, more significant than the narrower logical questions that are properly foundational", the latter are "amenable to solution, whereas solutions to the broader questions may depend upon further advances in mathematics itself, advances which we cannot as yet foresee"[21]. But this assumption overlooks the fact that, owing to Gödel's incompleteness theorems and related results, no logical question that is properly foundational has been solved by the philosophy of mathematics. Moreover, neither is there any evidence that the solutions to the broader questions may depend upon further advances in mathematics itself. As Wittgenstein put it, "even 500 years ago a philosophy of mathematics was possible, a philosophy of what mathematics was then"[22].

The reflection on mathematics also requires a detailed knowledge of the history of mathematics. Neglecting this has led the philosophy of mathematics to consider mathematics as a static building, based on linear relations of logical dependence between *a priori* determined axioms and theorems. On the contrary, the history of mathematics shows that mathematics is a dynamic process, which often develops through tortuous and tormented paths not determined *a priori*, and proceeds through false starts and standstills, routine periods and sudden turnings. This has prevented the philosophy of mathematics from accounting not only

[17] Dummett 1998, p. 124.
[18] *Ibid.*
[19] George-Velleman 2002, p. 2.
[20] Dummett 1998, p. 124.
[21] Mayberry 1994, p. 19.
[22] Wittgenstein 1978, V, § 52.

for mathematical discovery but also for the real processes of mathematical justification.

As to the idea that the philosophy of mathematics requires detailed knowledge of mathematical logic, not so much as part of the object of study as serving as a tool of inquiry, it risks being empty. For, although the philosophy of mathematics has carried out an intense study of the foundations of mathematics using mathematical logic as a tool of inquiry, the hard core of mathematics has turned out to be impervious to what is found there. Thus the aim of clarifying the foundations of mathematics has lost momentum. Even supporters of mathematical logic, like Simpson, acknowledge that "foundations of mathematics is now out of fashion. Today, most of the leading mathematicians are ignorant of foundations", and "foundations of mathematics is out of favor even among mathematical logicians"[23]. Indeed, the mainstream of mathematical logic has abandoned foundations to become a conventional albeit somewhat marginal branch of mathematics.

The idea that mathematical logic is the tool of inquiry of the philosophy of mathematics has its roots in the distrust towards the approach to philosophical problems of the philosophical tradition. This distrust has led to viewing the history of mathematical logic as a persistent struggle to free the subject from the grip of philosophy.

This distrust emerges among the first practitioners of the art, for example Russell, who maintains that "philosophy, from the earliest times, has made greater claims, and achieved fewer results, than any other branch of learning"[24]. Indeed, "so meagre was the logical apparatus that all the hypotheses philosophers could imagine were found to be inconsistent with facts"[25]. Nonetheless, "the time has now arrived when this unsatisfactory state of things can be brought to an end"[26]. This is made possible by mathematical logic, which has "introduced the same kind of advance into philosophy as Galilei introduced into physics"[27]. Mathematical logic "gives the method of research in philosophy, just as mathematics gives the method in physics. And as physics" finally "became a science through Galileo's fresh observation of facts and subsequent mathematical manipulation, so philosophy, in our own day, is becoming scientific through the simultaneous acquisition of new facts and logical methods"[28].

Statements of this kind are recurrent in the philosophy of mathematics. For example, Łukasiewicz claims that "philosophy must be reconstructed from its very foundations; it should take its inspiration from

---

[23] Simpson 1999, p. *vii*.
[24] Russell 1999, p. 13.
[25] *Ibid.*, p. 243.
[26] *Ibid.*, p. 13.
[27] *Ibid.*, pp. 68-69.
[28] *Ibid.*, p. 243.

scientific method and be based on the new logic"[29]. Beth maintains that "the lack of an adequate formal logic has strongly hampered the development of a systematic philosophy. Therefore", although reflection on the philosophy of the past will remain one of the elements of future philosophy, "an adequate formal logic will be a second element of future philosophy"[30]. Kreisel claims that the approach to philosophical problems of the philosophical tradition is viable only "at an early stage, when we know too little about the phenomenon involved and about our knowledge of it in order to ask sensible specific questions"[31]. Such an approach must be replaced by one based on mathematical logic, which is "a tool in the philosophy of mathematics; just as other mathematics, for example the theory of partial differential equations, is a tool in what used to be called natural philosophy"[32].

Distrust of the approach of the philosophical tradition and the urge to replace it by one based on mathematical logic are two basic features of the philosophy of mathematics which, on account of their very fruitlessness, have led to its progressive impoverishment and decline. This decline has become increasingly marked since the discovery of Gödel's incompleteness theorems, so much so that Mac Lane claimed that the philosophy of mathematics is "a subject dormant since about 1931"[33].

6. According to the dominant view, mathematics is theorem proving because it "is a collection of proofs. This is true no matter what standpoint one assumes about mathematics – platonism, anti-platonism, intuitionism, formalism, nominalism, etc."[34]. Perhaps "in 'doing mathematics' proving theorems isn't everything, but it's way ahead of whatever is in second place"[35]. Of course, "the activity of mathematics is not just randomly writing down formal proofs for random theorems", because "the choices of axioms, of problems, of research directions, are influenced by a variety of considerations – practical, artistic, mystical", but the latter are "really non-mathematical"[36]. Therefore they are not a concern of the philosophy of mathematics.

The view expressed in this book is instead that mathematics is problem solving. That does not mean that mathematics is only problem solving. First one must pose problems, then one can refine them, exhibit them, dismiss them or even dissolve them. But problem solving is the core of mathematical activity, so it seems justified to maintain that it is an essen-

---

29 Lukasiewicz 1970, p. 112.
30 Beth 1957, pp. 8-9.
31 Kreisel 1984, p. 82.
32 Kreisel 1967, p. 201.
33 Mac Lane 1981, p. 462.
34 Takeuti 1987, p. 1.
35 Franks 1989b, p. 12.
36 Monk 1976, p. 4.

tial feature of mathematics. Problem solving, however, does not concern single separated problems nor leads to a final solution. For each solution generates new problems, and depends on the solutions found for these new problems. Thus, no solution is final but is always subject to further reconsideration.

7. According to the dominant view, the method of mathematics is the axiomatic method. For "proof must begin from axioms that are not themselves proved"[37]. In fact, "to prove a proposition, you start from some first principles, derive some results from those axioms, then, using those axioms and results, push on to prove other results"[38]. The axiomatic method is available for the whole of mathematics, because "all mathematical theories, when sufficiently developed, are capable of axiomatization"[39]. Moreover, the axiomatic method provides "a strategy both for finding and remembering proofs", because "relatively few properties, Bourbaki's few, so-called basic structures, have been found adequate for similar strategies in a very broad domain of mathemathics", although "the use of axiomatic analysis as a proof strategy does not seem to be well known to people writing on heuristics, like Polya"[40]. Since the method of mathematics is the axiomatic method, "mathematics and science are intellectual undertakings which are complementary but opposed, distinguished by the direction of their view"[41]. For "the former proceeds forwards, from hypotheses to conclusions: i.e., from axioms to the theorems derivable from them. The latter proceeds backwards, from conclusions to premisses: i.e., from experimental data to physical laws from which they can be drawn"[42].

   The view expressed in this book is instead that the method of mathematics is the analytic method, a method which, unlike the axiomatic method, does not start from axioms which are given once and for all and are used to prove any theorem, nor does it proceed forwards from axioms to theorems, but proceeds backwards from problems to hypotheses. Thus proof does not begin from axioms that are not themselves proved. Unlike axioms, hypotheses are not given from the start, but are the very goal of the investigation. They are never definitive, but liable to be replaced by other hypotheses, and are introduced to solve specific problems[43].

---

[37] Maddy 1990, p. 144.
[38] Leary 2000, p. 48.
[39] Dummett 1991, p. 305.
[40] Kreisel-MacIntyre 1982, pp. 232-233.
[41] Odifreddi 2001, p. 233.
[42] *Ibid.*
[43] The analytic method is meant here not in the sense of Aristotle or Pappus but in the sense of Hippocrates of Chios and Plato. On this distinction see Cellucci 1998a.

The idea that, to prove a proposition, you start from some first principles, derive some results from those axioms, then, using those axioms and results, push on to prove other results, contrasts with mathematical experience which shows that in mathematics one first formulates problems, then looks for hypotheses to solve them. Thus one does not proceed, as in the axiomatic method, from axioms to theorems but proceeds, as in the analytic method, from problems to hypotheses. As Hamming points out, in mathematics deriving theorems from axioms "does not correspond to simple observation. If the Pythagorean theorem were found to not follow from postulates, we would again search for a way to alter the postulates until it was true. Euclid's postulates came from the Pythagorean theorem, not the other way"[44].

Similarly, the idea that the axiomatic method is available for the whole of mathematics because all mathematical theories, when sufficiently developed, are capable of axiomatization, contrasts with mathematical experience, which shows that axiomatization does not naturally apply to all parts of mathematics. Some of them are not suitable for axiomatization, and exist as collections of solved or unsolved problems of a certain kind. This is true, for example, of number theory and of much of the theory of partial differential equations.

The idea that the axiomatic method provides a strategy both for finding and remembering proofs also contrasts with mathematical experience, which shows that proofs based on the axiomatic method often appear to be found only by a stroke of luck, and seem artificial and difficult to understand. Showing only the final outcome of the investigation, established in a way that is completely different from how it was first obtained, such proofs hide the actual mathematical process, thus contributing to make mathematics a difficult subject.

Similarly, the idea that, since the method of mathematics is the axiomatic method, mathematics and science are intellectual undertakings which are complementary but opposed, distinguished by the direction of their view, contrasts with mathematical experience, which shows that mathematics, like other sciences, proceeds backwards from conclusions to premises, i.e. from problems to hypotheses which provide conditions for their solution. This is adequately accounted for by the analytic method, which assimilates mathematics to other sciences, and in particular assimilates the concept of mathematical proof to the concepts of proof of other sciences.

The limitations of the axiomatic method are acknowledged by several mathematicians. For example, Lang stresses that "axiomatization is what one does last, it's rubbish", it is merely "the hygiene of mathematics"[45]. Giusti states that "setting out axioms is never the starting point, but is

[44] Hamming 1980, p. 87.
[45] Lang 1985, p. 19.

rather the arrival point of a theory", and "the occasions where one started from the axioms are rather the exception than the rule"[46]. Hersh points out that, "in developing and understanding a subject, axioms come late", and even if "sometimes someone tries to invent a new branch of mathematics by making up some axioms and going from there", still "such efforts rarely achieve recognition or permanence. Examples, problems, and solutions come first. Later come axiom sets on which the existing theory can be 'based'. The view that mathematics is in essence derivations from axioms is backward. In fact, it's wrong"[47].

The limitations of the axiomatic method are also acknowledged by some supporters of the dominant view, like Mayberry, who recognizes that "no axiomatic theory, formal or informal, of first or of higher order can logically play a foundational role in mathematics"[48]. For "it is obvious that you cannot use the axiomatic method to explain what the axiomatic method is"[49]. Since any theory put forward "as the foundation of mathematics must supply a convincing account of axiomatic definition, it cannot, on pain of circularity, itself be presented by means of an axiomatic definition"[50].

8. According to the dominant view, the logic of mathematics is deductive logic. For theorems "are justified by deductive inference"[51]. In fact, "deductive inference patently plays a salient part in mathematics. The correct observation that the discovery of a theorem does not usually proceed in accordance with the strict rules of deduction has no force: a proof has to be set out in sufficient detail to convince readers, and, indeed, its author, of its deductive cogency"[52]. Admittedly, "deduction is only one component in mathematical reasoning understood in the broad sense of all the intellectual work that goes on when solving a mathematical problem. But this does not mean that the notion of deduction is not the key concept for understanding validity in mathematics, or that the distinction between discovery and justification loses its theoretical importance"[53]. For, "when it comes to explaining the remarkable phenomenon that work on a mathematical problem may end in a result that everyone finds definitive and conclusive, the notion of deduction is a central one"[54]. Indeed, "mathematics has a methodology unique among all the sciences. It is the only discipline in which

[46] Giusti 1999, p. 20.
[47] Hersh 1997, p. 6.
[48] Mayberry 1994, p. 34.
[49] *Ibid.*, p. 35.
[50] *Ibid.*
[51] Maddy 1984, p. 49.
[52] Dummett 1991, p. 305.
[53] Prawitz 1998, p. 332.
[54] *Ibid.*

deductive logic is the sole arbiter of truth. As a result mathematical truths established at the time of Euclid are still held valid today and are still taught. No other science can make this claim"[55].

The view expressed in this book is instead that the logic of mathematics is not deductive logic but a broader logic, dealing with non-deductive (inductive, analogical, metaphorical, metonymical, etc.) inferences in addition to deductive inferences. It is by non-deductive inferences that one finds the hypotheses by which mathematical problems are solved. The logic of mathematics is not, therefore, that studied by mathematical logic, which is simply a branch of mathematics, but consists of a set of non-deductive methods and techniques in addition to deductive methods and techniques, and hence is not a theory but a set of tools.

To claim that the logic of mathematics is deductive logic because theorems are justified by deductive inference, restricts mathematical experience to ways of reasoning found only in textbooks of mathematical logic, and neglects those that are really used in mathematical activity. Moreover, it does not account for the real nature of mathematics, because mathematical reasoning is based mainly on non-deductive inferences, not on deductive inferences, which play a somewhat restricted role within it. Contrary to widespread misunderstanding, mathematics is never deductive in the making, since mathematicians first state problems, then find hypotheses for their solution by non-deductive inferences. As even some supporters of the dominant view, like Halmos, acknowledge, mathematics "is never deductive in its creation. The mathematician at work makes vague guesses, visualizes broad generalizations, and jumps to unwarranted conclusions. He arranges and rearranges his ideas, and he becomes convinced of their truth long before he can write down a logical proof"[56]. The "deductive stage, writing the result down, and writing down its rigorous proof are relatively trivial once the real insight arrives; it is more like the draftsman's work, not the architect's"[57]. Furthermore, to claim that the logic of mathematics is deductive logic clashes with the results of the neurosciences, which show that the human brain is very inefficient even in moderately long chains of deductive inferences.

Similarly, to claim that, when it comes to explaining the remarkable phenomenon that work on a mathematical problem may end in a result that everyone finds definitive and conclusive, the notion of deduction is a central one, overlooks the fact that, according to the dominant view, several Euclid's proofs are flawed. Thus, in this view, the fact that everyone finds Euclid's results definitive and conclusive cannot depend on Euclid's proofs. The same applies to contemporary mathematics, where

[55] Franks 1989a, p. 68.
[56] Halmos 1968, p. 380.
[57] *Ibid.*

several published proofs are flawed – somewhat surprising if mathematics is the rigorous deduction of theorems from axioms.

Moreover, to claim that mathematics has a methodology unique among all the sciences because it is the only discipline in which deductive logic is the sole arbiter of truth, begs the question since it assumes that the logic of mathematics is deductive logic. On the contrary, the broader logic on which mathematics is based does not distinguish but rather assimilates the methodology of mathematics to that of the other sciences, which is based on inferences of the very same kind.

9. According to the dominant view, mathematical discovery is an irrational process based on intuition, not on logic. For "some intervention of intuition issuing from the unconscious is necessary at least to initiate the logical work"[58]. The activity "of a creating brain has never had any rational explanation, neither in mathematics nor in other fields"[59]. In particular, the discovery of axioms has nothing to do with logic, because "there is no hope, there is, as it were, a leap in the dark, a bet at any new axiom", so "we are no longer in the domain of science but in that of poetry"[60]. Generally, "the creative and intuitive aspects of mathematical work evade logical encapsulation"[61]. The "mathematician at work relies on surprisingly vague intuitions and proceeds by fumbling fits and starts with all too frequent reversals. In this picture the actual historical and individual processes of mathematical discovery appear haphazard and illogical"[62]. The role of intuition in mathematical discovery is decisive "in most researchers, who are often put on the track that will lead them to their goal by an albeit confused intuition of the mathematical phenomena studied by them"[63].

The view expressed in this book is instead that mathematics is a rational activity at any stage, including the most important one: discovery. Intuition does not provide an adequate explanation as to how we reach new hypotheses, so, either we must give up any explanation thus withdrawing into irrationalism, or we must provide an explanation, but then cannot appeal to intuition.

In fact, there is no need to appeal to intuition. Since ancient times, many have recognised not only that mathematical discovery is a rational process, but also that a method exists for it, namely the analytic method. This method gave great heuristic power to ancient mathematicians in solving geometrical problems, and has had a decisive role in the new

[58] Hadamard 1945, p. 112.
[59] Dieudonné 1988, p. 38.
[60] Girard 1989, p. 169.
[61] Feferman 1998, p. 178.
[62] *Ibid.*, p. 77.
[63] Dieudonné 1948, pp. 544-545.

developments of mathematics and physics since the beginning of the modern era. Within the analytic method, logic plays an essential role in the discovery of hypotheses, provided of course that logic is taken to include non-deductive inferences, unlike in the limited and somewhat parochial dominant view.

Not only is there no need to appeal to intuition, but Pascal even sets the mathematical mind against the intuitive mind. For he claims that "there are two kinds of mind, one mathematical, and the other what one might call the intuitive. The first takes a slow, firm and inflexible view, but the latter has flexibility of thought which it applies simultaneously to the diverse lovable parts of that which it loves"[64].

10. According to the dominant view, in addition to mathematical discovery, mathematical justification too is based on intuition. For, if one assumes that the method of mathematics is the axiomatic method, then justifying mathematics amounts to justifying the certainty of its axioms, and their certainty is directly or indirectly based on intuition. Directly, when through intuition "the axioms force themselves upon us as being true"[65]. Indirectly, when "we apply contentual" – and hence intuitive – "inference, in particular, to the proof of the consistency of the axioms"[66]. Thus, "accounting for intuitive 'knowledge' in mathematics is the basic problem of mathematical epistemology"[67].

The view expressed in this book is instead that justification is not based on intuition but on the fact that the hypotheses used in mathematics are plausible, i.e., compatible with the existing knowledge, in the sense that, if one compares the reasons for and against the hypotheses, the reasons for prevail. It is often claimed that 'plausible' has a subjective, psychological connotation, so that it is almost equivalent to 'rhetorically persuasive', hence plausible arguments are of little interest in mathematics. But 'plausible', in the sense explained above, has nothing subjective or psychological about it.

To assess whether a given hypothesis is plausible, one examines the reasons for and against it. This examination is carried out using facts which confirm the hypothesis or refute it, where these facts belong to the existing knowledge. Admittedly, such an assessment is fallible, because one's choice of facts may be inadequate, and moreover the existing knowledge is not static but develops continuously, each new development providing further elements for assessing the hypothesis, which may lead to its rejection. But this procedure is neither subjective nor psychological. On the

[64] Pascal 1913, p. 50.
[65] Gödel 1986-, II, p. 268.
[66] Hilbert 1996, p. 1132.
[67] Hersh 1997, p. 65.

contrary, a justification of mathematics based on intuition is subjective and psychological.

11. According to the dominant view, mathematics is a body of truths – indeed a body of absolutely certain and hence irrefutable truths. For mathematics is "the paradigm of certain and final knowledge: not fixed, to be sure, but a steadily accumulating coherent body of truths obtained by successive deduction from the most evident truths. By the intricate combination and recombination of elementary steps one is led incontrovertibly from what is trivial and unremarkable to what can be nontrivial and surprising"[68]. This derives from the fact that, "while a physical hypothesis can only be verified to the accuracy and the interpretation of the best work in the laboratory", a mathematical truth is established by a proof based on the axiomatic method, which "has the highest degree of certainty possible for man"[69]. Indeed, "there is at present no viable alternative to axiomatic presentation if the truth of a mathematical statement is to be established beyond reasonable doubt"[70]. While physical hypotheses come and go, none is definitive, and so "in physics nothing is completely certain", mathematics, as based on the axiomatic method, "lasts an eternity"[71].

    The view expressed in this book is instead that mathematics is a body of knowledge but contains no truths. Speaking of truth is not necessary in mathematics, just as it is not necessary in the natural sciences, and is not necessary anywhere except perhaps in theology and in lovers' quarrels. Assuming that mathematics is a body of truths leads to an inextricable muddle, which results in self-defeating statements such as: It is legitimate "to argue from 'this theory has properties we like' to 'this theory is true'" [72]. On this basis Frege could have argued that, since his ideography had the property he liked of reducing arithmetic to logic, his ideography was true, only to be belied by Russell's paradox.

    That mathematics is not a body of truths does not mean that it has no objective content. It only means that, as with any other science, mathematics does not consist of truths but only of plausibile statements, i.e., statements compatible with existing knowledge. The objectivity of mathematics does not depend on its being a body of truths but on its being a body of plausible statements.

    Moreover, the idea that mathematics is a body of absolutely certain and hence irrefutable truths, overlooks the fact that we cannot be sure of the current proofs of our theorems. For, by Gödel's incompleteness theorems and related results, we cannot be sure of the hypotheses on which they are based.

[68] Feferman 1998, p. 77.
[69] Jaffe 1997, p. 135.
[70] Rota 1997, p. 142.
[71] Jaffe 1997, pp. 139-140.
[72] Maddy 1997, p. 163.

As regards certainty, mathematics has no privilege and is as risky as any other human creation. Mathematical knowledge is not absolutely certain, but only plausible, i.e. compatible with the existing knowledge, and plausibility does not grant certainty, because the existing knowledge is not absolutely certain, but only plausible. For centuries mathematics was considered a body of absolutely certain truths, but now this is increasingly perceived as an illusion. Uncertainty and doubt have replaced the self-complacent certainty of the past. As some supporters of the dominant view, like Leary, also acknowledge, by Gödel's incompleteness theorems and related results, "mathematics, which had reigned for centuries as the embodiment of certainty, had lost that role"[73].

12. According to the dominant view, the question of the applicability of mathematics to the physical sciences is inessential for the philosophy of mathematics. Mathematics is "a unified undertaking which we have reason to study as it is, and the study of the actual methods of mathematics, which includes pure mathematics, quickly reveals that modern mathematics also has goals of its own, apart from its role in science"[74]. Admittedly, "it is a wonderful thing when a branch of mathematics suddenly becomes relevant to new discoveries in another science; both fields benefit enormously. But many (maybe most) areas of mathematics will never be so fortunate. Yet most mathematicians feel their work is no less valid and no less important than mathematics that has found utility in other sciences. For them it is enough to experience and share the beauty of a new theorem. New mathematical knowledge" is "an end in itself"[75].

    The view expressed in this book is instead that the question of the applicability of mathematics to the physical sciences is important for the reflection on mathematics. While, on the one hand, mathematics is continuous with philosophy, on the other hand it is also continuous with the physical sciences, and many of its developments, even in pure mathematics, are inextricably linked to the physical sciences.

13. According to the dominant view, mathematics is based only on conceptual thought. For mathematics "is the purest product of conceptual thought, which is a feature of human life that both pervasively structures it and sets it apart from all else"[76]. Mathematics is "unconstrained by experience", enters "the world touched only by the hand of reflection", and is "justified by pure ratiocination, perceptual observation via any of our five sensory modalities being neither necessary nor even relevant"[77].

---

[73] Leary 2000, p. 3.
[74] Maddy 1997, p. 205, footnote 15.
[75] Franks 1989b, p. 13.
[76] George-Velleman 2002, p. 1.
[77] *Ibid*

Thus, from the standpoint of the philosophy of mathematics, it is inessential to study questions concerning perception, or, more generally, "such questions as 'What brain, or neural activity, or cognitive architecture makes mathematical thought possible?' "[78]. These questions focus on "phenomena that are really extraneous to the nature of mathematical thought itself", i.e., "the neural states that somehow carry thought", whereas "philosophers, by contrast, are interested in the nature of those thoughts themselves, in the content carried by the neural vehicles"[79].

The view expressed in this book is instead that mathematics is based not only on conceptual thought but also on perception, which plays an important role in it, for example, in diagrams. Thus, from the viewpoint of the reflection on mathematics, it is important to study questions concerning perception and more generally the brain, the neural activity or the cognitive architecture which make mathematical thought possible. Mathematics, after all, is a human activity, and the only mathematics humans can do is what their brain, neural activity and cognitive architecture enable them to do. Therefore, what mathematics is essentially depends on what the human brain, neural activity or cognitive architectures are.

The idea that mathematics is based only on conceptual thought, and indeed is the purest product of conceptual thought, neglects the fact that the ability to distinguish shape, position and number is not restricted to humankind but is shared by several other forms of animal life. This ability is vital to these forms of life: they could not have survived without it. Mathematics, therefore, is not a feature of human life that distinguishes it from all the rest, but has its roots in certain basic abilities belonging both to humans and to several other forms of animal life, and is part of the process of adapting to the environment.

This brings to an end our examination of the main differences between the view expressed in this book and the dominant view. Not that there are no further differences, but those considered above will suffice to show to what extent the two views differ.

The arguments sketched above provide reasons for rejecting the dominant view. In short, the rejection is motivated by the fact that the dominant view does not explain how mathematical problems arise and are solved. Rather, it presents mathematics as an artificial construction, which does not reflect its important aspects, and omits those features which make mathematics a vital discipline. Thus the dominant view does not account for the richness, multifariousness, dynamism and flexibility of mathematical experience.

---

[78] George-Velleman 2002, p. 2.
[79] *Ibid.*

In showing the limitations of the dominant view, in this book I do not describe it in all its historical and conceptual articulations, which would require considerably more space than is available in a single book. I only present as much as is necessary to show that it is untenable.

For statements of the dominant view, the interested reader may wish to consult, in addition to introductory texts[80], the primary sources, many of which are readily available[81]. As to the different ways in which the reflection on mathematics has been carried out in the history of philosophy, he may wish to consult, in addition to introductory texts[82], the primary sources, from Plato to Mill, most of which are also readily available[83].

Partial challenges to the dominant view have been put forward by Pólya, Lakatos, Hersh and others[84]. The position stated in this book is, however, somewhat more radical, and perhaps more consequential.

For instance, unlike Pólya, I do not claim that "the first rule of discovery is to have brains and good luck", nor that the "the second rule of discovery is to sit tight and wait till you get a bright idea"[85]. Nor do I distinguish between, on the one hand, mathematics in a finished form, viewed as "purely demonstrative, consisting of proofs only", and, on the other hand, "mathematics in the making", which "resembles any other human knowledge in the making"[86]. Moreover, I do not claim that axiomatic reasoning, characteristic of mathematics in finished form, "is safe, beyond controversy, and final", unlike conjectural reasoning, characteristic of mathematics in the making, which "is hazardous, controversial, and provisional"[87]. Nor do I maintain that axiomatic reasoning is for the mathematician "his profession and the distinctive mark of his science"[88]. These views have prevented Pólya from developing a full alternative to the dominant view.

The view expounded in this book is a development of that presented in my earlier publications[89]. The reader might wish to consult them for matters which are not discussed or are discussed only too briefly here.

In this book I do not consider all philosophical questions concerning mathematics, even less all philosophical questions concerning knowledge,

---

80 See, for example, Giaquinto 2002, Shapiro 2000.
81 See Benacerraf-Putnam 1983, Ewald 1996, Hart 1996, Jacquette 2002, Mancosu 1998, van Heijenoort 1977.
82 See, for example, Barbin-Caveing 1996.
83 A basic choice can be found in Baum 1973.
84 See, for example, Tymoczko 1998.
85 Pólya 1948, p. 158.
86 Pólya 1954, I, p. *vi*.
87 *Ibid.*, I, p. *v*.
88 *Ibid.*, I, p. *vi*.
89 See, for example, Cellucci 1998a, 1998b, 2000, 2002b.

because that would require far more space than is available. To my mind, however, the questions discussed here should be dealt with in any investigation concerning the nature of mathematics.

The book consists of a number of short chapters, each of which can be read independently of the others, although its full meaning will emerge only within the context of the whole book. To illustrate my view, I often use fairly simple mathematical examples, which can be presented briefly and do not require elaborate preliminary explanations. Nonetheless, their simplicity does not detract from their exemplarity.

Since my view differs radically from the dominant view, which has exerted its supremacy for so long as to be now mistaken for common sense, I do not expect readers to agree with me immediately. I only ask that you to try and find counterarguments, and carefully assess whether they would stand up to the objections which could be raised against them from the viewpoint of this book.

References

Barbin, Evelyne, an1veing, Maurice (eds.), 1996. *Les philosophes et les mathématiques*, Ellipses, Paris.

Baum, Robert J. (ed.), 1973. *Philosophy and mathematics. From Plato to present*, Freeman, Cooper & Co., San Francisco, Ca.

Benacerraf, Paul, and Putnam, Hilary (eds.), 1983. *Philosophy of mathematics. Selected readings*, Cambridge University Press, Cambridge. [First published 1964].

Beth, Evert Willem, 1957. *La crise de la raison et la logique*, Gauthier-Villars, Paris, and Nauwelaerts, Louvain.

Cellucci, Carlo, 1998a. *Le ragioni della logica*, Laterza, Bari. [Reviewed by Donald Gillies, *Philosophia Mathematica*, vol. 7 (1999), pp. 213-222].

Cellucci, Carlo, 1998b. 'The scope of logic: deduction, abduction, analogy', *Theoria*, vol. 64, pp. 217-242.

Cellucci, Carlo, 2000. 'The growth of mathematical knowledge: an open world view', in Emily R. Grosholz and Herbert Breger (eds.), *The growth of mathematical knowledge*, Kluwer, Dordrecht, pp. 153-176.

Cellucci, Carlo, 2002a. *Filosofia e matematica*, Laterza, Bari. [Reviewed by Donald Gillies in *Philosophia Mathematica*, vol. 11 (2003), pp. 246-253].

Cellucci, Carlo, 2002b. 'La naturalizzazione della logica e della matematica', in Paolo Parrini (ed.), *Conoscenza e cognizione. Tra filosofia e scienza cognitiva*, Guerini, Milan, pp. 21-35.

Dieudonné, Jean, 1948. 'Les methodes axiomatiques modernes et les fondements des mathematiques', in François Le Lionnais (ed.), *Les grands courants de la pensée mathématique*, Blanchard, Paris, pp. 543-555.

Dieudonné, Jean, 1988. *Pour l'honneur de l'esprit humain: les mathématiques aujourd'hui*, Hachette, Paris.

Dummett, Michael, 1991. *Frege. Philosophy of Mathematics*, Duckworth, London.

Dummett, Michael, 1998. 'The philosophy of mathematics', in Anthony C. Grayling (ed.), *Philosophy 2. Further through the subject*, Oxford University Press, Oxford, pp. 122-196.

Dummett, Michael, 2001. *La natura e il futuro della filosofia*, il melangolo, Genova.
Ewald, William (ed.), 1996. *From Kant to Hilbert. A source book in the foundations of mathematics*, Oxford University Press, Oxford.
Feferman, Solomon, 1998. *In the light of logic*, Oxford University Press, Oxford.
Franks, John. 1989a. 'Review of James Gleick, 1987. *Chaos. Making a new science*, Viking Penguin, New York 1987', *The Mathematical Intelligencer*, vol. 11, no. 1, pp. 65-69.
Franks, John, 1989b. 'Comments on the responses to my review of Chaos', *The Mathematical Intelligencer*, vol. 11, no. 3, pp. 12-13.
Frege, Gottlob, 1967. 'Begriffsschrift, a formula language, modeled upon that of arithmetic, for pure thought', in Jean van Heijenoort (ed.), *From Frege to Gödel. A source book in mathematical logic, 1879-1931*, Harvard University Press, Cambridge, Mass., pp. 5-82.
George, Alexander, and Velleman, Daniel J.: 2002, *Philosophies of mathematics*, Blackwell, Oxford.
Giaquinto, Marcus, 2002. *The search for certainty. A philosophical account of foundations of mathematics*, Oxford University Press, Oxford [Reviewed by Carlo Cellucci, *European Journal of Philosophy*, vol. 11 (2003), pp. 420-423].
Girard, Jean-Yves, 1989. 'Le champ du signe ou la faillite du réductionnisme', in Ernest Nagel, James R. Newman, Kurt Gödel and Jean-Yves Girard, *Le théorème de Gödel*, Editions du Seuil, Paris, pp. 147-171.
Giusti, Enrico, 1999. *Ipotesi sulla natura degli oggetti matematici*, Bollati Boringhieri, Torino.
Gödel, Kurt, 1986-. *Collected works*, ed. by Solomon Feferman with the help of John William Dawson, Stephen Cole Kleene, Gregory H. Moore, Robert M. Solovay and Jean van Heijenoort, Oxford University Press, Oxford.
Hadamard, Jacques, 1945. *An essay on the psychology of invention in the mathematical field*, Princeton University Press, Princeton.
Halmos, Paul, 1968. 'Mathematics as a creative art', *American Scientist*, vol. 56, pp. 375-389.
Hamming, Richard Wesley, 1980. 'The unreasonable effectiveness of mathematics', *The American Mathematical Monthly*, vol. 87, pp. 81-90.
Hart, Wilbur Dyre (ed.), 1996. *The philosophy of mathematics*, Oxford University Press, Oxford.
Hersh, Reuben, 1997. *What is mathematics, really?*, Oxford University Press, Oxford.
Hilbert, David, 1996. 'The new grounding of mathematics. First report', in William Ewald (ed.), *From Kant to Hilbert. A source book in the foundations of mathematics*, Oxford University Press, Oxford, pp. 1115-1134.
Jacquette, Dale (ed.), 2002. *Philosophy of mathematics. An anthology*, Blackwell, Oxford.
Jaffe, Arthur, 1997. 'Proof and the evolution of mathematics', *Synthese*, vol. 111, pp. 133-146.
Körner, Stephan, 1960. *The philosophy of mathematics*, Hutchinson, London.
Kreisel, Georg, 1967. 'Mathematical logic: what has it done for the philosophy of mathematics?', in Ralph Schoenman (ed.), *Bertrand Russell, philosopher of the century*, Allen & Unwin, London, pp. 201-272.
Kreisel, Georg, 1984. 'Frege's foundations and intuitionistic logic', *The Monist*, vol. 67, pp. 72-91.
Kreisel, Georg, and MacIntyre, Angus, 1982. 'Constructive logic versus algebraization I', in Anne Sjerp Troelstra and Dirk van Dalen (eds.), *The L.E.J. Brouwer Centenary Colloquium*, North-Holland, Amsterdam, pp. 217-260.

Lang, Serge, 1985. *The beauty of doing mathematics. Three public dialogues*, Springer-Verlag, Berlin.
Leary, Christopher C., 2000. *A friendly introduction to mathematical logic*, Prentice Hall, Upper Saddle River, NJ.
Lehman, Hugh, 1979. *Introduction to the philosophy of mathematics*, Blackwell, Oxford.
Locke, John, 1975. *An essay concerning human understanding*, ed. by Peter H. Nidditch, Oxford University Press, Oxford. [First published 1690].
Lukasiewicz, Jan, 1970. *Selected works*, ed. by Ludwig Borkowski, North-Holland, Amsterdam.
Mac Lane, Saunders, 1981. 'Mathematical models: a sketch for the philosophy of mathematics', *The American Mathematical Monthly*, vol. 88, pp. 462-472.
Maddy, Penelope, 1984. 'Mathematical epistemology: What is the question', *The Monist*, vol. 67, pp. 46-55.
Maddy, Penelope, 1990. *Realism in mathematics*, Oxford University Press, Oxford. [Reviewed by Carlo Cellucci, *Physis*, vol. 34 (1997), pp. 418-426].
Maddy, Penelope, 1997. *Naturalism in mathematics*, Oxford University Press, Oxford.
Mancosu, Paolo (ed.), 1998. *From Brouwer to Hilbert. The debate on the foundations of mathematics in the 1920s*, Oxford University Press, Oxford.
Mayberry, John, 1994. 'What is required of a foundation for mathematics?', *Philosophia Mathematica*, vol. 2, pp. 16-35.
Monk, James Donald, 1976. *Mathematical logic*, Springer-Verlag, Berlin.
Odifreddi, Piergiorgio, 2001. *C'era una volta un paradosso. Storie di illusioni e verità rovesciate*, Einaudi, Torino.
Pascal, Blaise, 1913. 'Discours sur les passions de l'amour', in *Oeuvres Complètes*, vol. 2, Hachette, Paris, pp. 49-57.
Pólya, George, 1948. *How to solve it. A new aspect of mathematical method*, Princeton University Press, Princeton. [First published 1945].
Pólya, George, 1954. *Mathematics and plausible reasoning*, Princeton University Press, Princeton.
Prawitz, Dag, 1998. 'Comments on the papers', *Theoria*, vol. 64, pp. 283-337.
Rota, Gian-Carlo, 1997. *Indiscrete thoughts*, Birkhäuser, Boston.
Russell, Bertrand, 1999. *Our knowledge of the external world*, Routledge, London. [First published 1914].
Shapiro, Stewart, 2000. *Thinking about mathematics. The philosophy of mathematics*, Oxford University Press, Oxford.
Simpson, Stephen George, 1999. *Subsystems of second order arithmetic*, Springer-Verlag, Berlin.
Takeuti, Gaisi, 1987. *Proof Ttheory*, North-Holland, Amsterdam. [First published 1975].
Tymoczko, Thomas, 1998. *New directions in the philosophy of mathematics*, Princeton University Press, Princeton. [First published 1986].
van Heijenoort, Jean (ed.), 1967. *From Frege to Gödel. A source book in mathematical logic, 1879-1931*, Harvard University Press, Cambridge, Mass.
Wagner, Steven, 1982. 'Arithmetical fiction', *Pacific Philosophical Quarterly*, vol. 63, pp. 255-269.
Wittgenstein, Ludwig, 1978. *Remarks on the Foundations of Mathematics*, ed. By Georg Henrik von Wright, Rush Rees and Gertrude Elisabeth Margaret Anscombe, Blackwell, Oxford. [First published 1956].

3

On Proof and Progress in Mathematics

WILLIAM P. THURSTON

This essay on the nature of proof and progress in mathematics was stimulated by the article of Jaffe and Quinn, "Theoretical Mathematics: Toward a cultural synthesis of mathematics and theoretical physics". Their article raises interesting issues that mathematicians should pay more attention to, but it also perpetuates some widely held beliefs and attitudes that need to be questioned and examined.

The article had one paragraph portraying some of my work in a way that diverges from my experience, and it also diverges from the observations of people in the field whom I've discussed it with as a reality check.

After some reflection, it seemed to me that what Jaffe and Quinn wrote was an example of the phenomenon that people see what they are tuned to see. Their portrayal of my work resulted from projecting the sociology of mathematics onto a one-dimensional scale (speculation versus rigor) that ignores many basic phenomena.

Responses to the Jaffe-Quinn article have been invited from a number of mathematicians, and I expect it to receive plenty of specific analysis and criticism from others. Therefore, I will concentrate in this essay on the positive rather than on the contranegative. I will describe my view of the process of mathematics, referring only occasionally to Jaffe and Quinn by way of comparison.

In attempting to peel back layers of assumptions, it is important to try to begin with the right questions:

1. What is it that mathematicians accomplish?

There are many issues buried in this question, which I have tried to phrase in a way that does not presuppose the nature of the answer.

It would not be good to start, for example, with the question

How do mathematicians prove theorems?

This question introduces an interesting topic, but to start with it would be to project two hidden assumptions:

(1) that there is uniform, objective and firmly established theory and practice of mathematical proof, and
(2) that progress made by mathematicians consists of proving theorems.

It is worthwhile to examine these hypotheses, rather than to accept them as obvious and proceed from there.

The question is not even

How do mathematicians make progress in mathematics?

Rather, as a more explicit (and leading) form of the question, I prefer

How do mathematicians advance human understanding of mathematics?

This question brings to the fore something that is fundamental and pervasive: that what we are doing is finding ways for *people* to understand and think about mathematics.

The rapid advance of computers has helped dramatize this point, because computers and people are very different. For instance, when Appel and Haken completed a proof of the 4-color map theorem using a massive automatic computation, it evoked much controversy. I interpret the controversy as having little to do with doubt people had as to the veracity of the theorem or the correctness of the proof. Rather, it reflected a continuing desire for *human understanding* of a proof, in addition to knowledge that the theorem is true.

On a more everyday level, it is common for people first starting to grapple with computers to make large-scale computations of things they might have done on a smaller scale by hand. They might print out a table of the first 10,000 primes, only to find that their printout isn't something they really wanted after all. They discover by this kind of experience that what they really want is usually not some collection of "answers"—what they want is *understanding*.

It may sound almost circular to say that what mathematicians are accomplishing is to advance human understanding of mathematics. I will not try to resolve this by discussing what mathematics is, because it would take us far afield. Mathematicians generally feel that they know what mathematics is, but find it difficult to give a good direct definition. It is interesting to try. For me, "the theory of formal patterns" has come the closest, but to discuss this would be a whole essay in itself.

Could the difficulty in giving a good direct definition of mathematics be an essential one, indicating that mathematics has an essential recursive quality? Along these lines we might say that mathematics is the smallest subject satisfying the following:

- Mathematics includes the natural numbers and plane and solid geometry.
- Mathematics is that which mathematicians study.
- Mathematicians are those humans who advance human understanding of mathematics.

In other words, as mathematics advances, we incorporate it into our thinking. As our thinking becomes more sophisticated, we generate new mathematical concepts and new mathematical structures: the subject matter of mathematics changes to reflect how we think.

If what we are doing is constructing better ways of thinking, then psychological and social dimensions are essential to a good model for mathematical progress. These dimensions are absent from the popular model. In caricature, the popular model holds that

D. mathematicians start from a few basic mathematical structures and a collection of axioms "given" about these structures, that

T. there are various important questions to be answered about these structures that can be stated as formal mathematical propositions, and

P. the task of the mathematician is to seek a deductive pathway from the axioms to the propositions or to their denials.

We might call this the definition-theorem-proof (DTP) model of mathematics.

A clear difficulty with the DTP model is that it doesn't explain the source of the questions. Jaffe and Quinn discuss speculation (which they inappropriately label "theoretical mathematics") as an important additional ingredient. Speculation consists of making conjectures, raising questions, and making intelligent guesses and heuristic arguments about what is probably true.

Jaffe and Quinn's DSTP model still fails to address some basic issues. We are not trying to meet some abstract production quota of definitions, theorems and proofs. The measure of our success is whether what we do enables *people* to understand and think more clearly and effectively about mathematics.

Therefore, we need to ask ourselves:

2. How do people understand mathematics?

This is a very hard question. Understanding is an individual and internal matter that is hard to be fully aware of, hard to understand and often hard to communicate. We can only touch on it lightly here.

People have very different ways of understanding particular pieces of mathematics. To illustrate this, it is best to take an example that practicing mathematicians understand in multiple ways, but that we see our students struggling with. The derivative of a function fits well. The derivative can be thought of as:

(1) Infinitesimal: the ratio of the infinitesimal change in the value of a function to the infinitesimal change in a function.

(2) Symbolic: the derivative of x^n is nx^{n-1}, the derivative of $\sin(x)$ is $\cos(x)$, the derivative of $f \circ g$ is $f' \circ g * g'$, *etc.*

(3) Logical: $f'(x) = d$ if and only if for every ε there is a δ such that when $0 < |\Delta x| < \delta$,

$$\left|\frac{f(x+\Delta x)-f(x)}{\Delta x} - d\right| < \delta.$$

(4) Geometric: the derivative is the slope of a line tangent to the graph of the function, if the graph has a tangent.
(5) Rate: the instantaneous speed of $f(t)$, when t is time.
(6) Approximation: The derivative of a function is the best linear approximation to the function near a point.
(7) Microscopic: The derivative of a function is the limit of what you get by looking at it under a microscope of higher and higher power.

This is a list of different ways of *thinking about* or *conceiving of* the derivative, rather than a list of different *logical definitions*. Unless great efforts are made to maintain the tone and flavor of the original human insights, the differences start to evaporate as soon as the mental concepts are translated into precise, formal and explicit definitions.

I can remember absorbing each of these concepts as something new and interesting, and spending a good deal of mental time and effort digesting and practicing with each, reconciling it with the others. I also remember coming back to revisit these different concepts later with added meaning and understanding.

The list continues; there is no reason for it ever to stop. A sample entry further down the list may help illustrate this. We may think we know all there is to say about a certain subject, but new insights are around the corner. Furthermore, one person's clear mental image is another person's intimidation:

37. The derivative of a real-valued function f in a domain D is the Lagrangian section of the cotangent bundle $T^*(D)$ that gives the connection form for the unique flat connection on the trivial $\mathbf{R}$-bundle $D \times \mathbf{R}$ for which the graph of f is parallel.

These differences are not just a curiosity. Human thinking and understanding do not work on a single track, like a computer with a single central processing unit. Our brains and minds seem to be organized into a variety of separate, powerful facilities. These facilities work together loosely, "talking" to each other at high levels rather than at low levels of organization.

Here are some major divisions that are important for mathematical thinking:

(1) Human language. We have powerful special-purpose facilities for speaking and understanding human language, which also tie in to reading and writing. Our linguistic facility is an important tool for thinking, not just for communication. A crude example is the quadratic formula which people may remember as a little chant, "ex equals minus bee plus or minus the square root of bee squared minus four ay see all over two ay." The

mathematical language of symbols is closely tied to our human language facility. The fragment of mathematical symbolese available to most calculus students has only one verb, "=". That's why students use it when they're in need of a verb. Almost anyone who has taught calculus in the U.S. has seen students instinctively write "$x^3 = 3x^2$" and the like.

(2) Vision, spatial sense, kinesthetic (motion) sense. People have very powerful facilities for taking in information visually or kinesthetically, and thinking with their spatial sense. On the other hand, they do not have a very good built-in facility for inverse vision, that is, turning an internal spatial understanding back into a two-dimensional image. Consequently, mathematicians usually have fewer and poorer figures in their papers and books than in their heads.

An interesting phenomenon in spatial thinking is that scale makes a big difference. We can think about little objects in our hands, or we can think of bigger human-sized structures that we scan, or we can think of spatial structures that encompass us and that we move around in. We tend to think more effectively with spatial imagery on a larger scale: it's as if our brains take larger things more seriously and can devote more resources to them.

(3) Logic and deduction. We have some built-in ways of reasoning and putting things together associated with how we make logical deductions: cause and effect (related to implication), contradiction or negation, *etc.*

Mathematicians apparently don't generally rely on the formal rules of deduction as they are thinking. Rather, they hold a fair bit of logical structure of a proof in their heads, breaking proofs into intermediate results so that they don't have to hold too much logic at once. In fact, it is common for excellent mathematicians not even to know the standard formal usage of quantifiers (for all and there exists), yet all mathematicians certainly perform the reasoning that they encode.

It's interesting that although "or", "and" and "implies" have identical formal usage, we think of "or" and "and" as conjunctions and "implies" as a verb.

(4) Intuition, association, metaphor. People have amazing facilities for sensing something without knowing where it comes from (intuition); for sensing that some phenomenon or situation or object is like something else (association); and for building and testing connections and comparisons, holding two things in mind at the same time (metaphor). These facilities are quite important for mathematics. Personally, I put a lot of effort into "listening" to my intuitions and associations, and building them into metaphors and connections. This involves a kind of simultaneous quieting and focusing of my mind. Words, logic, and detailed pictures rattling around can inhibit intuitions and associations.

(5) Stimulus-response. This is often emphasized in schools; for instance, if you see 3927×253, you write one number above the other and draw a

line underneath, *etc*. This is also important for research mathematics: seeing a diagram of a knot, I might write down a presentation for the fundamental group of its complement by a procedure that is similar in feel to the multiplication algorithm.

(6) Process and time. We have a facility for thinking about processes or sequences of actions that can often be used to good effect in mathematical reasoning. One way to think of a function is as an action, a process, that takes the domain to the range. This is particularly valuable when composing functions. Another use of this facility is in remembering proofs: people often remember a proof as a process consisting of several steps. In topology, the notion of a homotopy is most often thought of as a process taking time. Mathematically, time is no different from one more spatial dimension, but since humans interact with it in a quite different way, it is psychologically very different.

3. How is mathematical understanding communicated?

The transfer of understanding from one person to another is not automatic. It is hard and tricky. Therefore, to analyze human understanding of mathematics, it is important to consider **who** understands **what**, and **when**.

Mathematicians have developed habits of communication that are often dysfunctional. Organizers of colloquium talks everywhere exhort speakers to explain things in elementary terms. Nonetheless, most of the audience at an average colloquium talk gets little of value from it. Perhaps they are lost within the first 5 minutes, yet sit silently through the remaining 55 minutes. Or perhaps they quickly lose interest because the speaker plunges into technical details without presenting any reason to investigate them. At the end of the talk, the few mathematicians who are close to the field of the speaker ask a question or two to avoid embarrassment.

This pattern is similar to what often holds in classrooms, where we go through the motions of saying for the record what we think the students "ought" to learn, while the students are trying to grapple with the more fundamental issues of learning our language and guessing at our mental models. Books compensate by giving samples of how to solve every type of homework problem. Professors compensate by giving homework and tests that are much easier than the material "covered" in the course, and then grading the homework and tests on a scale that requires little understanding. We assume that the problem is with the students rather than with communication: that the students either just don't have what it takes, or else just don't care.

Outsiders are amazed at this phenomenon, but within the mathematical community, we dismiss it with shrugs.

Much of the difficulty has to do with the language and culture of mathematics, which is divided into subfields. Basic concepts used every day within one subfield are often foreign to another subfield. Mathematicians give up on trying to understand the basic concepts even from neighboring subfields, unless they were clued in as graduate students.

In contrast, communication works very well within the subfields of mathematics. Within a subfield, people develop a body of common knowledge and known techniques. By informal contact, people learn to understand and copy each other's ways of thinking, so that ideas can be explained clearly and easily.

Mathematical knowledge can be transmitted amazingly fast within a subfield. When a significant theorem is proved, it often (but not always) happens that the solution can be communicated in a matter of minutes from one person to another within the subfield. The same proof would be communicated and generally understood in an hour talk to members of the subfield. It would be the subject of a 15- or 20-page paper, which could be read and understood in a few hours or perhaps days by members of the subfield.

Why is there such a big expansion from the informal discussion to the talk to the paper? One-on-one, people use wide channels of communication that go far beyond formal mathematical language. They use gestures, they draw pictures and diagrams, they make sound effects and use body language. Communication is more likely to be two-way, so that people can concentrate on what needs the most attention. With these channels of communication, they are in a much better position to convey what's going on, not just in their logical and linguistic facilities, but in their other mental facilities as well.

In talks, people are more inhibited and more formal. Mathematical audiences are often not very good at asking the questions that are on most people's minds, and speakers often have an unrealistic preset outline that inhibits them from addressing questions even when they are asked.

In papers, people are still more formal. Writers translate their ideas into symbols and logic, and readers try to translate back.

Why is there such a discrepancy between communication within a subfield and communication outside of subfields, not to mention communication outside mathematics?

Mathematics in some sense has a common language: a language of symbols, technical definitions, computations, and logic. This language efficiently conveys some, but not all, modes of mathematical thinking. Mathematicians learn to translate certain things almost unconsciously from one mental mode to the other, so that some statements quickly become clear. Different mathematicians study papers in different ways, but when I read a mathematical paper in a field in which I'm conversant, I concentrate on the thoughts that are between the lines. I might look over several paragraphs or strings of equations and think to myself "Oh yeah, they're putting in enough rigamarole to carry such-and-such idea." When the idea is clear, the formal setup is usually unnecessary and redundant—I often feel that I could write it out myself more easily than figuring out what the authors actually wrote. It's like a new toaster that comes with a 16-page manual. If you already understand toasters and if the toaster looks like previous toasters you've encountered, you might just plug it in and see if it works, rather than first reading all the details in the manual.

People familiar with ways of doing things in a subfield recognize various patterns of statements or formulas as idioms or circumlocution for certain concepts or mental images. But to people not already familiar with what's going on the same patterns are not very illuminating; they are often even misleading. The language is not alive except to those who use it.

I'd like to make an important remark here: there are some mathematicians who are conversant with the ways of thinking in more than one subfield, sometimes in quite a number of subfields. Some mathematicians learn the jargon of several subfields as graduate students, some people are just quick at picking up foreign mathematical language and culture, and some people are in mathematical centers where they are exposed to many subfields. People who are comfortable in more than one subfield can often have a very positive influence, serving as bridges, and helping different groups of mathematicians learn from each other. But people knowledgeable in multiple fields can also have a negative effect, by intimidating others, and by helping to validate and maintain the whole system of generally poor communication. For example, one effect often takes place during colloquium talks, where one or two widely knowledgeable people sitting in the front row may serve as the speaker's mental guide to the audience.

There is another effect caused by the big differences between how we think about mathematics and how we write it. A group of mathematicians interacting with each other can keep a collection of mathematical ideas alive for a period of years, even though the recorded version of their mathematical work differs from their actual thinking, having much greater emphasis on language, symbols, logic and formalism. But as new batches of mathematicians learn about the subject they tend to interpret what they read and hear more literally, so that the more easily recorded and communicated formalism and machinery tend to gradually take over from other modes of thinking.

There are two counters to this trend, so that mathematics does not become entirely mired down in formalism. First, younger generations of mathematicians are continually discovering and rediscovering insights on their own, thus reinjecting diverse modes of human thought into mathematics.

Second, mathematicians sometimes invent names and hit on unifying definitions that replace technical circumlocutions and give good handles for insights. Names like "group" to replace "a system of substitutions satisfying ...", and "manifold" to replace

> We can't give coordinates to parametrize all the solutions to our equations simultaneously, but in the neighborhood of any particular solution we can introduce coordinates

$(f_1(u_1, u_2, u_3), f_2(u_1, u_2, u_3), f_3(u_1, u_2, u_3), f_4(u_1, u_2, u_3), f_5(u_1, u_2, u_3))$

where at least one of the ten determinants

...[ten 3 × 3 determinants of matrices of partial derivatives]...

is not zero

may or may not have represented advances in insight among experts, but they greatly facilitate the *communication* of insights.

We mathematicians need to put far greater effort into communicating mathematical *ideas*. To accomplish this, we need to pay much more attention to communicating not just our definitions, theorems, and proofs, but also our ways of thinking. We need to appreciate the value of different ways of thinking about the same mathematical structure.

We need to focus far more energy on understanding and explaining the basic mental infrastructure of mathematics—with consequently less energy on the most recent results. This entails developing mathematical language that is effective for the radical purpose of conveying ideas to people who don't already know them.

Part of this communication is through proofs.

4. What is a proof?

When I started as a graduate student at Berkeley, I had trouble imagining how I could "prove" a new and interesting mathematical theorem. I didn't really understand what a "proof" was.

By going to seminars, reading papers, and talking to other graduate students, I gradually began to catch on. Within any field, there are certain theorems and certain techniques that are generally known and generally accepted. When you write a paper, you refer to these without proof. You look at other papers in the field, and you see what facts they quote without proof, and what they cite in their bibliography. You learn from other people some idea of the proofs. Then you're free to quote the same theorem and cite the same citations. You don't necessarily have to read the full papers or books that are in your bibliography. Many of the things that are generally known are things for which there may be no known written source. As long as people in the field are comfortable that the idea works, it doesn't need to have a formal written source.

At first I was highly suspicious of this process. I would doubt whether a certain idea was really established. But I found that I could ask people, and they could produce explanations and proofs, or else refer me to other people or to written sources that would give explanations and proofs. There were published theorems that were generally known to be false, or where the proofs were generally known to be incomplete. Mathematical knowledge and understanding were embedded in the minds and in the social fabric of the community of people thinking about a particular topic. This knowledge was supported by written documents, but the written documents were not really primary.

I think this pattern varies quite a bit from field to field. I was interested in geometric areas of mathematics, where it is often pretty hard to have a document that reflects well the way people actually think. In more algebraic or symbolic fields, this is not necessarily so, and I have the impression that in some areas documents are much closer to carrying the life of the field. But in any field, there is a strong social standard of validity and truth. Andrew Wiles's proof of Fermat's Last Theorem is a good illustration of this, in a field which is very algebraic. The experts quickly came to believe that his proof was basically correct on the basis of high-level ideas, long before details could be checked. This proof will receive a great deal of scrutiny and checking compared to most mathematical proofs; but no matter how the process of verification plays out, it helps illustrate how mathematics evolves by rather organic psychological and social processes.

When people are doing mathematics, the flow of ideas and the social standard of validity is much more reliable than formal documents. People are usually not very good in checking *formal correctness* of proofs, but they are quite good at detecting potential weaknesses or flaws in proofs.

To avoid misinterpretation, I'd like to emphasize two things I am *not* saying. First, I am *not* advocating any weakening of our community standard of proof; I am trying to describe how the process really works. Careful proofs that will stand up to scrutiny are very important. I think the process of proof on the whole works pretty well in the mathematical community. The kind of change I would advocate is that mathematicians take more care with their proofs, making them really clear and as simple as possible so that if any weakness is present it will be easy to detect. Second, I am *not* criticizing the mathematical study of formal proofs, nor am I criticizing people who put energy into making mathematical arguments more explicit and more formal. These are both useful activities that shed new insights on mathematics.

I have spent a fair amount of effort during periods of my career exploring mathematical questions by computer. In view of that experience, I was astonished to see the statement of Jaffe and Quinn that mathematics is extremely slow and arduous, and that it is arguably the most disciplined of all human activities. The standard of correctness and completeness necessary to get a computer program to work at all is a couple of orders of magnitude higher than the mathematical community's standard of valid proofs. Nonetheless, large computer programs, even when they have been very carefully written and very carefully tested, always seem to have bugs.

I think that mathematics is one of the most intellectually gratifying of human activities. Because we have a high standard for clear and convincing thinking and because we place a high value on listening to and trying to understand each other, we don't engage in interminable arguments and endless redoing of our mathematics. We are prepared to be convinced by others. Intellectually, mathematics moves very quickly. Entire mathematical landscapes change and change again in amazing ways during a single career.

When one considers how hard it is to write a computer program even approaching the intellectual scope of a good mathematical paper, and how much greater time and effort have to be put into it to make it "almost" formally correct, it is preposterous to claim that mathematics as we practice it is anywhere near formally correct.

Mathematics as we practice it is much more formally complete and precise than other sciences, but it is much less formally complete and precise for its content than computer programs. The difference has to do not just with the amount of effort: the kind of effort is qualitatively different. In large computer programs, a tremendous proportion of effort must be spent on myriad compatibility issues: making sure that all definitions are consistent, developing "good" data structures that have useful but not cumbersome generality, deciding on the "right" generality for functions, *etc.* The proportion of energy spent on the working part of a large program, as distinguished from the bookkeeping part, is surprisingly small. Because of compatibility issues that almost inevitably escalate out of hand because the "right" definitions change as generality and functionality are added, computer programs usually need to be rewritten frequently, often from scratch.

A very similar kind of effort would have to go into mathematics to make it formally correct and complete. It is not that formal correctness is prohibitively difficult on a small scale—it's that there are many possible choices of formalization on small scales that translate to huge numbers of interdependent choices in the large. It is quite hard to make these choices compatible; to do so would certainly entail going back and rewriting from scratch all old mathematical papers whose results we depend on. It is also quite hard to come up with good technical choices for formal definitions that will be valid in the variety of ways that mathematicians want to use them and that will anticipate future extensions of mathematics. If we were to continue to cooperate, much of our time would be spent with international standards commissions to establish uniform definitions and resolve huge controversies.

Mathematicians can and do fill in gaps, correct errors, and supply more detail and more careful scholarship when they are called on or motivated to do so. Our system is quite good at producing reliable theorems that can be solidly backed up. It's just that the reliability does not primarily come from mathematicians formally checking formal arguments; it comes from mathematicians thinking carefully and critically about mathematical ideas.

On the most fundamental level, the foundations of mathematics are much shakier than the mathematics that we do. Most mathematicians adhere to foundational principles that are known to be polite fictions. For example, it is a theorem that there does not exist any way to ever actually construct or even define a well-ordering of the real numbers. There is considerable evidence (but no proof) that we can get away with these polite fictions without being caught out, but that doesn't make them right. Set theorists construct many alternate and mutually contradictory "mathematical universes" such that if one is consistent, the others are too. This leaves very little confidence that one or the other is the right choice or the natural choice. Gödel's incompleteness theorem implies that there can be no formal system that is consistent, yet powerful enough to serve as a basis for all of the mathematics that we do.

In contrast to humans, computers are good at performing formal processes. There are people working hard on the project of actually formalizing parts of mathematics by computer, with actual formally correct formal deductions. I think this is a very big but very worthwhile project, and I am confident that we will learn a lot from it. The process will help simplify and clarify mathematics. In not too many years, I expect that we will have interactive computer programs that can help people compile significant chunks of formally complete and correct mathematics (based on a few perhaps shaky but at least explicit assumptions), and that they will become part of the standard mathematician's working environment.

However, we should recognize that the humanly understandable and humanly checkable proofs that we actually do are what is most important to us, and that they are quite different from formal proofs. For the present, formal proofs are out of reach and mostly irrelevant: we have good human processes for checking mathematical validity.

5. What motivates people to do mathematics?

There is a real joy in doing mathematics, in learning ways of thinking that explain and organize and simplify. One can feel this joy discovering new mathematics, rediscovering old mathematics, learning a way of thinking from a person or text, or finding a new way to explain or to view an old mathematical structure.

This inner motivation might lead us to think that we do mathematics solely for its own sake. That's not true: the social setting is extremely important. We are inspired by other people, we seek appreciation by other people, and we like to help other people solve their mathematical problems. What we enjoy changes in response to other people. Social interaction occurs through face-to-face meetings. It also occurs through written and electronic correspondence, preprints, and journal articles. One effect of this highly social system of mathematics is the tendency of mathematicians to follow fads. For the purpose of producing new mathematical theorems this is probably not very efficient: we'd seem to be better off having mathematicians cover the intellectual field much more evenly. But most mathematicians don't like to be lonely, and they have trouble staying excited about a subject, even if they are personally making progress, unless they have colleagues who share their excitement.

In addition to our inner motivation and our informal social motivation for doing mathematics, we are driven by considerations of economics and status. Mathematicians, like other academics, do a lot of judging and being judged. Starting with grades, and continuing through letters of recommendation, hiring decisions, promotion decisions, referees reports, invitations to speak, prizes, ... we are involved in many ratings, in a fiercely competitive system.

Jaffe and Quinn analyze the motivation to do mathematics in terms of a common currency that many mathematicians believe in: credit for theorems.

I think that our strong communal emphasis on theorem-credits has a negative effect on mathematical progress. If what we are accomplishing is advancing human understanding of mathematics, then we would be much better off recognizing and valuing a far broader range of activity. The people who see the way to proving theorems are doing it in the context of a mathematical community; they are not doing it on their own. They depend on understanding of mathematics that they glean from other mathematicians. Once a theorem has been proven, the mathematical community depends on the social network to distribute the ideas to people who might use them further—the print medium is far too obscure and cumbersome.

Even if one takes the narrow view that what we are producing is theorems, the team is important. Soccer can serve as a metaphor. There might only be one or two goals during a soccer game, made by one or two persons. That does not mean that the efforts of all the others are wasted. We do not judge players on a soccer team only by whether they personally make a goal; we judge the team by its function as a team.

In mathematics, it often happens that a group of mathematicians advances with a certain collection of ideas. There are theorems in the path of these advances that will almost inevitably be proven by one person or another. Sometimes the group of mathematicians can even anticipate what these theorems are likely to be. It is much harder to predict who will actually prove the theorem, although there are usually a few "point people" who are more likely to score. However, they are in a position to prove those theorems because of the collective efforts of the team. The team has a further function, in absorbing and making use of the theorems once they are proven. Even if one person could prove all the theorems in the path single-handedly, they are wasted if nobody else learns them.

There is an interesting phenomenon concerning the "point" people. It regularly happens that someone who was in the middle of a pack proves a theorem that receives wide recognition as being significant. Their status in the community—their pecking order—rises immediately and dramatically. When this happens, they usually become much more productive as a center of ideas and a source of theorems. Why? First, there is a large increase in self-esteem, and an accompanying increase in productivity. Second, when their status increases, people are more in the center of the network of ideas—others take them more seriously. Finally and perhaps most importantly, a mathematical breakthrough usually represents a new way of thinking, and effective ways of thinking can usually be applied in more than one situation.

This phenomenon convinces me that the entire mathematical community would become much more productive if we open our eyes to the real values in what we are doing. Jaffe and Quinn propose a system of recognized roles divided into "speculation" and "proving". Such a division only perpetuates the myth that our progress is measured in units of standard theorems deduced. This is a bit like the fallacy of the person who makes a printout of the first 10,000 primes. What we are producing is human understanding. We have many different ways to understand and many different processes that contribute to our understanding. We will be more satisfied, more productive and happier if we recognize and focus on this.

6. Some personal experiences

Since this essay grew out of reflection on the misfit between my experiences and the description of Jaffe and Quinn's, I will discuss two personal experiences, including the one they alluded to.

I feel some awkwardness in this, because I do have regrets about aspects of my career: if I were to do things over again with the benefit of my present insights about myself and about the process of mathematics, there is a lot that I would hope to do differently. I hope that by describing these experiences rather openly as I remember and understand them, I can help others understand the process better and learn in advance.

First I will discuss briefly the theory of foliations, which was my first subject, starting when I was a graduate student. (It doesn't matter here whether you know what foliations are.)

At that time, foliations had become a big center of attention among geometric topologists, dynamical systems people, and differential geometers. I fairly rapidly proved some dramatic theorems. I proved a classification theorem for foliations, giving a necessary and sufficient condition for a manifold to admit a foliation. I proved a number of other significant theorems. I wrote respectable papers and published at least the most important theorems. It was hard to find the time to write to keep up with what I could prove, and I built up a backlog.

An interesting phenomenon occurred. Within a couple of years, a dramatic evacuation of the field started to take place. I heard from a number of mathematicians that they were giving or receiving advice not to go into foliations—they were saying that Thurston was cleaning it out. People told me (not as a complaint, but as a compliment) that I was killing the field. Graduate students stopped studying foliations, and fairly soon, I turned to other interests as well.

I do not think that the evacuation occurred because the territory was intellectually exhausted—there were (and still are) many interesting questions that remain and that are probably approachable. Since those years, there have been interesting developments carried out by the few people who stayed in the field or who entered the field, and there have also been important developments in neighboring areas that I think would have been much accelerated had mathematicians continued to pursue foliation theory vigorously.

Today, I think there are few mathematicians who understand anything approaching the state of the art of foliations as it lived at that time, although there are some parts of the theory of foliations, including developments since that time, that are still thriving.

I believe that two ecological effects were much more important in putting a damper on the subject than any exhaustion of intellectual resources that occurred.

First, the results I proved (as well as some important results of other people) were documented in a conventional, formidable mathematician's style. They depended heavily on readers who shared certain background and certain insights. The theory of foliations was a young, opportunistic subfield, and the background was not standardized. I did not hesitate to draw on any of the mathematics I had learned from others. The papers I wrote did not (and could not) spend much time explaining the background culture. They documented top-level reasoning and conclusions that I often had achieved after much reflection and effort. I also threw out prize cryptic tidbits of insight, such as "the Godbillon-Vey invariant measures the helical wobble of a foliation", that remained mysterious to most mathematicans who read them. This created a high entry barrier: I think many graduate students and mathematicians were discouraged that it was hard to learn and understand the proofs of key theorems.

Second is the issue of what is in it for other people in the subfield. When I started working on foliations, I had the conception that what people wanted was to know the answers. I thought that what they sought was a collection of powerful proven theorems that might be applied to answer further mathematical questions. But that's only one part of the story. More than the knowledge, people want *personal understanding*. And in our credit-driven system, they also want and need *theorem-credits.*

I'll skip ahead a few years, to the subject that Jaffe and Quinn alluded to, when I began studying 3-dimensional manifolds and their relationship to hyperbolic geometry. (Again, it matters little if you know what this is about.) I gradually built up over a number of years a certain intuition for hyperbolic three-manifolds, with a repertoire of constructions, examples and proofs. (This process actually started when I was an undergraduate, and was strongly bolstered by applications to foliations.) After a while, I conjectured or speculated that all three-manifolds have a certain geometric structure; this conjecture eventually became known as the geometrization conjecture. About two or three years later, I proved the geometrization theorem for Haken manifolds. It was a hard theorem, and I spent a tremendous amount of effort thinking about it. When I completed the proof, I spent a lot more effort checking the proof, searching for difficulties and testing it against independent information.

I'd like to spell out more what I mean when I say I proved this theorem. It meant that I had a clear and complete flow of ideas, including details, that withstood a great deal of scrutiny by myself and by others. Mathematicians have many different styles of thought. My style is not one of making broad sweeping but careless generalities, which are merely hints or inspirations: I make clear mental models, and I think things through. My proofs have turned out to be quite reliable. I have not had trouble backing up claims or producing details for things I have proven. I am good in detecting flaws in my own reasoning as well as in the reasoning of others.

However, there is sometimes a huge expansion factor in translating from the encoding in my own thinking to something that can be conveyed to someone else. My mathematical education was rather independent and idiosyncratic, where for a number of years I learned things on my own, developing personal mental models for how to think about mathematics. This has often been a big advantage for me in thinking about mathematics, because it's easy to pick up later the standard mental models shared by groups of mathematicians. This means that some concepts that I use freely and naturally in my personal thinking are foreign to most mathematicians I talk to. My personal mental models and structures are similar in character to the kinds of models groups of mathematicians share—but they are often different models. At the time of the formulation of the geometrization conjecture, my understanding of hyperbolic geometry was a good example. A random continuing example is an understanding of finite topological spaces, an oddball topic that can lend good insight to a variety of questions but that is generally not worth developing in any one case because there are standard circumlocutions that avoid it.

Neither the geometrization conjecture nor its proof for Haken manifolds was in the path of any group of mathematicians at the time—it went against the trends in topology for the preceding 30 years, and it took people by surprise. To most topologists at the time, hyperbolic geometry was an arcane side branch of mathematics, although there were other groups of mathematicians such as differential geometers who did understand it from certain points of view. It took topologists a while just to understand what the geometrization conjecture meant, what it was good for, and why it was relevant.

At the same time, I started writing notes on the geometry and topology of 3-manifolds, in conjunction with the graduate course I was teaching. I distributed them to a few people, and before long many others from around the world were writing for copies. The mailing list grew to about 1200 people to whom I was sending notes every couple of months. I tried to communicate my real thoughts in these notes. People ran many seminars based on my notes, and I got lots of feedback. Overwhelmingly, the feedback ran something like "Your notes are really inspiring and beautiful, but I have to tell you that we spent 3 weeks in our seminar working out the details of §*n.n*. More explanation would sure help."

I also gave many presentations to groups of mathematicians about the ideas of studying 3-manifolds from the point of view of geometry, and about the proof of the geometrization conjecture for Haken manifolds. At the beginning, this subject was foreign to almost everyone. It was hard to communicate—the infrastructure was in my head, not in the mathematical community. There were several mathematical theories that fed into the cluster of ideas: three-manifold topology, Kleinian groups, dynamical systems, geometric topology, discrete subgroups of Lie groups, foliations, Teichmüller spaces, pseudo-Anosov diffeomorphisms, geometric group theory, as well as hyperbolic geometry.

We held an AMS summer workshop at Bowdoin in 1980, where many mathematicans in the subfields of low-dimensional topology, dynamical systems and Kleinian groups came.

It was an interesting experience exchanging cultures. It became dramatically clear how much proofs depend on the audience. We prove things in a social context and address them to a certain audience. Parts of this proof I could communicate in two minutes to the topologists, but the analysts would need an hour lecture before they would begin to understand it. Similarly, there were some things that could be said in two minutes to the analysts that would take an hour before the topologists would begin to get it. And there were many other parts of the proof which should take two minutes in the abstract, but that none of the audience at the time had the mental infrastructure to get in less than an hour.

At that time, there was practically no infrastructure and practically no context for this theorem, so the expansion from how an idea was keyed in my head to what I had to say to get it across, not to mention how much energy the audience had to devote to understand it, was very dramatic.

In reaction to my experience with foliations and in response to social pressures, I concentrated most of my attention on developing and presenting the infrastructure in what I wrote and in what I talked to people about. I explained the details to the few people who were "up" for it. I wrote some papers giving the substantive parts of the proof of the geometrization theorem for Haken manifolds—for these papers, I got almost no feedback. Similarly, few people actually worked through the harder and deeper sections of my notes until much later.

The result has been that now quite a number of mathematicians have what was dramatically lacking in the beginning: a working understanding of the concepts and the infrastructure that are natural for this subject. There has been and there continues to be a great deal of thriving mathematical activity. By concentrating on building the infrastructure and explaining and publishing definitions and ways of thinking but being slow in stating or in publishing proofs of all the "theorems" I knew how to prove, I left room for many other people to pick up credit. There has been room for people to discover and publish other proofs of the geometrization theorem. These proofs helped develop mathematical concepts which are quite interesting in themselves, and lead to further mathematics.

What mathematicians most wanted and needed from me was to learn my ways of thinking, and not in fact to learn my proof of the geometrization conjecture for Haken manifolds. It is unlikely that the proof of the general geometrization conjecture will consist of pushing the same proof further.

A further issue is that people sometimes need or want an accepted and validated result not in order to learn it, but so that they can quote it and rely on it.

Mathematicians were actually very quick to accept my proof, and to start quoting it and using it based on what documentation there was, based on their experience and belief in me, and based on acceptance by opinions of experts with whom I spent a lot of time communicating the proof. The theorem now is documented, through published sources authored by me and by others, so most people feel secure in quoting it; people in the field certainly have not challenged me about its validity, or expressed to me a need for details that are not available.

Not all proofs have an identical role in the logical scaffolding we are building for mathematics. This particular proof probably has only temporary logical value, although it has a high motivational value in helping support a certain vision for the structure of 3-manifolds. The full geometrization conjecture is still a conjecture. It has been proven for many cases, and is supported by a great deal of computer evidence as well, but it has not been proven in generality. I am convinced that the general proof will be discovered; I hope before too many more years. At that point, proofs of special cases are likely to become obsolete.

Meanwhile, people who want to use the geometric technology are better off to start off with the assumption "Let M^3 be a manifold that admits a geometric decomposition," since this is more general than "Let M^3 be a Haken manifold." People who don't want to use the technology or who are suspicious of it can avoid it. Even when a theorem about Haken manifolds can be proven using geometric techniques, there is a high value in finding purely topological techniques to prove it.

In this episode (which still continues) I think I have managed to avoid the two worst possible outcomes: either for me not to let on that I discovered what I discovered and proved what I proved, keeping it to myself (perhaps with the hope of proving the Poincaré conjecture), or for me to present an unassailable and hard-to-learn theory with no practitioners to keep it alive and to make it grow.

I can easily name regrets about my career. I have not published as much as I should. There are a number of mathematical projects in addition to the geometrization theorem for Haken manifolds that I have not delivered well or at all to the mathematical public. When I concentrated more on developing the infrastructure rather than the top-level theorems in the geometric theory of 3-manifolds, I became somewhat disengaged as the subject continued to evolve; and I have not actively or effectively promoted the field or the careers of the excellent people in it. (But some degree of disengagement seems to me an almost inevitable by-product of the mentoring of graduate students and others: in order to really turn genuine research directions over to others, it's necessary to really let go and stop oneself from thinking about them very hard.)

On the other hand, I have been busy and productive, in many different activities. Our system does not create extra time for people like me to spend on writing and research; instead, it inundates us with many requests and opportunities for extra work, and my gut reaction has been to say 'yes' to many of these requests and opportunities. I have put a lot of effort into non-credit-producing activities that I value just as I value proving theorems: mathematical politics, revision of my notes into a book with a high standard of communication, exploration of computing in mathematics, mathematical education, development of new forms for communication of mathematics through the Geometry Center (such as our first experiment, the "Not Knot" video), directing MSRI, *etc.*

I think that what I have done has not maximized my "credits". I have been in a position not to feel a strong need to compete for more credits. Indeed, I began to feel strong challenges from other things besides proving new theorems.

I do think that my actions have done well in stimulating mathematics.

4

The Informal Logic of Mathematical Proof

ANDREW ABERDEIN

The proof of mathematical theorems is central to mathematical practice and to much recent debate about the nature of mathematics: as Paul Erdös once remarked, 'a mathematician is a machine for turning coffee into theorems' [9, p. 7]. This paper is an attempt to introduce a new perspective on the argumentation characteristic of mathematical proof. I shall argue that this account, an application of informal logic to mathematics, helps to clarify and resolve several important philosophical difficulties.

It might be objected that formal, deductive logic tells us everything we need to know about mathematical argumentation. I shall leave it to others [14, for example] to address this concern in detail. However, even the protagonists of explicit reductionist programmes—such as logicists in the philosophy of mathematics and the formal theorem proving community in computer science—would readily concede that their work is not an attempt to capture actual mathematical practice. Having said that, mathematical argumentation is certainly not inductive either. Mathematical proofs do not involve inference from particular observations to general laws. A satisfactory account of mathematical argumentation must include deductive inference, even if it is not exhausted by it. It must be complementary, rather than hostile, to formal logic. My contention is that a suitable candidate has already been developed independently: informal logic.

Informal logic is concerned with all aspects of inference, including those which cannot be captured by logical form. It is an ancient subject, but has been a degenerating research programme for a long time. Since the nineteenth century it has been overshadowed by the growth of formal logic. More fundamentally, it has suffered by identification with the simplistic enumeration of fallacies, without any indication of the circumstances in which they are illegitimate. Since most fallacies can be exemplified in some contexts by persuasive, indeed valid, arguments, this approach is of limited use. In recent decades more interesting theories have been developed. I shall look at two of the most influential, and discuss their usefulness for the analysis of mathematical proof.

1. Toulmin's pattern of argument

One of the first modern accounts of argumentation is that developed in Stephen Toulmin's *The Uses of Argument* [18]. Toulmin offers a general account of the layout of an argument, as a claim (C) derived from data (D), in respect of a warrant (W). Warrants are general hypothetical statements of the form 'Given D, one may take it that C' [18, p. 99]. Hence the laws of logic provide a warrant for deductive inferences. However, the pattern is intended to be more general, and provides for different, weaker warrants, although these would not permit us to ascribe the same degree of certainty to C. This is recognized by the inclusion of a modal qualifier (Q), such as 'necessarily', 'probably', 'presumably',..., in the pattern. If the warrant is defeasible, we may also specify the conditions (R) under which it may be rebutted. Finally, the argument may turn on the backing (B) which can be provided for W. Toulmin's claim is that the general structure of a disparate variety of arguments may be represented as in Figure 1.

Interpreting the letters as above, this diagram may be read as follows: "Given D, we can (modulo Q) claim C, since W (on account of B), unless R". In Toulmin's vintage example: "Given that Harry was born in Bermuda, we can presumably claim that he is British, since anyone born in Bermuda will generally be British (on account of various statutes ...), unless he's a naturalized American, or his parents were aliens, or ...". In simpler examples B, Q and R may not all be present, but D, W and C are taken to be essential to any argument, hence the description of this model as the DWC pattern. Toulmin stresses the field dependency of the canons of good argument: what counts as convincing may vary substantially between the law court, the laboratory and the debating chamber. In particular, what counts as acceptable backing will turn significantly on the field in which the argument is conducted [18, p. 104].

Toulmin's work has been very influential in the study of argument, despite an initially chilly reception amongst philosophers and logicians.[1] It was quickly adopted by communication theorists, after the publication in 1960 of a celebrated paper by Wayne Brockriede and Douglas Ehninger [4]. Toulmin's account of argumentation is now the dominant model in this field. More recently, his work has been widely studied by computer scientists attempting to model natural argumentation [13, for example]. One purpose of Toulmin's critique of logic is to dispute the utility of formal logic for the analysis of any argumentational discourse *other* than mathematics: he regards mathematical proof as one of the few success stories for the formal logic tradition. Nevertheless, mathematical proofs can be subsumed under

---

[1] 'Unanimous ... condemnation' according to van Eemeren & *al.* [7, p. 164], who survey the book's reviewers.

the DWC pattern, where the warrant is backed by various axioms, rules of inference and mathematical techniques providing grounds for supposing the claim to be necessary, given the data. Toulmin provides an example in a later collaborative work [19, p. 89], by reconstructing Theaetetus's proof that there are exactly five polyhedra. The data and warrant consist of various facts about the platonic solids, the warrant is backed by the axioms, postulates and definitions of three-dimensional Euclidean geometry, and the modal qualifier 'with strict geometrical necessity' admits of no rebuttal or exception within the bounds of Euclidean geometry.

2 Applying Toulmin to mathematics

The significance of Toulmin's work for mathematical proof is explored at greater length in what I believe to be the only study so far of the application of informal logic to mathematics, a paper written in Catalan by Jesús Alcolea Banegas [1].[2] Alcolea makes use of a further distinction of Toulmin's, introduced in [19]: that between *regular* and *critical* arguments. This distinction echoes Thomas Kuhn's contrast between normal and revolutionary science: a regular argument is an argument within a field which appeals to the already well-established warrants characteristic of the field, whereas a critical argument is an argument used to challenge prevailing ideas, focusing attention on the assumptions which provide a backing for the warrants of regular arguments. Critical arguments must therefore appeal to different warrants. Mathematical proofs are regular arguments, although they may give rise to critical arguments if they are especially interesting or controversial. Conversely, metamathematical debates are critical arguments, but they often provide new opportunities for proofs, that is, regular arguments.

Alcolea uses Toulmin's layout to reconstruct one regular and one critical argument from mathematics. The critical argument, the debate over the admissibility of the axiom of choice, is the more fully developed and persuasive of Alcolea's case studies. It is perhaps not too surprising that critical arguments in mathematics are similar to critical arguments in the other sciences, since ultimately they are not arguments *in* mathematics, but arguments *about* mathematics, that is to say they are metamathematical. However, my concern is primarily with the argumentation of mathematics itself, rather than that of metamathematics. Hence I shall concentrate on Alcolea's example of a regular argument: Kenneth Appel and Wolfgang Haken's proof of the four colour conjecture.[3] He reconstructs the central argument of the proof as a derivation from the data D_1–D_3

---

[2] I am grateful to Miguel Gimenez of the University of Edinburgh for translating this paper.

[3] For further detail of the proof see [2], [25] or [12].

(D_1) Any planar map can be coloured with five colours.
(D_2) There are some maps for which three colours are insufficient.
(D_3) A computer has analysed every type of planar map and verified that each of them is 4-colorable.

of the claim C, that

(C) Four colours suffice to colour any planar map.

by employment of the warrant W, which has backing B

(W) The computer has been properly programmed and its hardware has no defects.
(B) Technology and computer programming are sufficiently reliable. [1, pp. 142f.]

He regards this as making clear that, since the warrant is not wholly mathematical, the proof must leave open the possibility of 'a specific counterexample, that is to say, a particular map that cannot be coloured with four colours might still exist' [1, p. 143].[4]

This example demonstrates both the strengths and the dangers of this approach. To complete Toulmin's layout we are obliged to make explicit not merely the premises and the conclusion, but also the nature of the support which the former is supposed to lend the latter. Thus the focus of Appel and Haken's critics, the heterodox deployment of a computer in a mathematical proof, is made glaringly obvious. However, it is premature to draw from this surface dissimilarity the inference that Appel and Haken's result is less convincing than other mathematical proofs. A closer reading of Alcolea's reconstruction may clarify this point. Premises D_1 and D_2 have conventional mathematical proofs, as Alcolea points out. (D_1 is not strictly relevant to the derivation of C, although its proof originated techniques which were instrumental to Appel and Haken's work.) D_3 is a very concise summary of the central results of Appel and Haken's work. It may help to spell out the details at greater length.

There are two essential ideas behind the Appel and Haken proof: unavoidability and reducibility. An unavoidable set is a set of configurations, that is countries or groups of adjacent countries, at least one of which must be present in any planar map. For example, all such maps must contain either a two-sided, a three-sided, a four-sided or a five-sided country, so these configurations constitute an unavoidable set. A configuration is reducible if any map containing it may be shown to be four-colorable. Two-sided, three-sided, and four-sided countries are all reducible. To prove the four colour theorem it suffices to exhibit an unavoidable set of reducible configurations.

[4] '... un contraexemple específic, és a dir, que es trobe un mapa particular que no puga colarar-se amb quatre colors'

Alfred Kempe, who introduced the concepts of unavoidability and reducibility, was believed to have proved the four colour theorem in 1879 by showing that five-sided countries were also reducible [11]. However, in 1890 a flaw was discovered in his reasoning: the five-sided country is not reducible, hence a larger unavoidable set is required if all its configurations are to be reducible. Appel and Haken used a computer to search for such a set, eventually discovering one with 1,482 members. The unavoidability of this set could be demonstrated by hand, but the reducibility of all its members would be far too protracted a task for human verification. Subsequent independent searches have turned up other unavoidable sets. The smallest to date is a set of 633 reducible configurations found by Neil Robertson, Daniel Sanders, Paul Seymour and Robin Thomas in 1994.[5] Verifying the reducibility of these configurations still requires a computer.

So for there to be a non-four-colourable planar map, as Alcolea suggests, Appel and Haken (and their successors) must have erred either in the identification of the unavoidable set, or in the demonstration of the reducibility of its member configurations. Since the former step can be verified by conventional methods, the computer can only be suspected of error in demonstrating reducibility. Two sorts of computer error should be distinguished: a mistake may be made in the programming, or a fault may arise in the computer itself (the hardware or firmware). The former error would arise due to a human failure to correctly represent the mathematical algorithms which the computer was programmed to implement. This sort of mistake does not seem to be interestingly different from the traditional type of mathematical mistake, such as that made by Kempe in his attempt to prove the four colour conjecture. The second sort of error is genuinely new. However, it would seem to be profoundly unlikely.

Computer hardware can exhibit persistent faults, some of which can be hard to detect.[6] However, the potential risks of such faults can be minimized by running the program on many different machines. One might still worry about Appel and Haken's programs, since they were written in machine code and would therefore be implemented in more or less exactly the same manner on any computer capable of running them, perhaps falling foul of the same bug each time. This sort of checking might be suspected of being no better than buying two copies of the same newspaper to check the veracity of its reporting.[7] However, the same reducibility results were achieved independently, using different programs, as part of the refereeing process for Appel and Haken's work. Moreover, the more recent programs of Robertson & *al.* were written in higher level languages, as are the programs employed in most

[5] For details of their publications, see [25, p. 244].
[6] For example, the notorious Pentium FDIV bug.
[7] As Wittgenstein once remarked in a different context: [26, §265].

other computer-assisted proofs. The existence of different compilers and different computer platforms ensures that these programs can be implemented in many intrinsically different ways, reducing the likelihood of hardware or firmware induced error to the astronomical.

Thus we may derive an alternative reconstruction of Appel and Haken's argument: "Given that (D_4) the elements of the set U are reducible, we can (Q) almost certainly claim that (C) four colours suffice to colour any planar map, since (W) U is an unavoidable set (on account of (B) conventional mathematical techniques), unless (R) there has been an error in either (i) our mathematical reasoning, or (ii) the hardware or firmware of all the computers on which the algorithm establishing D_4 has been run." If, in addition, we observe that (i) appears to be orders of magnitude more likely than (ii), then C would seem to be in much less doubt than it did in the light of Alcolea's reconstruction. The purpose of the preceding has been not so much to rescue the four colour conjecture from Alcolea's critique (although few if any graph theorists would accept that a counterexample is possible), but to show up the limitations of Toulmin's pattern as a descriptive technique. As other critics have pointed out, reconstructing an argument along Toulmin's lines 'forces us to rip propositions out of context' [24, p. 318]. The degree of abstraction necessary to use the diagram at all can make different, incompatible, reconstructions possible, leaving the suspicion that any such reconstruction may involve considerable (and unquantified) distortion.

3 Walton's new dialectic

There has been significant progress in informal logic since the publication of *The uses of argument*. One milestone was the publication of Charles Hamblin's *Fallacies* [8] in 1970. This demonstrated the inadequacies of much of traditional fallacy theory and, by way of remedy, proposed an influential dialectical model of argumentation. Further impetus has come from the recent work of communication theorists such as Frans van Eemeren and Rob Grootendorst [6]. One contemporary logician who shows the influence of both traditions is Douglas Walton.[8] The focus of his work is the dialectical context of argument. Walton distinguishes between 'inference', defined as a set of propositions, one of which is warranted by the others, 'reasoning', defined as a chain of inferences, and 'argument', defined as a dialogue employing reasoning. This dialectical component entails that arguments require more than one arguer: at the very least there must be an assumed audience, capable in principle of answering back.

[8] Walton has published a great number of works on informal logic. [22] provides an overview of the general method common to many of them.

Winston Churchill once praised the argumentational skills of the celebrated barrister and politician F. E. Smith, 1st Earl of Birkenhead, by stressing their suitability to context: 'The bludgeon for the platform; the rapier for a personal dispute; the entangling net and unexpected trident for the Courts of Law; and a jug of clear spring water for an anxious perplexed conclave' [5, p. 176]. Toulmin also stresses the domain specificity of good practice in argument. What is distinctive about Walton's analysis is the attempt to characterize dialectical context in terms of general features which are not themselves domain specific. Without pretending to have an exhaustive classification of argumentational dialogue, he is able to use these features to draw several important distinctions. The principal features with which he is concerned are the 'initial situation' and the 'main goal' of the dialogue. The initial situation describes the circumstances which give rise to the dialogue, in particular the differing commitments of the interlocutors. The main goal is the collective outcome sought by both (all) participants, which may be distinct from their individual goals.

If we simplify the situation by permitting each discussant to regard some crucial proposition as either true, false or unknown, four possibilities emerge. Either (0) the discussants agree that the proposition is true (or that it is false), in which case there is no dispute; or (1) one of them takes it to be true and the other false, in which case they will be in direct conflict with each other; or (2) they both regard it as unknown, which may result in a dialogue as they attempt to find out whether it is true or false; or (3) one of them believes the proposition to be true (or false) but the other does not know which it is. Thus we may distinguish three types of initial situation from which an argumentational dialogue may arise: a conflict, an open problem, or an unsatisfactory spread of information. A conflict may produce several different types of dialogue depending on how complete a resolution is sought. For a stable outcome one interlocutor must persuade the other, but, even if such persuasion is impossible they may still seek to negotiate a practical compromise on which future action could be based. Or they may aim merely to clear the air by expressing their contrasting opinions, without hoping to do more than merely agree to disagree: a quarrel. These three goals—stable resolution, practical settlement and provisional accommodation—can also be applied to the other two initial situations, although not all three will be exemplified in each case. So open problems can lead to stable resolutions, or if this is not achievable, to practical settlement. However, provisional accommodation should not be necessary if the problem is genuinely open, since neither discussant will be committed to any specific view. Where the dialogue arises merely from the ignorance of one party then a stable resolution should always be achievable, obviating the other goals. The interplay of these different types of initial situation and main goal thus allows Walton to identify six principal types of dialogue, Persuasion, Negotiation, Eristic, Inquiry, Deliberation and Information Seeking, which may be represented diagrammatically as in Table 1. The contrasting

TABLE 1: Walton & Krabbe's 'Systematic survey of dialogue types' [23, p. 80]

Main Goal	Initial Situation		
	Conflict	Open Problem	Unsatisfactory Spread of Information
Stable Agreement/ Resolution	Persuasion	Inquiry	Information Seeking
Practical Settlement/ Decision (Not) to Act	Negotiation	Deliberation	
Reaching a (Provisional) Accommodation	Eristic		

properties of these different types of dialogue are set out in Table 2. This table also states the individual goals of the interlocutors typical to each type, and includes two derivative types: the debate, a mixture of persuasion and eristic dialogue, and the pedagogical dialogue, a subtype of the information seeking dialogue. Many other familiar argumentational contexts may be represented in terms of Walton's six basic types of dialogue by such hybridization and subdivision.[9]

TABLE 2: Walton's types of dialogue [21, p. 605]

Type of Dialogue	Initial Situation	Individual Goals of Participants	Collective Goal of Dialogue	Benefits
Persuasion	Difference of opinion	Persuade other party	Resolve difference of opinion	Understand positions
Inquiry	Ignorance	Contribute findings	Prove or disprove conjecture	Obtain knowledge
Deliberation	Contemplation of future consequences	Promote personal goals	Act on a thoughtful basis	Formulate personal priorities
Negotiation	Conflict of interest	Maximize gains (self-interest)	Settlement (without undue inequity)	Harmony
Information-Seeking	One party lacks information	Obtain information	Transfer of knowledge	Help in goal activity
Quarrel (Eristic)	Personal conflict	Verbally hit out at and humiliate opponent	Reveal deeper conflict	Vent emotions
Debate	Adversarial	Persuade third party	Air strongest arguments for both sides	Spread information
Pedagogical	Ignorance of one party	Teaching and learning	Teaching and knowledge	Reserve transfer

[9] See Table 3.1 in [23, p. 66] for some further examples.

It is central to Walton's work that the legitimacy of an argument should be assessed in the context of its use: what is appropriate in a quarrel may be inappropriate in an inquiry, and so forth. Although some forms of argument are never legitimate (or never illegitimate), most are appropriate if and only if they are "in the right place". For example, threats are inappropriate as a form of persuasion, but they can be essential in negotiation. In an impressive sequence of books, Walton has analyzed a wide variety of fallacious or otherwise illicit argumentation as the deployment of strategies which are sometimes admissible in contexts in which they are inadmissible. However, Walton has not directly addressed mathematical argumentation. In the next section I shall set out to explore how well his system may be adapted to this purpose.

4 Applying Walton to mathematics

In what context (or contexts) do mathematical proofs occur? The obvious answer is that mathematical proof is a special case of inquiry. Indeed, Walton states that the collective goal of inquiry is to 'prove or disprove [a] conjecture'. An inquiry dialogue proceeds from an open problem to a stable agreement. That is to say from an initial situation of mutual ignorance, or at least lack of commitment for or against the proposition at issue, to a main goal of shared endorsement or rejection of the proposition. This reflects a standard way of reading mathematical proofs: the prover begins from a position of open-mindedness towards the conjecture, shared with his audience. He then derives the conjecture from results upon which they both agree, by methods which they both accept.

But this is not the only sort of dialogue in which a mathematical proof may be set out. As William Thurston has remarked, mathematicians 'prove things in a certain context and address them to a certain audience' [17, p. 175]. Indeed, crucially, there are several different audiences for any mathematical proof, with different goals. Satisfying the goals of one audience need not satisfy those of the others. For example, a proof may be read by:

- Journal referees, who have a professional obligation to play devil's advocate;
- Professional mathematicians in the same field, who may be expected to quickly identify the new idea(s) that the proof contains, grasping them with only a few cues, but who may already have a strong commitment to the falsehood of the conjecture;
- Professional mathematicians in other (presumably neighbouring) fields, who will need more careful and protracted exposition;
- Students and prospective future researchers in the field, who could be put off by too technical an appearance, or by the impression that all the important results have been achieved;

- Posterity, or in more mercenary terms, funding bodies: proof priority can be instrumental in establishing cudos with both.

This list suggests that the initial situation of a proof dialogue cannot always be characterized as mutual open-mindedness. Firstly, in some cases, the relationship between the prover and his audience will be one of conflict. If the conjecture is a controversial one, its prover will have to convince those who are committed to an incompatible view. And if an article is refereed thoroughly, the referees will be obliged to adopt an adversarial attitude, irrespective of their private views.

Secondly, as the later items indicate, proofs have a pedagogic purpose. Thurston relates his contrasting experiences in two fields to which he made substantial contributions. As a young mathematician, he proved many results in foliation theory using powerful new methods. However, his proofs were of a highly technical nature and did little to explain to the audience how they too might exploit the new techniques. As a result, the field evacuated: other mathematicians were afraid that by the time they had mastered Thurston's methods he would have proved all the important results. In later work, on Haken manifolds, he adopted a different approach. By concentrating on proving results which provided an infrastructure for the field, in a fashion which allowed others to acquire his methods, he was able to develop a community of mathematicians who could pursue the field further than he could alone. The price for this altruism was that he could not take all the credit for the major results. Proofs which succeed in the context Thurston advocates proceed from an initial situation closer to Walton's 'unsatisfactory spread of information'. This implies that information seeking is another context in which mathematical proofs may be articulated. Of course, the information which is being sought is not merely the conjecture being proved, but also the methods used to prove it.

An unsatisfactory spread of information, unlike a conflict or an open problem, is an intrinsically asymmetrical situation. We have seen that proofs can arise in dialogues wherein the prover possesses information sought by his interlocutors. Might there be circumstances in which we should describe as a proof a dialogue in which the prover is the information seeker? This is the question considered by Thomas Tymoczko [20, p. 71] and Yehuda Rav [14], to somewhat different ends. Tymoczko considers a community of Martian mathematicians who have amongst their number an unparalleled mathematical genius, Simon. Simon proves many important results, but states others without proof. Such is his prestige, that "Simon says" becomes accepted as a form of proof amongst the Martians. Rav considers a fantastical machine, Pythiagora, capable of answering mathematical questions instantaneously and infallibly. Both thought experiments consider the admission of a dialogue with an inscrutable but far better informed interlocutor as a possible method of proof.

In both cases we are invited to reject this admission, although, interestingly, for different reasons. Rav sees his scenario as suggesting that proof cannot be purely epistemic: if it were then Pythiagora would give us all that we needed, but Rav suggests that we would continue to seek conventional proofs for their other explanatory merits. He concludes that Pythiagora could not give us proof. Tymoczko draws an analogy between his thought experiment and the use of computers in proofs such as that of the four colour theorem. He argues that there is no formal difference between claims backed by computer and claims backed by Simon: they are both appeals to authority. The difference is that the computer can be a warranted authority. Hence, on Tymoczko's admittedly controversial reading, computer assisted proof is an information seeking dialogue between the prover and the computer.

So far we have seen that the initial situation of a proof dialogue can vary from that of an inquiry. What of the main goal—must this be restricted to stable resolution? Some recent commentators have felt the need for a less rigorous form of mathematics, with a goal closer to Walton's practical settlement. Arthur Jaffe and Frank Quinn [10] introduced the much discussed, if confusingly named, concept of 'theoretical mathematics'. They envisage a division of labour, analogous to that between theoretical and experimental physics, between conjectural or speculative mathematics and rigorous mathematics. Where traditional, rigorous mathematicians have theorems and proofs, theoretical mathematicians make do with 'conjectures' and 'supporting arguments'. This echoes an earlier suggestion by Edward Swart [16] that we should refrain from accepting as theorems results which depend upon lengthy arguments, whether by hand or computer, of which we cannot yet be wholly certain. He suggests that 'these additional entities could be called agnograms, meaning theoremlike statements that we have verified as best we can but whose truth is not known with the kind of assurance that we attach to theorems and about which we must thus remain, to some extent, agnostic' [16, p. 705]. In both cases the hope is that further progress will make good the shortfall: neither Jaffe and Quinn's conjectures nor Swart's agnograms are intended as replacements for rigorously proved theorems.

More radical critics of the accepted standards of mathematical rigour suggest that practical settlement can be a goal of proof and not merely of lesser, analogous activities. For instance, Doron Zeilberger [27] envisages a future of semi-rigorous (and ultimately non-rigorous) mathematics in which the ready availability by computer of near certainty reduces the pursuit of absolute certainty to a low resource allocation priority. Hence he predicts that a mathematical abstract of the future could read "We show, in a certain precise sense, that the Goldbach conjecture is true with probability larger than 0.99999, and that its complete truth could be determined with a budget of $10 billion"

[27, p. 980]. 'Proofs' of this sort explicitly eschew stable resolution for practical settlement. Thus Zeilberger is arguing that proofs could take the form of deliberation or negotiation.

The last of Walton's dialogue types is the eristic dialogue, in which no settlement is sought, merely a provisional accommodation in which the commitments of the parties are made explicit. This cannot be any sort of proof, since no conclusion is arrived at. But it is not completely without interest. A familiar diplomatic euphemism for a quarrel is "a full and frank exchange of views", and such activity does have genuine merit. Similarly, even failed mathematical proofs can be of use, especially if they clarify previously imprecise concepts, as we saw with Kempe's attempted proof of the four colour conjecture [11]. This process has something in common, if not with a quarrel, at least with a debate, which we saw to be a related type of dialogue.

To take stock, we have seen that most of Walton's dialogue types are reflected to some degree in mathematical proof. Table 3, an adaptation of Table 2, sets out the difference between the various types of proof dialogue introduced.

TABLE 3: Some types of proof dialogue

Type of Dialogue	Initial Situation	Main Goal	Goal of Prover Prover	Goal of Interlocutor
Proof as Inquiry	Open-mindedness	Prove or disprove conjecture	Contribute to outcome	Obtain knowledge
Proof as Persuasion	Difference of opinion	Resolve difference of opinion with rigour	Persuade interlocutor	Persuade prover
Proof as Information Seeking (Pedagogical)	Interlocutor lacks information	Transfer of knowledge	Disseminate knowledge of results & methods	Obtain knowledge
'Proof' as Information-Seeking (*e.g.* Tymoczko)	Prover lacks information	Transfer of knowledge	Obtain information	Presumably inscrutable
'Proof' as Deliberation (*e.g.* Swart)	Open-mindedness	Reach a provisional conclusion	Contribute to outcome	Obtain warranted belief
'Proof' as Negotiation (*e.g.* Zeilberger)	Difference of opinion	Exchange resources for a provisional conclusion	Contribute to outcome	Maximize value of exchange
'Proof' as Eristic/Debate	Irreconcilable difference of opinion	Reveal deeper conflict	Clarify position	Clarify position

5 Proof dialogues

In this last section I shall explore how the classification of proof dialogues may help to clarify many of the problems that have arisen in the philosophical debate over the nature of mathematical proof. We can see that proofs may occur in several distinct types of dialogue, even if we do not count the suspect cases (the entries for Table 3 where 'proof' is in scare quotes). An ideal proof will succeed within inquiry, persuasion and pedagogic proof dialogues. Suboptimal proofs may fail to achieve the goals of at least one of these dialogue types. In some cases, this may be an acceptable, perhaps inevitable, shortcoming; in others it would fatally compromise the argument's claim to be accepted as a proof.

As Thurston's experience with foliation theory demonstrated, not every proof succeeds pedagogically. Proofs in newly explored areas are often hard to follow, and there are some results which have notoriously resisted all attempts at clarification or simplification.[10] Yet, if these proofs succeed in inquiry and persuasion dialogues, we have no hesitation in accepting them. Conversely, there are some 'proofs' which have a heuristic usefulness in education, but which would not convince a more seasoned audience. Pedagogic success is neither necessary nor sufficient for proof status—but it is a desirable property, nonetheless.

An argument might convince a neutral audience, but fail to persuade a determined sceptic. Just this happened to Andrew Wiles's first attempt at a proof of the Fermat conjecture: the initial audience were convinced, but the argument ran into trouble when exposed to determined criticism from its referees. Such a case might be seen as success within an inquiry proof dialogue, followed by failure in a persuasion proof dialogue. A similar story could be told about Kempe's 'proof' of the four colour conjecture: a result which received far less scrutiny than Wiles's work, and was thereby widely accepted for eleven years. On the other hand, if even the sceptics are convinced, then an open-minded audience should follow suit. Thus, on the conventional understanding of mathematical rigour, success within both inquiry and persuasion proof dialogues is necessary for an argument to count as a proof.

We saw in the last section how a variety of differently motivated departures from the prevailing standards of mathematical rigour may be understood as shifts to different types of proof dialogue. Indeed, one of Walton's principal concerns in his analysis of natural argumentation is the identification of shifts from one type of dialogue to another. Such shifts can take a variety of forms: either gradual or abrupt, and either replacing the former type of dialogue or embedding the new type within the old. These processes are an

[10] For example, von Staudt's proof of the equivalence of analytic and synthetic projective geometry has retained its difficulty for nearly two centuries. See [15, pp. 193 f.] for a discussion.

essential and productive aspect of argumentation, but they are also open to abuse. Similar warnings apply to shifts towards less rigorous types of proof dialogue.

Many of the concerns which the critics of these forms of argumentation have advanced may be understood as an anxiety about illicit shifts of proof dialogue type. For example, the published discussion of mathematical conjecture is something which Jaffe and Quinn welcome: their concern is that such material not be mistaken for theorem-proving. Although some of their critics interpreted their advocacy of 'theoretical mathematics' as a radical move, their primary goal was a conservative one: to maintain a sharp demarcation between rigorous and speculative work. Their 'measures to ensure "truth in advertising"' [10, p. 10] are precisely calculated to prevent illicit shifts between inquiry and deliberation proof dialogues.[11] A similar story could be told about Tymoczko or Swart's discussion of methods they see as falling short of conventional rigour. Zeilberger is advocating the abandonment of rigour, but he recognizes at least a temporary imperative to separate rigorous from 'semi-rigorous' mathematics.

As Toulmin & *al.* remark 'it has never been customary for philosophers to pay much attention to the *rhetoric* of mathematical debate' [19, p. 89]. The goal of this article has been to exhibit some of the benefits that may accrue from a similarly uncustomary interest in the *dialectic* of mathematical debate—a dialectic which informal logic can do much to illuminate.

References

1. JESÚS ALCOLEA BANEGAS, 1998, *L'argumentació en matemàtiques*, in ***XIIè Congrés Valencià de Filosofia*** (E. Casaban i Moya, editor), València, pp. 135–147.
2. KENNETH APPEL & WOLFGANG HAKEN, 1978, *The four colour problem*, reprinted 2002 in ***The philosophy of mathematics: An anthology*** (D. Jacquette, editor), Blackwell, Oxford, pp. 193–208.
3. J. BORWEIN, P. BORWEIN, R. GIRGENSOHN & S. PARNES, 1995, *Experimental mathematics: A discussion*, ***CECM preprint no. 95:032***, http://www.cecm.sfu.ca/preprints/1995pp.html.
4. WAYNE BROCKRIEDE & DOUGLAS EHNINGER, 1960, *Toulmin on argument: An interpretation and application* ***Quarterly journal of speech***, **46**, pp. 44–53.
5. WINSTON CHURCHILL, 1937, ***Great contemporaries***, Thornton Butterworth, London.
6. FRANS VAN EEMEREN & ROB GROOTENDORST, 1992, ***Argumentation, communication and fallacies***, Lawrence Erlbaum Associates, Hillsdale, N.J.
7. FRANS VAN EEMEREN, ROB GROOTENDORST & TJARK KRUIGER, 1987, ***Handbook of argumentation theory: A critical survey of classical backgrounds and modern studies***, Foris, Dordrecht.
8. CHARLES HAMBLIN, 1970, ***Fallacies***, Methuen, London.

[11] Indeed, misleading advertisements are one of Walton's principal examples of an illicit dialogue shift. See, for example, [22, p. 206 ff.].

9. PAUL HOFFMAN, 1998, ***The man who loved only numbers***, Fourth Estate, London.
10. ARTHUR JAFFE & FRANK QUINN, 1993, *"Theoretical mathematics": Toward a cultural synthesis of mathematics and theoretical physics*, ***Bulletin of the American Mathematical Society***, **29**, pp. 1–13.
11. ALFRED KEMPE, 1879, *On the geographical problem of the four colours*, ***American journal of mathematics***, **2**, pp. 193–200.
12. DONALD MACKENZIE, 1999, *Slaying the Kraken: The sociohistory of a mathematical proof*, ***Social studies of science***, **29**, pp. 7–60.
13. SUSAN NEWMAN & CATHERINE MARSHALL, 1992, *Pushing Toulmin too far: Learning from an argument representation scheme*, ***Xerox PARC technical report no. SSL-92-45***, http://www.csdl.tamu.edu/~marshall/toulmin.pdf.
14. YEHUDA RAV, 1999, *Why do we prove theorems?*, ***Philosophia Mathematica***, **7**, pp. 5–41.
15. GIAN-CARLO ROTA, 1997, *The phenomenology of mathematical proof*, ***Synthese***, **111**, pp. 183–196.
16. EDWARD SWART, 1980, *The philosophical implications of the four-color problem*, ***The American mathematical monthly***, **87**, pp. 697–707.
17. WILLIAM THURSTON, 1994, *On proof and progress in mathematics*, ***Bulletin of the American Mathematical Society***, **30**, pp. 161–171.
18. STEPHEN TOULMIN, 1958, ***The uses of argument***, Cambridge University Press, Cambridge.
19. STEPHEN TOULMIN, RICHARD RIEKE & ALLAN JANIK, 1979, ***An introduction to reasoning***, Macmillan, London.
20. THOMAS TYMOCZKO, 1979, *The four-color problem and its philosophical significance*, ***Journal of philosophy***, **76**, pp. 57–83.
21. DOUGLAS WALTON, 1997, *How can logic best be applied to arguments?* ***Logic journal of the IGPL***, **5**, pp. 603–614.
22. DOUGLAS WALTON, 1998, ***The new dialectic: Conversational contexts of argument***, Toronto University Press, Toronto.
23. DOUGLAS WALTON & ERIK KRABBE, 1995, ***Commitment in dialogue: Basic concepts of interpersonal reasoning***, SUNY Press, Albany, N.Y..
24. CHARLES WILLARD, 1976, *On the utility of descriptive diagrams for the analysis and criticism of arguments*, ***Communication monographs***, **43**, pp. 308–319.
25. ROBIN WILSON, 2002, ***Four colours suffice: How the map problem was solved***, Allen Lane, London.
26. LUDWIG WITTGENSTEIN, 1953, ***Philosophical Investigations***, Blackwell, Oxford.
27. DORON ZEILBERGER, 1993, *Theorems for a price: Tomorrow's semi-rigorous mathematical culture*, ***Notices of the American Mathematical Society***, **40**, pp. 978–981.

5

Philosophical Problems of Mathematics in the Light of Evolutionary Epistemology

YEHUDA RAV

Introduction

When one speaks of the foundations of mathematics or of its foundational problems, it's important to remember that mathematics is not an edifice which risks collapse unless it is seated on solid and eternal foundations that are supplied by some logical, philosophical, or extra-mathematical construction. Rather, mathematics ought to be viewed as an ever-expanding mansion floating in space, with new links constantly growing between previously separated compartments, while other chambers atrophy for lack of interested or interesting habitants. The foundations of mathematics also grow, change, and further interconnect with diverse branches of mathematics as well as with other fields of knowledge. Mathematics flourishes on open and thorny problems, and foundational problems are no exception. Such problems arose already in antiquity, but the rapid advance in the second half of the nineteenth century toward higher levels of abstraction and the recourse to the actual infinite by Dedekind[1] and Cantor all pressed for an intense concern with foundational questions. The discovery of irrational numbers, the use of negative numbers (from the Latin *negare,* literally, "to deny," or "to refuse"), the introduction of imaginary numbers, the invention of the infinitesimal calculus and the (incoherent) calculations with divergent series, and so forth, each of these novelties precipitated at their time uncertainties and resulted in methodological reflections. But starting with the creation of non-Euclidean geometries[2] and culminating in Cantor's theory of transfinite numbers, the *rate* at which new foundational problems presented themselves grew to the point of causing in some quarters a sense of crisis-hence the talk of a foundational crisis at the beginning of this century.

The philosophy of mathematics is basically concerned with systematic reflection about the nature of mathematics, its methodological problems, its relations to reality, and its applicability. Certain foundational inquiries, philosophical at the outset, were eventually internalized. Thus, the impetus resulting from philosophically motivated researches produced spectacular developments in the field of logic, with their ultimate absorption within mathematics proper. Today, the various descendants of foundational work, such as proof theory, axiomatic set theory, recursion theory, and so on, are part and parcel of the mainstream of mathematical research. This does not mean that the philosophy of mathematics has or ought to have withered away. On the contrary. Nowadays, many voices hail a renaissance in the philosophy of mathematics and acclaim its new vigor. Note also the current dynamic preoccupation by biologists and philosophers alike with foundational problems of biology. (A special journal, *Biology and Philosophy,* was created in 1986 to serve as a common forum.) By contrast, the mathematical community is rather insular, and most mathematicians now have a tendency to spurn philosophical reflections. Yet without philosophy we remain just stone heapers: "Tu peux certes raisonner sur l'arrangement des pierres du temple, tu ne toucheras point l'essentiel qui échappe aux pierres." ("You can certainly reason about the arrangement of the stones of the temple, but you'll never grasp its essence which lies beyond the stones," (my translation; SaintExupéry, 1948:256).

It is significant to notice that in the current literature on the philosophy of mathematics there is a marked shift towards an analysis of *mathematical practice* (cf. Feferman, 1985; Hersh, 1979; Kitcher, 1983; Kreisel, 1973; Resnik, 1975, 1981, 1982; Resnik and Kusher, 1987; Shapiro, 1983; Steiner, 1978a, 198Th, 1983; Van Bendegem, 1987). This is most refreshing, for it is high time that the philosophy of mathematics liberates itself from ever enacting the worn-out tetralogy of Platonism, logicism, intuitionism, and formalism. As Quine (1980:14) has pointed out, the traditional schools of the philosophy of mathematics have their roots in the medieval doctrines of realism, conceptualism and nominalism. Whereas the quarrel about universals and ontology *had* its meaning and significance within the context of medieval Christian culture, it is an intellectual scandal that some philosophers of mathematics can still discuss whether whole numbers exist or not. It was an interesting question to compare mathematical "objects" with physical objects as long as the latter concept was believed to be unambiguous. But, with the advent of quantum mechanics, the very concept of a physical object became more problematic than any mathematical concept.3 In a nutshell, philosophy too has its paradigms, and a fertile philosophy of mathematics, like any other "philosophy of," must be solidly oriented towards the practice of its particular discipline and keep contact with actual currents in the philosophy of science. The purpose of this essay is to explore one such current in the philosophy of science, namely, evolutionary epistemology, with the tacit aim of hopefully obtaining some new insights concerning the nature of mathematical knowledge. This is not a reductionist program. But the search for new insights seem more fruitful than treading forever on the quicksand of neo-scholasticism and its offshoots. I concur with Wittgenstein that "a philosophical work consists essentially of elucidations" (1983:77).

The Main Tenets of Evolutionary Epistemology

Evolutionary Epistemology (EE) was independently conceived by Lorenz, a biologist; Campbell, a psychologist; and Volimer, a physicist and philosopher. Though its origins can be traced to nineteenth-century evolutionary thinkers, EE received its initial formulation by Lorenz (1941) in a little-noticed paper on Kant. Christened in 1974 by Campbell and systematically developed in a book by Voilmer in 1975, evolutionary epistemology has quickly become a topic of numerous papers and books (see Campbell, Hayes, and Callebaut, 1987). In the opening paragraph of an essay in honor of Sir Karl Popper, where the term evolutionary epistemology appears for the first time, Campbell states:

An evolutionary epistemology would be at minimum an epistemology taking cognizance of and compatible with man's status as a product of biological and social evolution. In the present essay it is also argued that evolution-even in its biological aspects-is a knowledge process, and that the natural-selection paradigm for such knowledge increments can be generalized to other epistemic activities, such as learning, thought, and science (Campbell, 1974:413).

My aim is to add mathematics to that list. Riedl characterizes evolutionary epistemology as follows:

In contrast to the various philosophical epistemologies, evolutionary epistemology attempts to investigate the mechanism of cognition from the point of view of its phylogeny. It is mainly distinguished from the traditional position in that it adopts a point of view outside the subject and examines different cognitive mechanisms comparatively. It is thus able to present objectively a series of problems [including the problems of traditional epistemologies, not soluble on the level of reason alone but soluble from the phylogenetic point of view]. (Reidel, 1984:220,1988:287.)

In an extensive survey article, Bradie (1986) introduced a distinction between two interrelated but distinct programs that go under the name of evolutionary epistemology. One one hand, there is an "attempt to account for the characteristics of cognitive mechanisms in animals and humans by a straightforward extension of the biological theory of evolution to those aspects or traits of animals which are the biological substrates of cognitive activity, e.g., their brains, sensory systems, motor systems, etc." (Bradie 1986:403). Bradie refers to this as the "Evolutionary Epistemology Mechanism program" (EEM). On the other hand, the EE Theory program, EET, "attempts to account for the evolution of ideas, scientific theories and culture in general by using models and metaphors drawn from evolutionary biology.' Both programs have their roots in 19th century biology and social philosophy, in the work of Darwin, Spencer and others" (Bradie, 1986:403). Popper is generally considered to be the main representative of the EET program, though Popper himself would not call himself an evolutionary epistemologist.5 The great impetus to the EE Mechanisms program came from the work of Konrad Lorenz and his school of ethology. Through extensive studies of the behavior of animals in their natural habitat, Lorenz has deepened our understanding of the interplay between genetically determined and learned behavioral patterns. To Lorenz, the evolution of the cognitive apparatus is not different in kind from the evolution of organs. The same evolutionary mechanisms account for both. As Lorenz puts it in a famous passage:

Just as the hoof of the horse, this central nervous apparatus stumbles over unforeseen changes in its task. But just as the hoof of the horse is adapted to the ground of the steppe which it copes with, so our central nervous apparatus for organizing the image of the world is adapted to the real world with which man has to cope. Just like any organ, this apparatus has attained its expedient species-preserving form through this coping of real with the real during its genealogical evolution, lasting many eons (Lorenz, 1983:124).

In the fascinating 1941 paper already mentioned, Lorenz reinterpreted the Kantian categories of cognition in the light of evolutionary biology. By passing from Kant's *prescriptive epistemology* to an evolutionary *descriptive epistemology,* the category of *a priori cognition is* reinterpreted as the individual's inborn (a priori) *cognitive mechanisms* that have evolved on the basis of the species' a posteriori confrontation with the environment. In short, the phylogenetically a posteriori became the ontogenetically a priori. In the words of Lorenz, "The categories and modes of perception of man's cognitive apparatus are the natural products of phylogeny and thus adapted to the parameters of external reality in the same way, and for the same reason, as the horse's hooves are adapted to the prairie, or the fish's fins to the water" (1977:37,1985:57).

Any epistemology worthy of its name must start from some postulate of realism: that there exists a real world with some organizational regularities. "In a chaotic world not only knowledge, but even organisms would be impossible, hence non-existent" (Volimer, 1983:29). But the world includes also the reflecting individual. Whereas the idealist, to paraphrase Lorenz, looks only into the mirror and turns a back to reality, the realist looks only outwardly and is not aware of being a mirror of reality. Each ignores the fact that the mirror also has a nonreflecting side that is part and parcel of reality and consists of the physiological apparatus that has evolved in adaptation to the real world. This is the subject of Lorenz's remarkable book *Behind the Mirror*. Yet reality is not given to immediate and direct inspection. "Reality is veiled," to use the deft expression of d'Espagnat. But the veil can progressively be *transluminated, so* to speak, by conceptual modeling and experimentation. This is the credo of the working scientist. Evolutionary epistemology posits a minimal ontology, known under the name of *hypothetical realism,* following a term coined and defined by Campbell as follows:

My general orientation I shall call hypothetical realism. An "external" world is hypothesized in general, and specific entities and processes are hypothesized in particular, and the observable implications of these hypotheses (or hypostatizations, or reifications) are sought out for verification. No part of the hypotheses has any "justification" or validity prior to, or other than through, the testing of these implications. Both in specific and in general they are always to some degree tentative (Campbell, 1959:156).

The reader is referred to the treatises by Voilmer (1987a, 1985; 1986) for a systematic discussion of evolutionary epistemology. See also Ursua (1986) and Volimer's (1984) survey article. Subsequently, I will also draw on the insights furnished by the genetic (or developmental) epistemology of Piaget and his school (which I consider part of the EE Mechanisms program), as well as on the work of Oeser (1987, Oeser and Seitelberger 1988).

Some Perennial Questions in the Philosophy of Mathematics

The Hungarian mathematician Alfréd Rényi has written a delightful little book entitled *Dialogues on Mathematics*.' The first is a Socratic dialogue on the nature of mathematics, touching on some central themes in the philosophy of mathematics. From the following excerpts, *1* shall extract the topics of our subsequent discussion.

SOCRATES What things does a mathematician study? ... Would you say that these things exist? ... Then tell me, if there were no mathematicians, would there be prime numbers, and if so, where would they be?

SOCRATES Having established that mathematicians are concerned with things that do not exist in reality, but only in their thoughts, let us examine the statement of Theaitetos, which you mentioned, that mathematics gives us more trustworthy knowledge than does any other branch of science.

HIPPOCRATES ... [Im reality you never find two things which are exactly the same; ... but one may be sure that the two diagonals of a rectangle are exactly equal ... Heraditus ... said that everything which exists is constantly changing, but that sure knowledge is only possible about things that never change, for instance, the odd and the even, the straight line and the circle.

SOCRATES ... We have much more certain knowledge about persons who exist only in our imagination, for example, about characters in a play, than about living persons ... The situation is exactly the same in mathematics.

HIPPOCRATES But what is the use of knowledge of non-existing things such as that which mathematics offers?

SOCRATES: . . . *How* to explain that, as often happens, mathematicians living far apart from each other and having no contact, independently discover the same truth? I never heard of two poets writing the same poem ... It seems that the object of [mathematicians'] study has some sort of existence which is independent of their person.

SOCRATES: But tell me, the mathematician who finds new truth, does he discover it or invent it?

HIPPOCRATES The main aim of the mathematician is to explore the secrets and riddles of the sea of. *These exist independently of the mathematician, though not from humanity as a whole* [italics mine].

SOCRATES: We have not yet answered the question: what is the use of exploring the wonderful sea of human thought?

SOCRATES: *If you want to be a mathematician, you must realize you will be working mostly for the future* [italics mine]. Now, let us return to the main question. We saw that knowledge about another world of thought, about things which do not exist in the usual sense of the word, can be used in everyday life to answer questions about the real world. Is this not surprising?

HIPPOCRATES More than that, it is incomprehensible. It is really a miracle.

HIPPOCRATES [B]ut I do not see any similarity between the real world and the imaginary world of mathematics.

HIPPOCRATES Do you want to say that the world of mathematics is a reflected image of the real world in the mirror of our thinking?

SOCRATES: [D]o you think that someone who has never counted real objects can understand the abstract notion of number? The child arrives at the notion of a sphere through experience with round objects like balls. Mankind developed all fundamental notions of mathematics in a similar way. These notions are crystallized from a knowledge of the real world, and thus it is not surprising but quite natural that they bear the marks of their origin, as children do of their parents. And exactly as children when they grow up become the supporters of their parents, so any branch of mathematics, if it is sufficiently developed, becomes a useful tool in exploring the real world,

HIPPOCRATES Now we have found that the world of mathematics is nothing else but a reflection in our mind of the real world,

SOCRATES: ... *I* tell you [that] the answer is not yet complete.

SOCRATES: *We have kept too close to the simile of the reflected image. A simile is like a bow-if you stretch it too far, it snaps* [italics mine].

Schematically, the key issues that emerge from the dialogue are the following:

1. *Ontology*. In what sense can one say that mathematical "objects" exist?' If discovered, what does it mean to say that mathematical propositions are true independently of the knowing subject(s) and *prior* to their discovery?
2. *Epistemology*. How do we come to know "mathematical truth" and why is mathematical knowledge considered to be certain and apodictic?
3. *Applicability*. Why is mathematical knowledge applicable to reality?
4. *Psychosociology*. If invented, how can different individuals invent the "same" proposition? What is the role of society and culture?

It has been stressed by Körner *(1960)* and by Shapiro *(1983)* that problem *3 is* least adequately dealt with by *each* of the traditional philosophies of mathematics. As Shapiro rightly observes: "many of the reasons for engaging in philosophy at all make an account of the relationship between mathematics and culture a priority ... Any world view which does not provide such an account is incomplete at best" *(1983:524). To* answer this challenge, there are voices that try to revive Mill's long-buried empiricist philosophy of mathematics, notwithstanding the obvious fact that mathematical propositions are not founded on sense impressions nor could any ever be refuted by empirical observations. How are we supposed to derive from experience that every continuous function on a closed interval is Riemann integrable? A more shaded empiricism has been advocated by Kalmar and Lakatos. Their position was sharply criticized by Goodstein *(1970) and I fully agree with Goodstein's arguments (cf. the discussion in Lolli, 1982)*. In a different direction, Kòrner *(1965)* has sought an empiicist *justification* of mathematics via empirically verifiable propositions modulo translation of mathematical propositions into empirical ones. The problematics of translation apart, the knotty question of inductive justification "poppers" up again, and not much seems to be gained from this move. Though the road to empiricism is paved with good intentions, as with all such roads, the end point is the same. Yet the success of mathematics as a scientific tool is itself an empirical fact.

Moreover, empirical elements seem to be present in the more elementary parts of mathematics, and they are difficult to account for. To say that some mathematical concepts were formed by "abstraction" from experience only displaces the problem, for we still don't know how this process of abstraction is supposed to work. Besides, it is not the elementary part of mathematics that plays a fundamental role in the elaboration of scientific theories; rather, it is the totality of mathematics, with its most abstract concepts, that serves as a pool from which the scientist draws *conceptual schemes* for the elaboration of scientific theories. In order to account for this process, evolutionary epistemology starts from a minimal physical ontology, known as "hypothetical realism"; it just assumes the existence of an objective reality that is independent of our taking cognizance of it. Living beings, idealistic philosophers included, are of course part of objective reality. It is sufficient to assume that the world is nonchaotic; or, put positively, the world is assumed to possess organizational regularities. But I would not attribute to reality 'objective *relations*', 'quantitative *relations*', 'immutable *laws*', and so forth. All these are epistemic concepts and can only have a place within the frame of scientific theories. Some philosophers of mathematics have gone far beyond hypothetical realism and thereby skirt the pitfalls of both empiricism and Platonism, such as Ruzavin when he writes: "In complete conformity with the assertions of science, dialectical materialism considers mathematical objects as images, photographs, copies of the real quantitative relations and space forms of the world which surrounds us" (1977:193). But we are never told how, for instance, Urysohn's metrization theorem of topological spaces could reflect objective reality. Are we supposed to assume that through its pre-image in objective reality, Urysohn's theorem was already true before anybody ever thought of topological spaces? Such a position is nothing but Platonism demystified, and it would further imply that every mathematical problem is decidable independently of any underlying theoretical framework.

To reiterate: mathematics and objective reality are related, but the relationship is extremely complex, and no magic formula can replace patient epistemological analysis. We turn now to the task of indicating a direction for such an analysis from the point of view of evolutionary epistemology (cf. Vollmer, 1983).

MATHEMATICS AND REALITY

Many consider it a miracle-as Rényi had Hippocrates say-that mathematics is applicable to questions of the real world. In a famous article, Wigner expressed himself in a similar way: "the enormous usefulness of mathematics in the natural sciences is something bordering on the mysterious and... there is no rational explanation for it" (Wigner, 1960:2) It is hard to believe that our reasoning power was brought, by Darwin's process of natural selection, to the perfection which it seems to possess.

> With due respect to the awe of the great physicist, *there is a rational explanation for the usefulness of mathematics* and it is the task of any epistemology to furnish one. Curiously, we'll find its *empirical basis* in the very evolutionary process which puzzled Wigner. Here is the theory that I propose.

The core element, the depth structure of mathematics, incorporates cognitive mechanisms, which have evolved like other biological mechanisms, by confrontation with reality and which have become genetically fixed in the course of evolution. I shall refer to this core structure as the *logico-operational component* of mathematics. Upon this scaffold grew and continues to grow the *thematic component* of mathematics, which consists of the specific content of mathematics. This second level is culturally determined and originated, most likely, from ritual needs. (The ritual origin of mathematics has been discussed and documented by numerous authors (cf. Seidenberg, 1981; Carruccio, 1977:10; Michaels, 1978, and their respective bibliographies). Notice that ritual needs were practical needs, seen in the context of the prevailing cultures; hence there is no more doubt about the practical origin of mathematics Marchack (1972) has documented the presence of mathematical notations on bones dating to the Paleolithic of about thirty thousand years ago. This puts it twenty thousand years prior to the beginnings of agriculture; hence some mathematical knowledge was already available for the needs of land measurements, prediction of tides, and so forth. Given this remarkably long history, mathematics has been subjected to a lengthy cultural molding process akin to an environmental selection. Whereas the thematic component of mathematics is culturally transmitted and is in a continuous state of growth, the logico-operational component is based on genetically transmitted cognitive mechanisms and this is fixed. (This does not mean that the logico-operational level is ready for use at birth; it is still subject to an ontogenetic development.9 The genetic program is an *open program* (Mayr, 1974:651-52) that is materialized in the phenotype under the influence of internal and external factors and is realized by stages in the development of the individual).

> Let us look closer at the nature of cognitive mechanisms. Cognition is a fundamental physical process; in its simplest form it occurs at the molecular level when certain stereospecific configurations permit the aggregation of molecules into larger complexes. (There is no more anthropomorphism here in speaking of *molecular cognition* than in using the term *force* in physics.) As we move up the ladder of complexity, cognition plays a central role in prebiotic chemical evolution, and in the formation of self-replicating units. Here, in the evolution of macromolecules, "survival of the fittest" has a literal meaning: that *which fits, sticks* (chemically so!). That which doesn't fit, well, it just stays out of the game; it is "eliminated." These simple considerations should have a sobering effect when looking at more complicated evolutionary processes. The importance of cognition in the process of the self-organization of living matter cannot be overemphasized. Thus Maturana writes: *"Living systems are cognitive systems, and living as a process is a process of cognition" (1980:13)*. What I wish to stress here is that there is a continuum of cognitive mechanisms, from molecular cognition to cognitive acts of organisms, and that some of these fittings have become genetically fixed and are transmitted from generation to generation. Cognition is not a passive act on the part of an organism but a dynamic process realized in and through *action.* Lorenz *(1983:102)* has perceptively pointed out that the German word for reality, *Wirklichkeit, is* derived from the verb *wirken,* "to act upon." The evolution of cognitive mechanisms is the story of successive fittings of the organism's actions upon the internal and external environments.

It is remarkable how complicated and well adapted inborn behavioral patterns can be, as numerous studies by ethologists have shown.

Consider, for instance [writes Bonner], a solitary wasp. The female deposits her eggs in small cavities, adds some food, and seals off the chamber. Upon emergence the young wasp has never seen one of its own kind, yet it can walk, fly, eat, find a mate, mate, find prey, and perform a host of other complex behavioral patterns. This is all done without any learning from other individuals. It is awesome to realize that so many (and some of them complex) behavioral patterns can be determined by the genes. *(1980:40)*

Isn't this as remarkable as "that our reasoning power was brought, by Darwin's process of natural selection, to the perfection which it seems to possess?" (Recall the quote from Wigner.) From the rigid *single-choice behavior,* as in the case of the solitary wasp, through the evolution of *multiple-choice behavior 10* and up to our capacity of *planned actions,* all intermediate stages occur and often concur.

As behavior and sense organs became more complex [writes Simpson], perception of sensation from those organs obviously maintained a realistic relationship to the environment. To put it crudely but graphically, the monkey who did not have a realistic perception of the tree branch he jumped for was soon a dead monkey and therefore did not become one of our ancestors. Our perceptions do give true, even though not complete, representations of the outer world because that was and is a biological necessity, built into us by natural selection. If we were not so, we would not be here! We do now reach perceptions for which our ancestors had no need, for example, of X-rays or electric potentials, but we do so by translating them into modalities that are evolution-tested. (1963:84)

The nervous system is foremost a steering device for the internal and external coordination of activities. There is no such thing as an "illogical" biological coordination mechanism, else survival would not have been possible. "For survival," writes Oeser, "it is not the right images which count but the corresponding (re)actions" (1988:38). The coordinating activities of the nervous system proceed mostly on a subconscious level; we become aware of the hand that reaches out to catch a falling glass only at the end of the action. (It is estimated that from an input of 10 bits/sec, only 102 bits/sec reach consciousness.) Yet another crucial mechanism has evolved, known on the human level as *planned action.* It permits a choice of action or hypothetical reasoning: we can imagine, prior to acting, the possible outcome of an action and thereby minimize all risks. The survival value of anticipatory schemes is obvious. *When we form a representation for possible action, the nervous system apparently treats this representation as if it were a sensory input, hence processes it by the same logico-operational schemes as when dealing with an environmental situation* (cf. Shepard and Cooper, 1981, for some fascinating data). From a different perspective, Maturana and Varela express it this way: "all states of the nervous system are internal states, and the nervous system cannot make a distinction in its process of transformations between its internally and externally generated changes" (1980:131). Thus, the logical schemes in hypothetical representations are the same as the logical schemes in the coordination of actions, schemes that have been tested through eons of evolution and which by now are genetically fixed.

The preceding considerations have far reaching implications for mathematics. Under *"logico-mathematical schemes"* Piaget understands the cognitive schemes that concern groupings of physical objects, arranging them in order, comparing groupings, and so on. These basic premathematical schemes have a genetic envelope but mature by stages in the intellectual development of the individual. They are based on the equally genetically fixed *logico-operational schemes,* a term that I have introduced, as these schemes operate also on the nonhuman level. The logico-operational schemes form the basis of our logical thinking. As it is a fundamental property of the nervous system to function through recursive loops, any hypothetical representation that we form is dealt with by the same "logic" of coordination as in dealing with real life situations. Starting from the elementary logico-mathematical schemes, a hierarchy is established. Under the impetus of sociocultural factors, new mathematical concepts are progressively introduced, and each new layer fuses with the previous layers." in structuring new layers, the same cognitive mechanisms operate with respect to the previous layers as they operate with respect to an environmental input. This may explain, perhaps, why the working mathematician is so prone to Platonistic illusions. The sense of reality that one experiences in dealing with mathematical concepts stems in part from the fact that in all our hypothetical reasonings, the object of our reasoning is treated in the nervous system by cognitive mechanisms, which have evolved through interactions with external reality (see also the quotation from Borel in note 16).

To summarize: mathematics does not reflect reality. But our cognitive mechanisms have received their imprimatur, so to speak, through dealing with the world. The empirical component in mathematics manifests itself not on the thematic level, which is culturally determined, but through the logico-operational and logico-mathematical schemes. As the patterns and structures that mathematics consists of are molded by the logico-operational neural mechanisms, these abstract patterns and structures acquire the status of *potential* cognitive schemes for forming abstract hypothetical world pictures. Mathematics is a singularly rich *cognition pool* of humankind from which schemes can be drawn for formulating *theories* that deal with phenomena that lie outside the range of daily experience and, hence, for which ordinary language is inadequate. Mathematics is structured by cognitive mechanisms, which have evolved in confrontation with experience, and, in its turn, mathematics is a tool for structuring domains of indirect experience. But mathematics is more than just a tool. *Mathematics is a collective work of art that derives its objectivity through social interaction.* "A mathematician, like a painter or a poet, is a maker of patterns," wrote Hardy (1969:84). The metaphor of the weaver has been frequently evoked. But the mathematician is a weaver of a very special sort. When the weaver arrives at the loom, it is to find a fabric already spun by generations of previous weavers and whose beginnings lie beyond the horizons. Yet with the yarn of creative imagination, existing patterns are extended and sometimes modified. The weaver may only add a beautiful motif, or mend the web; at times, the weaver may care more about the possible use of the cloth. But the weaving hand, for whatever motive it may reach out for the shuttle, is the very *prehensile* organ that evolved as a grasping and branch clutching organ, and its coordinating actions have stood the test of an adaptive evolution." In the mathematician, the artisan and artist are united into an inseparable whole, a unity that reflects the uniqueness of humankind as *Homo artifex*.

The Trilemma of a Finitary Logic and Infinitary Mathematics

In 1902, *L'enseignement mathématique* launched an inquiry into the working methods of mathematicians. The questionnaire is reproduced (in English translation) as appendix I in Hadamard (1945). Of particular interest is question thirty, which, among others, Hadamard addressed to Einstein. (No date for the correspondence is given, but I situate it in the forties when Hadamard was at Columbia University). Question thirty reads as follows:

> It would be very helpful for the purpose of psychological investigation to know what internal or mental images, what kind of "internal world" mathematicians make use of, whether they are motor, auditory, visual, or mixed, depending on the subject which they are studying.

In his answer to Hadamard (1945, appendix 2:142-43), Einstein wrote:

(a) The words or the language, as they are written or spoken, do not seem to play any role in my mechanism of thought. The physical entities which seem to serve as elements in thought are certain signs and more or less clear images which can be "voluntarily" reproduced and combined
(b) The above mentioned elements are, in my case, of *visual and some of muscular type*. Conventional words or other signs have to be sought for laboriously only in a secondary stage, when the mentioned associative play is sufficiently established and can be reproduced at will [italics mine],

In the previous section I have discussed the core structures of mathematics which consist of the logico-operational schemes for the coordination of actions. Throughout the evolution of hominoids, the coordinating mechanisms of the hand and eye played a particularly important role, leading to the feasibility of extensive use of tools and, thereby, to further cortical developments. It is therefore not surprising that in dealing with concepts, where the same neural mechanisms are involved, visual and traces of kinesthetic elements manifest themselves in consciousness, as Einstein's testimonial confirms.

The world of our immediate actions is *finite*, and the neural mechanisms for anticipatory representations were forged through dealing with the finite.

Formal logic is not the source of our reasoning but only *codifies* parts of the reasoning processes. But whence comes the feeling of safety and confidence in the soundness of the schemes that formal logic incorporates? To an evolutioiary epistemologist, logic is not based on conventions; rather, we look for the biological substrata of the fundamental schemes of inference. Consider for instance *modus ponens*:

$$\frac{\begin{array}{c}A \rightarrow B\\ A\end{array}}{B}$$

If a sheep perceives only the muzzle of a wolf, it promptly flees for its life. Here, "muzzle -wolf" is "wired" into its nervous system. Hence the mere sight of a muzzle the muzzle of any wolf, not just the muzzle of a particular wolf-results in "inferring" the presence of a wolf. Needless to say, such inborn behavioral patterns are vital. For related examples, see Lorenz (1973) and Riedl (1979). The necessary character of logic, qua codified logico-operational schemes, thus receives a coherent explanation in view of its phylogenetic origin. It follows furthermore that *as far as logic is concerned, finitism does not need any further philosophical justification*. It is biologically imposed. The situation is different with respect to the thematic component of mathematics. Once the cultural step was taken in inventing number words and symbols which can be indefinitely extended, mathematics proper, as the science of the infinite, came into being. The story of the early philosophical groping with mathematical and possible physical infinity is well known. When at last full citizenship was conferred on the actual infinite-de facto by Kummer and Dedekind; de jure by Cantor and Zermelo-an intense preoccupation with foundational problems was set in motion.14 The first school to emerge was logicism à la Frege and Russell. "The logicistic thesis is," writes Church, "that logic and mathematics are related, not as two different subjects, but as earlier and later parts of the same subject, and indeed in such a way that mathematics can be obtained from pure logic without the introduction of additional primitives or additional assumptions" (Church, 1962:186). Had the logicist programme succeeded, then infinitary mathematics, a cultural product, would have received a finitary foundation in finitary, biologically based logic. But as early as 1902, Keyser already showed that mathematical induction required an axiom of infinity, and, finally, Russell had to concede that such an axiom (plus the axiom of reducibility) had to be added to his system. Thus, the actual infinite is the rock upon which logicism foundered. Still, the efforts of the logicist school were not in vain, as Church has pointed out: "it does not follow that logicism is barren of fruit. Two important things remain. One of these is the reduction of mathematical vocabulary to a surprisingly brief list of primitives, all belonging to the vocabulary of pure logic. The other is the basing of all existing mathematics on one comparatively simple unified system of axioms and rules of inference" (1962:186).15

The second attempt in finitist foundations for mathematics was undertaken by Hilbert in his famous program. It may not be inopportune to stress that Hilbert never maintained seriously that mathematics is devoid of content, and his oft-cited mot d'esprit that "mathematics is a game played according to certain simple rules with meaningless marks on paper" has regrettably resulted in unwarranted philosophical extrapolations. Hilbert's formalist program is a technique, a device, for proving the consistency of infinitary mathematics by finitistic means. In the very article in which he outlines his program, Hilbert said the following concerning Cantor's theory of transfinite numbers: "This appears to me the most admirable flower of the mathematical intellect and in general one of the highest achievements of purely rational human activity" (1967:373). A meaningless game? Hardly!

Through formalization of thematic mathematics, Hilbert proposed that "contentual inference [be] replaced by manipulation of signs according to rules" (1967:381). This *manipulation (manus,* literally "hand"), this *handling* of inscriptions in the manner one handles physical objects would be founded, from the perspective of evolutionary epistemology, on the safe *logico-operational schemes for dealing with the finite*. It was a magnificent program, and though in view of Gödel's incompleteness theorem it could not be carried out as originally conceived, its offshoot, proof theory, is a major flourishing branch of mathematical logic. Thus, the contributions of logicism and Hilbert's program are of lasting value. As to the original intent, we just have to accept that *one cannot catch an infinite fish with a finite net!* Thus there remain three alternatives:

1. Use an infinite net, say of size ε^0 (Gentzen)
2. Eat only synthetic fish (Brouwer)
3. Be undernourished and settle for small fish (strict finitism).

Chacun à son goût!

Invention Versus Discovery

"But tell me," asked Socrates in Rényi's dialogue, "the mathematician who finds new truth, does he discover it or invent it?" We all know that a time-honored way to animate an after-dinner philosophical discussion is to ask such a question. People agree that following common usage of language, Columbus did not invent America, nor did Beethoven discover the ninth symphony. But when a new drug has been synthesized we commonly speak of a discovery, though the molecule never existed anywhere prior to the creative act of its synthesizers. Hadamard, in the introduction to his book *The Psychology of Invention in the Mathematical Field,* observes that "there are plenty of examples of scientific results which are as much discoveries as inventions," and thus he prefers not to insist on the distinction between invention and discovery (1945:xi). Yet there are philosophers of mathematics who are committed to an essential distinction between discovery and invention. To the intuitionist, mathematical propositions are mental constructions and, as such, could not result from a discovery. The Platonist, on the other hand, believes "that mathematical reality lies outside us, that our function is to discover or *observe* it," as Hardy (1969:123) put it. The conventionalist, though for different reasons, would side with the intuitionist and consider mathematics to be invented. Apparently, neither logicism nor formalism is committed to a discovery/ invention dichotomy. Is the debate about invention versus discovery an idle issue, or can one use the commonsense distinction between the two terms in order to elucidate the distinct components in the growth of mathematical knowledge? Let us examine the issue through a standard example. I propose to argue that: (1) the *concept* of 'prime number' is an invention; (2) the *theorem* that there are infinitely many prime numbers is a discovery. (N.b., Euclid's formulation, book 9, prop. 20, reads: "Prime numbers are more than any assigned multitude of prime numbers.")

Why should the concept of prime number be considered an invention, a purely creative step that need not have been taken, while contrariwise it appears that an examination of the factorization properties of the natural numbers leads immediately to the "discovery" that some numbers are composite and others are not, and this looks like a simple "matter of fact"? Weren't the prime numbers already there, tucked away in the sequence of natural numbers prior to anyone noticing them? Now things are not that simple. First of all, the counting numbers, like other classificatory schemes, did not make a sudden appearance as an indefinitely extendable sequence. Some cultures never went beyond coining words for the first few whole numbers. There are even languages destitute of pure numeral words. But even in cultures with a highly developed arithmetic, like the ones in ancient Babylonia, Egypt, or China, the concept of prime number was absent. Mo (1982) has shown how mathematicians in ancient China, though lacking the concept of prime number, solved problems such as reduction of fractions to lowest terms, addition of fractions, and finding Pythagorean triplets. Could it reasonably be said that the Chinese just missed "discovering" the prime numbers, and that so did the Babylonians and the Egyptians, in spite of their highly developed mathematical culture extending over thousands of years? I don't think so. In retrospect it seems to us that there was some sort of necessity that the concept of prime number be stumbled upon. But this is a misleading impression. *Evolution, be it biological or cultural, is opportunistic.* Much of our modern mathematics would still stay intact if the concept of prime number were lacking, though number theory and, hence, portions of abstract algebra would be different. There are $2^{\aleph_0}$ subsets of N of which only $\aleph_0$ can be defined by any linguistic means. We neither discover nor invent any one of these subsets separately. But when the inventive step was taken in formulating the *concept* of prime numbers, one of the subsets of N was singled out (that is, to serve as a *model,* in modern terminology). Some historians of mathematics attribute to the Pythagoreans certain theorems involving primes, but it is more likely that the concept of prime number is of a later date. It is conceivable that there is a connection between cosmological reflections about the ultimate constituents of matter by the Greek atomists and thoughts about numerical atoms, that is, prime numbers. Whatever the tie may be, one thing is certain: the invention of mathematical concepts is tied to culture. As White (1956) affirmed contra Platonistic doctrines, the "locus of mathematical reality is cultural tradition." The evolution of mathematical concepts can be understood only in the appropriate sociocultural context.16

Let us note that concepts can be defined explicitly, as in the case of prime numbers, or implicitly, by a system of axioms, like the concept of a group. In either case it is an inventive act. Theorems, on the other hand, have more the character of a discovery, in the sense that one *discovers* a road linking different localities. Once certain concepts have been introduced and, so to speak, are already there, it is a matter of discovering their connection, and this is the function of proofs. To come back to the theorem that no finite set of primes can contain all the prime numbers, it has the character of a discovery when one establishes a *road map* (Goodstein, 1970) linking "set of primes," "number of elements," and so on, to yield a path to the conclusion. A proposed path may or may not be valid, beautiful, or interesting. But to say that a proof renders a proposition "true" is as metaphorical as when one claims to have found a "true" path. *It seems best to dispense altogether with the notion of mathematical truth.* (This has no bearing on the *technical* metamathematical notion of 'truth' in the sense of Tarski.) Gone is, then, too the outdated Aristotelian conception of 'true axioms'. (Think of Euclidean and non-Euclidean geometries.) Such a "no truth" view also resolves the infinite regress involved in the apparent flow of truth from axioms to theorems that Lakatos (1962) endeavored to eliminate by an untenable return to empiricism. The creative work of the mathematician consists of inventing concepts and developing methods permitting one to chart paths between concepts.17 This is how mathematics grows in response to internal and external *problems* and results in an edifice that is beautiful and useful at the same time.

Recapitulation and Concluding Remarks

The evolutionary point of view dominated this essay, both in its metaphorical as well as in its strict biological sense. I started with the view of mathematics as an evolving mansion, foundations included. In harmony with the current emphasis in the philosophy of mathematics on actual mathematical practice, one of my chief concerns was to elucidate the relationship between mathematics and external reality. Though I rejected empiricism as an inadequate philosophy of mathematics, I endeavored to account for the empirical components in mathematics whose presence is clearly felt but which are difficult to locate. Mathematics is a science of structures, of abstract patterns (cf. Resnik, 1981, 1982). It is a human creation, hence it is natural to look for biological as well as sociocultural factors that govern the genesis of mathematical knowledge. The success of mathematics as a cognitive tool leaves no doubt that some basic biological mechanisms are involved. The acquisition of knowledge by organisms, even in its simplest form, presupposes mechanisms that could have evolved only under environmental pressure. Evolutionary epistemology starts from the empirical fact that our cognitive apparatus is the result of evolution and holds that our world picture must be appropriate for dealing with the world, because otherwise survival would not have been possible (cf. Bollmer, 1975:102). Indeed, it is from the coordination of *actions* in dealing with the world that anticipatory schemes of action have evolved; these, in turn, are at the root of our logical thinking. Thus, the *phylogenetically* but not individually empirical element manifests itself in our logico-operational schemes of actions, which lie at the root of the elementary logico-mathematical operations as studied by Piaget. On the other hand, the *content* of mathematical theories is culturally determined, but the overall mathematical formation sits on the logico-operational scaffold. Mathematics is thus seen as a twotiered web: a *logico-operational level* based on cognitive mechanisms which have become fixed in adaptation to the world, and a *thematic level* determined by culture and social needs and hence in a continuous process of growth. This special double-tiered structure endows mathematics in addition to its artistic value with the function of a *cognition pool* which is singularly suitable beyond ordinary language for formulating scientific concepts and theories.

In the course of my discussion I also reassessed the rationale of logicism and Hilbert's program. Of the traditional philosophies of mathematics, only Platonism is completely incompatible with evolutionary epistemology. "How is it that the Platonistic conception of mathematical objects can be so convincing, so fruitful and yet so clearly false?" writes Paul Ernest in a review (1983).18 I disagree with Ernest on only one point: I do not think that Platonism is fruitful. As a matter of fact, Platonism has *negative effects* on research by blocking a dynamical and dialectic outlook. Just think of set theorists who keep looking for "the true axioms" of set theory, and the working mathematician who will not explore on equal footing the consequences of the negation of the continuum hypothesis as well as the consequences of the affirmation of the continuum hypothesis. For the same reason too many logicians still ignore paraconsistent and other "deviant" (!) logics. Like the biological theory of preformation-which is just another side of the same coin-Platonism has deep sociological and ideological roots. What Dobzhansky had to say about the preformist way of thinking applies *mutatis mutandis* to Platonism:

The idea that things are preformed, predestined, just waiting around the corner for their turn to appear, is pleasing and comforting to many people. Everything is destiny, fate. But to other people predestination is denial of freedom and novelty. They prefer to think that the flow of events in the world may be changed creatively, and that new things do arise. The influence of these two types of thinking is very clear in the development of biological theories. (1955:223)

And so it is in the philosophy of mathematics.

Starting with the misleading metaphor of mathematical truth, Platonists graft onto it the further misleading metaphor of mathematical objects as physical objects to which "truth" is supposed to apply. Metaphors are illuminating, but when metaphors are stacked one upon the other without end, the result is obscurity and finally obscurantism. I frankly confess that I am absolutely incapable of understanding what is meant by "ontological commitment" and the issue of the "existence of abstract objects," and I begin to suspect that the emperor wears no clothes. No, there are no preordained, predetermined mathematical "truths" that lie just out or up there. Evolutionary thinking teaches us otherwise.

Caminante, son tus huellas el camino y nada mas; caminante, no hay camino, se hace camino al andar. (Antonio Machado)

Walker, just your footsteps are the path and nothing more; walker, no path was there before, the path is made by act of walking.

NOTES

This is an expanded version of talks presented at the International Congress "Communication and Cognition. Applied Epistemology," Ghent, December 6-10, 1987; at the Logic Seminar of the Kurt-Gödel-Gesellschaft, Technical University, Vienna, May 30,1988; and at the Séminaire de Philosophie et Mathématique, Ecole Normale Supérieure, Paris, November 7, 1988. I thank the various organizers and participants for numerous stimulating conversations and discussions.

1. Concerning the role of Dedekind, often neglected in foundational discussions, see Edwards (1983). In his review of Edwards's paper, Dieudonné makes the following significant observation: "Dedekind broke entirely new ground in his free use of 'completed' infinite sets as single objects on which one could compute as with numbers, long before Cantor began his work on set theory" (Mathematical Reviews 84d:01028).
2. "With non-Euclidean geometry came into being a new state of mind which impressed its spirit of freedom on the whole development of modem mathematics" (Toth, 1986:90; this fascinating essay deals in considerable depth with the epistemological problem of non-Euclidean geometries).
3. From a current point of view, physical objects are considered as events or states that rest unaltered for a nonnegligible time interval. Though "event" and "state" *refer* to reality, in order to *speak* of them one needs the mathematical apparatus incorporated in physical theories. Thus one ends up again with mathematical concepts. Hence it is futile to look at mathematical concepts as *objects* in the manner of physical objects and then, to crown it all, relegate them to a Platonic abode. For a further discussion of ontological questions concerning physical objects, see Dalla Chiara, 1985; Dalla Chiara and Toraldo di Francia, 1982; Quine, 1976.

4. Volimer (1987a, 198Th) also stresses the difference between EE à la Lorenz as a *biological theory of the evolution of cognitive systems* and EE à la Popper as a theory of the *evolution of scientific ideas*.
5. Cf. Popper's disclaimer in Riedl et al., 1987:24. In Popper's philosophy, factual knowledge cannot serve as a basis for an epistemology, whereas evolutionary epistemology is committed to an "irresolvable nexus between empirical knowledge and metatheoretical reflections," following Vollmer. Moreover, the great strides of science in the last fifty years are due to ever-refined experimental techniques and technologies coupled with piecemeal modeling, rather than to the elaboration of grand theories. When one peeks into a modern research institute, one scarcely finds scientists engaging in a grandiose search for bold hypotheses and a frantic pursuit of refutations, but rather humbly approaching "nature with the view, indeed, of receiving information from it, not, however, in the character of a pupil, who listens to all that his master chooses to tell him, but in that of a judge, who compels the witness to reply to those questions which he himself thinks fit to propose" (Kant, 1934:10-11)
6. See Rényi, 1967. The booklet contains three dialogues: (1) "A Socratic Dialogue on Mathematics," whose protagonists are Socrates and Hippocrates; (2) "A Dialogue on the Applications of Mathematics," featuring Archimedes and Hieron; and (3) "A Dialogue on the Language of the Book of Nature," whose chief character is Galileo.
7. I have excerpted this material from pages 7-25 in A. Rényi, *Dialogues on Mathematics* (San Francisco: Holden Day, 1967). These excerpts are quoted with the kind permission of the publishers.
8. Aristotle, 1941 discussed the difficulties with the Platonist notion of mathematical *objects* and their existence. See *Metaphysics,* book 13, chap. 1-3, 1076a 33-s1078b 6.
9. We owe much of our understanding of the ontogenetic development of the various logico-mathematical schemes to the work of Piaget and his school. Compare MUller, 1987:102-6, for a succinct summary of Piaget's theory. Note that much though not all of Piaget's (onto)genetic epistemology is compatible with evolutionary epistemology. See the discussions by Apostel (1987) and Oeser (1988:40,165).
10. These terms are given by Bonner (1980). Of particular importance is Bonner's extension of the concept of 'culture', which he defines as follows: By culture I mean the transfer of information by behavorial means, most particularly by the process of teaching and learning, It is used in a sense that contrasts with the transmission of genetic information passed by direct inheritance of genes. The information passed in a cultural fashion accumulates in the form of knowledge and tradition, but the stress of this definition is on the mode of transmission of information, rather than its result. *In this* simple definition I have taken care not to limit it to man. (1980:10; italics mine)
11. It is a "fundamental principle of neuro-epistemology," writes Oeser, "that each new cognitive function results from an integration with previously formed and already existing functions" (1988:158).
12. The evolution of the hand as a prehensile organ not only enabled humans to grasp physical objects but led concomitantly to neural mechanisms enabling them to grasp relationships between objects. This is the path from prehension to comprehension, or in German, as Lorenz has pointed out, from *greifen* (to grasp), via *begreifen* (to understand), to *Begriff* (concept) (see Lorenz, 1973:192-94; Vollmer, 1975:104-5; Oeser and Seitelberger, 1988:159). From a neurophysiological point of view, notice the large area of the cortical maps of the hands (see Granit, 1977:64-65).

13. For a collection of most of the relevant passages in Aristotle, see Apostel (1952). Augustine had no qualms about the *actual infinite in* mathematics, to wit: "Every number is defined by its own unique character, so that no number is equal to any other. They are all unequal to one another and different, and the individual numbers are finite but *as a class they are infinite" (1984:496; my italics)*.
14. The "paradoxes" played only a minor role in this process and none in the case of Frege. For a discussion, compare Garciadiego, 1986 and the review by Corcoran in *Mathematical Reviews* 1988 (88a:01026).
15. However, there is no *unique* set theory with a unique underlying logic from which all presently known mathematics can be derived. (Just recall the numerous independence results and the needs of category theory.) Moreover, when one examines actual mathematical practice, the deficiencies of "standard logic" are apparent, as Corcoran (1973) has perspicaciously pointed out. Furthermore, cognitive psychologists and workers in artificial intelligence are keenly aware of the fact that our current schemes of formal logic are inapplicable for analyzing actual reasoning processes. (cf. Gardner, 1985:368-70 and the references cited therein). Much work needs to be done in developing a logic of actual reasoning.
16. For a further discussion, see White, 1956; Wilder, 1981. And Borel adds the following perceptive observation:
    [W]e tend to posit existence on all those things which belong to civilization or culture in that we share them with other people and can exchange thoughts about them. Something becomes objective (as opposed to "subjective") as soon as we are convinced that it exists in the minds of others in the same form as it does in ours, and that we can think about it and discuss together. Because the language of mathematics is so precise, it is ideally suited to defining concepts for which such a consensus exists. In my opinion, that is sufficient to provide us with a *feeling* of an objective existence, of a reality of mathematics (1983:13)
17. A radioscopy of mathematical proofs reveals their logical structure, and this aspect has traditionally been overemphasized at the expense of seeing the meat and flesh of proofs. The path between concepts not only has a logical part which serves to *convince;* such paths also establish interconnections which *modify* and *illuminate* complexes of mathematical ideas, and this is how proofs differ from derivations.
18. Similarly, Machover writes concerning Platonism: "The most remarkable thing about this utterly incredible philosophy is its success" (1983:4). And further down: "The clearest condemnation of Platonism is not so much its belief in the occult but its total inability to account for constructive mathematics" (1983:5).

References

Apostel, L.

1987 "Evolutionary Epistemology, Genetic Epistemology, History and Neurology." Pp. *311-26* in W. Callebaut and R. Pinxten *(1987)*.

Apostle, H. G.

1952 Aristotle's Philosophy of Mathematics. Chicago: University of Chicago Press.

Aristotle

1941 The Basic Works of Aristotle. New York: Random House.

Augustine

1984 Concerning the City of God against the Pagans, a new translation by H. Bettenson, Harmondsworth: Penguin.

Beth, E. J., and J. Piaget
1966 *Mathematical Epistemology and Psychology*. Dordrecht: D. Reidel.

Bonner, J. T.
1980 *The Evolution of Culture in Animals*. Princeton: Princeton University Press.

Borel, A.
1983 "Mathematics, Art and Science." *The Mathematical Intelligencer 5, 4:9-17.*

Bradie, M.
1986 "Assessing Evolutionary Epistemology." *Biology and Philosophy 1:401-59.*

Callebaut, W., and R. Pinxten, eds.
1987 *Evolutionary Epistemology: A Multiparadigm Program*. Dordrecht: D. Reidel.

Campbell, D. T.
1959 "Methodological Suggestions from a Comparative Psychology of Knowledge Processes." *Inquiry 2:152-82.*
1974 "Evolutionary epistemology." Pp. *413-63* in P. A. Schilpp (ed.), *The Philosophy of K. Popper,* pt. 1. La Salle: Open Court.

Campbell, D. 1., C. Heyes, and W. Callebaut
1987 "Evolutionary Epistemology Bibliography." Pp. *402-31* in Callebaut and R. Pinxten *(1987)*.

Carruccio, E.
1977 *Appunti di storia delle matematiche, della logica, della metamatematica.* Bologna: Pitagora Editrice.

Church, A.
1962 "Mathematics and Logic." Pp. *181-86* in E. Nagel, P. Suppes and A. Tarski, (eds.), *Logic, Methodology and Philosophy of Science*. Proceedings, *1960* International Congress on Logic, Methodology and Philosophy of Science, Stanford. Stanford: Stanford University Press.

Corcoran, J.
1973 "Gaps between Logical Theory and Mathematical Practice." Pp. *23-50* in M. Bunge (ed.), *The Methodological Unity of Science*. Dor-drecht: D. Reidel.

d'Espagnat, B.
1979 A la recherche du reel: Le regard d'un physicien. Paris: Gauthier-Villard d'Espagnat, B. and M. Paty
1980 "La physique et le réel". *Bulletin de la société francaise de philosophie 84:1-42.*

Dalla Chiara, M. L.
1985 "Some Foundational Problems in Mathematics Suggested by Physics." *Synthese 62, 2:303-15.*

Dalla Chiara, M. L. and G. T. di Francia
1982 "Consideraciones ontologicas sobre los objectos de la fisica moderna." *Analisis Filosofico 2:35-46.*

Dobzhansky, H. M.
1955 *Evolution, Genetics, and Man*. New York: John Wiley.

Edwards, H. M.
1983 "Dedekind's Invention of Ideals." *Bulletin of the London Mathematical Society. 15:8-17.*

Ernest, P.
1983 *Mathematical Reviews 83K(00010):4381*

Feferman, S.
1985 "Working Foundations." *Synthese 62,2:229-54.*

Garciadiego, A. R.
1986 "On Rewriting the History of the Foundations of Mathematics." *Historia Mathematica 13:38-41.*

Gardner, H.
1985 The Mind's New Science: A History of the Cognitive Revolution. New York: Basic Books.

Goodstein, R. L.
1970 "Empiricism in Mathematics." *Dialectica 23:50-57.*

Granit, R.
1977 *The Purposive Brain*. Cambridge, Mass.: MIT Press.

Hadamard, Jacques S.
1945 The Psychology of Invention in the Mathematical Field. Princeton: Princeton University Press.

Hardy, G. H.
1940 *A Mathematician's Apology*. Cambridge: Cambridge University Press.

Hersh, R.
1979 "Some proposals for reviving the philosophy of mathematics." *Advances in Mathematics 31:31-50.*

Hilbert, D.
1967 "On the Infinite." Pp. *369-92* in J. Van Heijenoort (ed.), *From Frege to Gode!.* Cambridge, Mass.: Harvard University Press.

Kant, E.
1934 *Critique of Pure Reason* second ed., translated by J. H. D. Meiklejohn. London: Everyman's Library.

Kaspar, R
1983 "Die biologischen Grundlagen der evolutionären Erkenntnistheotie." Pp. *125-45* in K. Lorenz et al. *(1983).*

Keyser, C. J.
1902 "Concerning the Axiom of Infinity and Mathematical Induction." *Bulletin of the American Mathematical Society. 9:424-34.*

Kitcher, P.
1981 "Mathematical Rigor-Who Needs It?" Nous *15:469-93.*
1983 *The Nature of Mathematical Knowledge.* New York: Oxford University

Körner, S.
1960 *The Philosophy of Mathematics.* London: Hutchins.
1956 "An Empiricist Justification of Mathematics." Pp. *222-27* in Y. Bar-Hillel (ed.), *Logic, Methodology and Philosophy of Science.* Proceedings, *1964,* International Congress of Logic, Methodology and Philosophy of Science, Jerusalem. Amsterdam: North-Holland.

Kreisel, G.
1973 "Perspectives in the Philosophy of Mathematics." Pp. *255-77* in P. Suppes (ed.), *Logic, Methodology and Philosophy of Science IV.* Amster-*dam, North-Holland.*

Lakatos, I.
1962 "Infinite Regress and the Foundations of Mathematics." *Aristotelian Society Proceedings 36:155-84.*

Lolli, G.
1982 "La dimostrazione in matematica: Analisi di un dibattido." *Bolletino* Unione Matematica Italiana Vol. 61, 1:197-216.

Lorenz, K.
1983 "Kants Lehre vom Apriorischen in Lichte gegenwärtiger Biologie." *Pp. 95-124* in Lorenz and Wuketits, *(1983).*

Lorenz, K. and F. M. Wuketits, eds.
1983 *Die Evolution des Denkens.* Munich: R. Pieper.

Machover, M.
1983 "Towards a new philosophy of mathematics." *British Journal for Philosophy of Science 34:1-11.*

Marshack, A.
1972 *The Roots of Civilization.* London: Weiderfeld & Nicolson.

Maturana, H. R.
1980 "Biology of Cognition." Pp. *5-57* in Maturana and Varela *(1980).*

Maturana, H. R. and F. J. Varela
1980 *Autopoiesis and Cognition.* Dordrecht: D. Reidel.

Mayr, E.
1974 "Behavioral Programs and Evolutionary Strategies." *American Scientist 62:650-659.*

Michaels, A.
1978 *Beweisverfahren in der vedischen Sakralgeometrie.* Wiesbaden: Franz Steiner.

Mo, S. K.
1982 "What Do We Do If We Do Not Have the Concept of Prime Numbers? A Note on the History of Mathematics in China." *Journal of Mathematical Research*

Expositions 2:183-87. In Chinese; English summary in *Mathematical Reviews 84i:01024*.

Muffler, H. M.
1987 *Evolution, Kognition und Sprache*. Berlin: Paul Parey.

Oeser, E.
1987 Psychozoikum: Evolution und Mechanismus der menschlichen Erkenntnis*fähigkeit*. Berlin: Paul Parey.

Oeser, E., and F. Seitelberger
1988 Gehirn, Bewusstsein und Erkenntnis. Berlin: Paul Parey.

Piaget, J.
1967 Biologie et Connaissance. Paris: Gallimard.
1970 *Genetic Epistemology*. New York: Columbia University Press.
1971 Biology and Knowledge: An Essay on the Relations between Organic Regu*lations and Cognitive Processes*. Chicago: University of Chicago Press. (English translation by Beatrix Walsh of Piaget, *1967)*.

Plotkin, H. C., ed.
1982 *Learning, Development and Culture: Essays in Evolutionary Epistemology*. New York: John Wiley.

Popper, K. R.
1972 *Objective Knowledge: An Evolutionary Approach*. Oxford: Oxford University Press.

Quine, W. V. O.
1976 "Whither Physical Objects?." Pp. *497-504* in R. S. Cohen, P. K. Feyerabend and M. W. Wartofsky (eds.), *Essays in Memory of Imre Lakatos*. Dordrecht: D. Reidel.
1980 "On What There Is." Pp. *1-19* in W. V. O. Quine, *From a Logical Point of View*. Cambridge, Mass.: Harvard University Press.

Rényi, A.
1967 *Dialogues on Mathematics*. San Francisco: Holden-Day.

Resnik, M. D.
1975 "Mathematical Knowledge and Pattern Cognition." *Canadian Journal for Philosophy 5:25-39*.
1981 "Mathematics as a Science of Patterns: Ontology and Reference." *NoQs 15:529-50*.
1982 "Mathematics as a Science of Patterns: Epistemology." Nous *16:95-105*.

Resnik, M. D., and D. Kushner
1987 "Explanation, Independence and Realism in Mathematics." *British Journal for Philosophy of Science 38:141-58*.

Riedl, R.
1979 *Biologic der Erkenntnis: Die stammgeschichtlichen Grundlagen der Vernunft*. Berlin: Paul Parey.

1984 *Biology of Knowledge: The Evolutionary Basis of Reason*. Chichester: John Wiley. (English translation by Paul Feulkes of Riedl, *1979.)*

Riedl, R. and F. M. Wuketits, eds.
1987 *Die evolutionäre Erkenntnistheorie*. Berlin: Paul Parey.

Ruzavin, G. I.
1977 *Die Natur der mathematischen Erkenntnis*. Berlin: Akademie Verlag.

Saint-Exupéry, A.
1948 *Citadel*. Paris: Gallimard.

Seidenberg, A.
1962 "The ritual origin of counting." *Archives of the History of the Exact Sciences 2:1-40.*
1977 "The origin of mathematics." *Archives of the History of the Exact Sciences 18:301-42.*
1981 "The ritual origin of the circle and square." *Archives of the History of the Exact Sciences 25:269-327.*

Shapiro, S.
1983 "Mathematics and Reality." *Philosophy of Science 50:523-548.*

Shepard, R. N., and L. A. Cooper
1981 *Mental Images and Their Transformations*. Cambridge, Mass.: MIT Press.

Simpson, G. G.
1963 "Biology and the Nature of Science." *Science* 139:81-88.

Steiner, M.
1978a"Mathematics, Explanation, and Scientific Knowledge." Nofls 12:17-28.
1978b"Mathematical Explanation." *Philosophical Studies,* 34:135-51.
1983 "Mathematical realism." *Nous* 17:363-85.

Toraldo di Francia, G.
1978 "What is a physical object?" *Scientia* 113:57-65.

Toth, I.
1986 "Mathematische Philosophie und hegelsche Dialektik." Pp. 89-182 in M. J. Petry, (ed.), *Hegel und die Naturwissenschaften*. Stuttgart: Frommann-Holzboog.

Ursua, N.
1986 "Conocimento y realidad: Aproximacion a una hipotesis." *Theoria* 2, 5-6;461-502.

Van Bendegem, J. P.
1987 "Fermat's Last Theorem Seen as an Exercise in Evolutionary Epistemology." Pp. 337-63 in W. Callebaut and R. Pinxten (1987).

Voilmer, G.

1984 "Mesocosm and Objective Knowledge." Pp. 69-121 in F. M. Wuketits (ed.), *Concepts and Approaches in Evolutionary Epistemology*. Dordrecht: D. Reidel.

1985 *Was kdnnen wir wissen? Vol. 1: Die Natur der Erkenntnis*. Stuttgart: S. Hirzel.

1986 *Was kdnnen wir wissen?* Vol. 2: *Die Erkenntnis der Natur*. Stuttgart: S. Hirzel.

1987a *Evolutionàre Erkenntnistheorie*. Stuttgart: S. Hirzel.

1987b "What Evolutionary Epistemology Is Not." Pp. 203-21 in W. Callebaut and R. Pinxten (1987).

White, L. A.

1956 "The Locus of Mathematical Reality: An Anthropological Footnote." Pp. 2348-64 in J. R. Newman (ed.), *The World of Mathematics, vol.* 4. New York: Simon and Schuster.

Wigner, E. P.

1960 "The Unreasonable Effectiveness of Mathematics in the Natural Sciences". *Communications in Pure and Applied Mathematics*. 13:1-14.

Wilder, R. L.

1981 *Mathematics as a Cultural System*. Oxford: Pergamon.

Wittgenstein, L.

1983 *Tractatus Logico-Philosophicus*. London: Routledge and Kegan Paul.

6

Toward a Semiotics of Mathematics

BRIAN ROTMAN

Preface

As the sign system whose grammar has determined the shape of Western culture's techno-scientific discourse since its inception, mathematics is implicated, at a deeply linguistic level, in any form of distinctively intellectual activity; indeed, the norms and guidelines of the 'rational' - valid argument, definitional clarity, coherent thought, lucid explication, unambiguous expression, logical transparency, objective reasoning - are located in their most extreme, focused, and highly cultivated form in mathematics. The question this essay addresses - what is the nature of mathematical language? should therefore be of interest to semioticians and philosophers as well as mathematicians.

There are, however, certain difficulties. inherent in trying to address such disparate types of readers at the same time which it would he disingenuous not to acknowledge at the outset.

Consider the mathematical reader. On the one hand it is no accident that Peirce, whose writings created the possibility of the present essay, was a mathematician: nor one that I have practiced as a mathematician; nor that Hilbert, Brouwer, and Frege - the authors of the accounts of mathematics I shall dispute - were mathematicians. Mathematics is cognitively difficult, technical, abstract, and (for many) defeatingly impersonal: one needs, it seems, to have been inside the dressing room in order to make much sense of the play. On the other hand, one cannot stay too long there if the play is not to disappear inside its own performance. In this respect mathematicians confronted with the nature of their subject arc no different from anybody else. The language that textual critics, for example, use to talk about criticism will be permeatcd by precisely those features - figures of ambiguity, polysemy, compression of meaning, subtlety and plurality of interpretation, rhetorical tropes, and so on - which these critics value in the texts they study; likewise mathematicians will create and respond to just those discussions of mathematics that ape what attracts them to their subject matter. Where textual critics literize their rnetalanguage. mathematicians mathcmatize theirs. And since for mathematicians the principal activity is proving new theorems, what they will ask of my description of their subject is: can it be the source of new mathematical material? Does it suggest new notational systems, definitions, assertions, proofs? Now it is certainly the case that the accounts offered by Frege. Brouwer, and Hilbert all satisfied this requirement: each put forward a program that engendered new matbematics: each acted in and wrote the play and, in doing so, gave a necessarily truncated and misleading account of mathematics. Thus, if a semiotic approach to mathematics can be made to yield theorems and be acceptable to mathematicians, it is unlikely to deliver the kind of exterior view of mathematics it promises; if it does not engender theorems, then mathematicians will be little interested in its project of redescribing their subject -'queen of sciences' – via an explanatory formalism that (for them) is in a pre-scientific stage of arguing about its own fundamental terms. Since the account I have given is not slanted toward the creation of new mathematics, the chances of interesting mathematicians - let alone making a significant impact on them - look slim.

With readers versed in semiotics the principal obstacle is getting them sufficiently behind the mathematical spectacle to make sense of the project without losing them in the stage machinery. To this end I've kept the presence of technical discussion down to the absolute minimum. If I have been successful in this then a certain dissatisfaction presents itself: the sheer *semiotic* skimpiness of the picture I offer. Rarely do I go beyond identifying an issue, clearing the ground, proposing a solution, and drawing a consequence or two. Thus, to take a single example, readers familiar with recent theories of narrative are unlikely to feel more than titillated by being asked to discover that the persuasive force of proofs, of formal arguments within the mathematical Code, are to be found in stories situated in the metaCode. They would want to know what sort of stories, how they relate to each other, what their genres are, whether they are culturally and historically invariant, how what they tell depends on the telling, and so on. To have attempted to enter into these questions would have entailed the very technical mathematical discussions I was trying to avoid. Semioticians, then, might well feel they have been served too thin a gruel. To them I can only say that beginnings are difficult and that if what I offer has any substance, then others - they themselves, perhaps - will be prompted to cook it into a more satisfying sort of semiotic soup.

Finally, there are analytic philosophers. Here the difficulty is not that of unfamiliarity with the mathematical issues. On the contrary, no one is morc familiar with them: the major thrust of twentieth-century analytic philosophy can be seen as a continuing responsc to the questions of reference, meaning, truth, naming, existence, and knowledge that emergcd from work in mathematics, logic, and metamathematics at the end the last century. And indeed, all the leading figures in the modern analytic tradition - from Frege, Russell, and Carnap to Quine, Wittgenstein, and Kripke - have directly addressed the question of mathematics in some way or other. The problem is rather that of incompatibility, a lack of engagement between forms of enquiry. My purpose has been to describe mathematics as a practice, as an ongoing cultural endeavor: and while it is unavoidable that any description I come up with will be riddled with unresolved philosophical issues, these are not - they cannot be if I am to get off the ground - my concern. Confronted, for example, with thc debates and counter-debates contained in the elaborate secondary and tertiary literature on Frege, my response has been to avoid them and regard Frege's thought from a certain kind of semiotic scratch. So that if entering these debates is the only route to the attention of analytic philosophers, then the probability of engaging such readers seems not very great.

Obviously, I hope that these fears arc exaggerated and misplaced, and that there will be readers from each of these three academic specialisms prepared to break through what are only, after all, disciplinary barriers.

Introduction

My purpose here is to initiate the project of giving a semiotic analysis of mathematical signs; a project which, though implicit in the repeated references to mathematics in Peirce's writings on signs, seems not so far to have been carried out. Why this is so - why mathematics, which is so obviously a candidate for semiotic attention, should have have received so little of it - will, I hope, emerge. Let me begin by presenting a certain obstacle, a difficulty of method, in the way of beginning the enterprise.

It is possible to distinguish, without being at all subtle about it, three axes or aspects of any discourse that might serve as an external starting point for a semiotic investigation of the code that underlies it. There is the refcrential aspect, which concerns itself with the code's secondarity, with the objects of discourse, the things that are supposedly talked about and referred to by the signs of the code; the formal aspect, whose focus is on the manner and form of the material means through which the discourse operates, its physical manifcstation as a medium; and the psychological aspect, whose interest is primarily in those interior meanings which the signs of the code answer to or invoke. While all three of these axes can be drawn schematically through any given code, it is nonetheless the case that some codes seem to present themselves as more obviously biased toward just one of them. Thus, so-called representational codes such as perspectival painting or realistically conceived literature and film come clothed in a certain kind of secondarity; before all else they seem to he 'about' some world external to themselves. Then there are those signifying systems, such as that of non-representational painting for example, where secondarity seems not to be in evidence, but where there is a highly palpable sensory dimension - a concrete visual order of signifiers - whose formal material status has a first claim on any semiotic account of these codes. And again, there are codes such as those of music and dance where what is of principal semiotic interest is how the dynamics of performance, of enacted gestures in space or time, are seen to be in the service of some prior psychological meanings assumed or addressed by the code.

With mathematics each of these external entry points into a semiotic account seems to be highly problematic: mathematics is an art that is practiced, not performed; its signs seem to be constructed - as we shall see - so as to sever their signifieds, what they are supposed to mean, from the real time and space within which their material signifiers occur; and the question of secondarity, of whether mathematics is 'about' anything, whether its signs have referents, whether they are signs of something outside themselves, is precisely what one would expect a semiotics of mathematics to be in the business of discussing. In short, mathematics can only offer one of these familiar semiotic handles on itself-the referential route through an external world, the formal route through material signifiers, the psychological route through prior meanings - at the risk of begging the very semiotic issues requiring investigation.

To clarify this last point and put these three routes in a wider perspective, let me anticipate a discussion that can only be given fully later in this essay, after a semiotic model of mathematics has been sketched. For a long time mathematicians, logicians, and philosophers who write on the foundations of mathematics have agreed that (to put things at their most basic) there are really only three serious responses - mutually antagonistic and incompatible - to the question 'what is mathematics?' The responses - formalism, intuitionism, and platonism - run very briefly as follows.

For the formalist, mathematics is a species of game, a determinate play of written marks that are transformed according to explicit unambiguous formal rules. Such marks are held to be without intention, mere physical inscriptions from which any attempt to signify, to mean, is absent: they operate like the pieces and moves in chess which, though they can be made to carry significance (rcpresenting strategies, for example), function independently of such - no doubt useful but inherently posterior, after the event - accretions of meaning. Formalism, in other words, reduces mathematical signs to material signifiers which are, in principle, without signifieds. In Hilbert's classic statement of the formalist credo, matheniatics consists of manipulating 'meaningless marks on paper'.

Intuitionists, in many ways the natural dialectical antagonists of formalists, deny that signifiers - whether written, spoken, or indeed in any other medium - play any constitutive role in mathematical activity. For intuitionism mathematics is a species of purely mental construction, a form of internal cerebral labor, performed privately and in solitude within the individual - but cognitively universal - mind of the mathematician. If formalism characterizes mathematics as the manipulation of physical signifiers in the visible, intersubjective space of writing, intuitionism (in Brouwer's formulation) sees it as the creation of immaterial signifieds within the Kantian - inner, a priori, intuited - category of time. And as the formalist reduces the signified to an inessential adjunct of the signifier, so the intuitionist privileges the signified and dismisses the signifier as a useful but theoretically unnecessary epiphenomenon. For Brouwer it was axiomatic that 'mathematics is a languageless activity'.

Last, and most important, since it is the orthodox position representing the view of all but a small minority of mathematicians, there is platonism. For platonists mathematics is neither a formal and meaningless game nor some kind of languageless mental construction, but a science, a public discipline concerned to discover and validate objective or logical truths. According to this conception mathematical assertions are true or false propositions, statements of fact about some definitc state of affairs, some objective reality, which exists independently of and prior to the mathematical act of investigating it. For Frege, whose logicist program is the principal source of twentieth-century platonism, mathematics seen in this way was nothing other than an extension of pure logic. For his successors there is a separation: mathematical assertions are facts - specifically, they describe the properties of abstract collections (sets) - while logic is merely a truth-preserving form of inference which provides the means of proving that these descriptions are 'true'. Clearly, then, to the platonist mathematics is a realist science, its symbols are symbols of certain real - pre-scientific - things, its assertions are consequently assertions about some determinate, objective subject matter, and its epistemology is framed in terms of what can be proved true concerning this reality,

The relevance of these accounts of mathematics to a semiotic project is twofold. First, to have persisted so long each must encapsulate, however partially, an important facet of what is felt to be intrinsic to mathematical activity. Certainly, in some undeniable but obscure way, mathematics seems at the same time to be a meaningless game, a subjective construction, and a source of objective truth. The difficulty is to extract these part truths: the three accounts seem locked in an impasse which cannot be escaped from within the common terms that have allowed them to impinge on each other. As with the scholastic impasse created by nominalism, conceptualism, and realism - a parallel made long ago by Quine - the impasse has to be transcended. A semiotics of mathematics cannot, then, be expected to offer a synoptic reconciliation of these views; rather, it must attempt to explain - from a semiotic perspective alien to all of them - how each is inadequate, illusory, and undeniably attractive. Second, to return to our earlier difficulty of where to begin, each of these pictures of mathematics, though it is not posed as such, takes a particular theory-laden view about what mathematical signs are and are not; so that, to avoid a sell-fulfilling circularity, no one of them can legitimately serve as a starting point for a semiotic investigation of mathematics. Thus, what we called the route through material signifiers is precisely the formalist obsession with marks, the psychological route through prior meaning comprehends intuitionisni, and the route through an external world of referents is what all forms of mathematical platonism require.

A semiotic model of mathematics

Where then can one start? Mathematics is an activity, a practice. If one observes its participants it would be perverse not to infer that for large stretches of time they are engaged in a process of *communicating* with themselves and each other; an inference prompted by the constant presence of standardly presented formal written texts (notes, textbooks, blackboard lectures, articles, digests, reviews, and the like) being read, written, and exchanged, and of informal signifying activities that occur when they talk, gesticulate, expound, make guesses, disagree, draw pictures, and so on. (The relation between the formal and informal modes of communication is an important and interesting one to which we shall return later; for the present I want to focus on the written mathematical text.)

Taking the participants' word for it that such texts are indeed items in a communicative network, our first response would be to try to 'read' them, to try to decode what they arc about and what sorts of things they are saying. Pursuing this, what we observe at once is that any mathematical text is written in a mixture of words, phrases, and locutions drawn from some recognizable natural language together with mathematical marks, signs, symbols, diagrams, and figures that (wc suppose) are being used in some systematic and previously agreed upon way. We will also notice that this mixture of natural and artificial signs is conventionally punctuated and divided up into what appear to he complete grammatical sentences; that is, syntactically self-contained units in which noun phrases ('all points on X', 'the number y', 'the first and second derivatives of', 'the theorem alpha', and so on) are systematically related to verbs ('count', 'consider', 'can be evaluated', 'prove', and so on) in what one takes to be the accepted sense of connecting an activity to an object.

Given the problem of 'objects'. and of all the issues of ontology, reference, 'truth', and secondarity that surface as soon as one tries to identify what mathematical particulars and entities such as numbers, points, lines, functions, relations, spaces, orderings, groups, sets, limits, morphisms, functors, and operators 'are', it would be sensible to defer discussion of the interpretation of nouns and ask questions about the 'activity' that makes up the remaining part of the sentence; that is, still trusting to grammar, to ask about verbs.

Linguistics makes a separation between verbs functioning in different grammatical moods; that is, between modes of sign use which arise, in the case of speech, from different roles which a speaker can select for himself and his hearer. The primary such distinction is between the indicative and the imperative.

The indicative mood has to do with asking for (interrogative case) or conveying (declarative case) information - 'the speaker of a clause which has selected the indicative plus declarative has selected for himself the role of informant and for his hearer the role of informed' (Berry 1975: 166). For mathematics, the indicative governs all those questions, assumptions, and statements of information - assertions, propositions, posits, theorems, hypotheses, axioms, conjectures, and problems - which either ask for, grant, or deliver some piece of mathematical content, some putative mathematical fact such as 'there are infinitely many prime numbers', 'all groups with 7 elements are abelian'. '5 + 11 + 3 = 11 + 3 + 5', 'there is a continuous curve with no tangent at any point', 'every even number can he written as the sum of two prime numbers', or, less obviously, those that might be said to convey metalingual information such as 'assertion A is provable', 'x is a counterexample to proposition P', 'definition D is legitimate', 'notational system N is inconsistent', and so on.

In normal parlance, the indicative bundles information, truth, and validity indiscriminately together; it being equivalent to say that an assertion is 'true', that it 'holds', that it is 'valid', that it is 'the case', that it is informationally 'correct', and so on. With mathematics it is necessary to be more discriminating: being 'true' (whatever that is ultimately to mean) is not the same attribute of an assertion as being valid (that is, capable of being proved): conversely, what is informationally correct is not always, even in principle, susceptible of mathematical proof. The indicative mood, it seems, is inextricably tied up with the notion of mathematical proof. But proof in turn involves the idea of an argument, a narrative structure of sentences, and sentences can be in the imperative rather than the indicative.

According to the standard grammatical description, 'the speaker of a clause which has chosen the imperative has selected for himself the role of controller and for his hearer the role of controlled. The speaker expects more than a purely verbal response. He expects some form of action' (Berry 1975: 166). Mathematics is so permeated by instructions for actions to be carried out, orders, commands, injunctions to be obeyed -'prove theorem T', 'subtract x from y', 'drop a perpendicular from point P onto line L', 'count the elements of set S', 'reverse the arrows in diagram D', 'consider an arbitrary polygon with k sides', and similarly for the activities specified by the verbs add', 'multiply', 'exhibit', 'find', 'enumerate', 'show', 'compute', 'demonstrate', 'define', 'eliminate', 'list', 'draw', 'complete', 'connect', 'assign', evaluate', 'integrate', 'specify', 'differentiate', 'adjoin', 'delete', 'iterate', 'order', 'complete', 'calculate', 'construct', etc. - that mathematical texts seem at times to be little more than sequences of instructions written in an entirely operational, exhortatory language.

Of course, mathematics is highly diverse, and the actions indicated even in this very incomplete list of verbs differ very widely. Thus, depending on their context **and** their domain of application (algebra. calculus. arithmetic, topology, and so on), they display radical differences in scope, fruitfulness, complexity, and logical character: some (like 'adjoin') might he finitary, others (like 'integrate') depend essentially on an infinite process; some (like 'count') apply solely to collections, others solely to functions or relations or diagrams, whilst others (like 'exhibit') apply to any mathematical entity; some can he repeated on the states or entities they produce, others cannot. To pursue these differences would require technical mathematical knowedge which would be out of place in the present project. It would also be beside the point: our focus on these verbs has to do not with the particular mathematical character of the actions they denote, but with differences between them - of an epistemological and semiotic kind - which are reflected in their grammatical status, and specifically in their use in the imperative mood.

Corresponding to the linguist's distinction between inclusive imperatives ('Let's go') and exclusive imperatives ('Go'), there seems to be a radical split between types of mathematical exhortation: inclusive commands marked by the verbs 'considcr', 'define', 'prove' and their synonyms - demand that speaker and hearer institute and inhabit a common world or that they share some specific argued conviction about an item in such a world; and exclusive commands - essentially the mathematical actions denoted by all other verbs - dictate that certain operations meaningful in an already shared world be executed.

Thus, for example, the imperative 'consider a Hausdorff space' is an injunction to establish a shared domain of Hausdorff spaces: it commands its recipient to introduce a standard, mutually agreed upon ensemble of signs - symbolized notions, definions, proofs, and particular cases that bring into play the ideas of topological neighborhood, limit point, a certain separability condition in such a way as to determine what it means to dwell in the world of such spaces. By contrast, an imperative like 'integrate the function f', for example, is mechanical and exclusive: it takes for granted that a shared frame (a world within the domain of calculus) has already been set, and asks that a specific operation relevant to this world be carried out on the function f. Likewise, the imperative 'define ...' (or equivalently, 'let us define . . .') dictates that certain sign uses be agreed upon as the shared givens for some particular universe of discourse. Again, an imperative of the type 'prove (or demonstrate or show) there are infinitely many prime numbers' requires its recipient to construct a certain kind of argument, a narrative whose persuasive force establishes a commonality between speaker and hearer with respect to the world of integers. By contrast, an imperative like 'multiply the integer x by its successor' is concerned not to establish commonality of any sort, but to effect a specific operation on numbers.

One can gloss the distinction between inclusive and exclusive cornmands by observing that the familiar natural language process of forming nouns from verbs - the gerund 'going' from the verb 'go' - is not available for verbs used in the mood of the inclusive imperative. Thus, while exclusive commands can always (with varying degrees of artificiality, to be sure) be made to yield legitimate mathematical objects -'add' gives rise to an 'adding' in the sense of the binary operation of addition, 'count' yields a 'counting' in the sense of a well-ordered binary relation of enumeration, and so on - such is not the case with inclusive commands: normal mathematical practice does not allow a 'defining' or a 'considering' or a 'proving' to be legitimate objects of mathematical discourse. One cannot, in other words, prove results about or consider or define a 'considering', a 'proving', or a 'defining' in the way that one can for an 'adding', a 'counting', etc. (An apparent exception to this occurs in meta mathematics, where certain sorts *of* definitions and proofs are themselves considered and defined, and have theorems proved about them; but metamathematics is still mathematics - it provides no violation of what is being elaborated here.)

The grammatical line we have been following formulates the imperative in terms of speakers who dictate and hearers who carry out actions. But what is to be understood by 'action' in relation to mathematical practice? What does the hearer-reader, recipient, addressee actually do in responding to an imperative? Mathematics can be an activity whose practice is silent and sedentary. The only things mathematicians can be supposed to do with any certainty are scribble and think; they read and write inscriptions which seem to be inescapably attached to systematically meaningful mental events. If this is so, then whatever actions they perform must be explicable in terms of a scribbling/thinking amalgam. It is conceivable, as we have seen, to deny any necessary amalgamation of these two terms and to construe mathematics purely as scribbling (as entirely physical and 'real': formalism's 'meaningless marks on paper') or purely as thinking (as entirely mental and 'imaginary': intuitionism's 'languageless activity'); but to adopt either of these polemical extremes is to foreclose on any semiotic project whatsoever, since each excludes interpreting mathematics as a business of using those signifier/signified couples one calls signs. Ultimately. then, our object has to be to articulate what mode of signifying, of scribbling/thinking, mathematical activity is, to explain how, within this mode, mathematical imperatives are discharged; and to identify who or what semiotic agency issues and obeys these imperatives.

Leaving aside scribbling for the moment, let us focus on mathematical 'thinking'. Consider the imperative 'consider a Hausdorfl space'. 'Consider' means view attentively, survey, examine, reflect, etc.; the visual imagery here being part of' a wider pattern of cognitive body metaphors such as understand, comprehend, defend, grasp, or get the feel of an idea or thesis. Therefore, any attempt to explicate mathematical thought is unlikely to escape the net of' such metaphors; indeed, to speak (as we did) of 'dwelling in a world of Hausdorff spaces' is metaphorically to equate mathematical thinking with physical exploration. Clearly, such worlds are imagined, and the actions that take place within these worlds are imagined actions. Someone has to be imagining worlds and actions, and something else has to be performing these imaginary actions. In other words, someone - some subjective agency - is imagining itself' to act. Seen in this way, mathematical thinking seems to have much in common with the making of self-reflective thought experiments. Such indeed was the conclusion Peirce arrived at:

> It is a familiar experience to every human being to wish for something quite beyond his present mcans, and to follow that wish by

> the question, 'Should I wish for that thing just the same, if I had ample means to gratify it?'. To answer that question, he searches his heart, and in so doing makes what I term an abstractive observation. He makes in his imagination a sort of skeleton diagram, or outline sketch of himself, considers what modifications the hypothetical state of things would require to be made in that picture, and then examines it, that is, observes what he has imagined, to see whether the same ardent desire is there to be discerned. By such a process, which is at bottom very much like mathematical reasoning, we can reach conclusions as to what *would* be true of signs in all cases...(Buehler 1940: 98)

Following the suggestion in Peirce's formulation, we are led to distinguish between sorts of mathematical agency: the one who imagines (what Peirce simply calls the 'self' who conducts a reflective observation), which we shall call the *Mathematician,* and the one who is imagined (the skeleton diagram and surrogate of this self), which we shall call the *Agent.* In terms of the distinction between imperatives, it is the Mathematician who carries out inclusive demands to 'consider' and 'define' certain worlds and to 'prove' theorems in relation to these, and it is his Agent who executes the actions within such fabricated worlds, such as 'count', 'integrate', and so on, demanded by exclusive imperatives.

At first glance, the relation between Mathematician and Agent seems no more than a version of that which occurs in the reading of a road map, in which one propels one's surrogate, a fingertip model of oncself, around the world of roads imaged by the lines of the map. Unfortunately. the parallel is misleading, since the point of a road map is to represent real roads - real in the sense of beirg entities which exist prior to and independently of the map, so that an imagined journey by an agent is conceived to be (at least in principle) realizable. With mathematics the existence of such priorly occurring 'real' worlds is, from a semiotic point of view, problematic; if mathematical signs are to he likened to maps, they are maps of purely imaginary territory.

In what semiotic sense is the Agent a skeleton diagram of the Mathematician? Our picture of the Mathematician is of a conscious - intentional, imagining subject who creates a fictional self, the Agent, and fictional worlds within which this self acts. But such creation cannot. of course, be effected as pure thinking: signifieds are inseparable from signifiers: in order to create fictions, the Mathematician scribbles.

Thus in response to the imperative 'add the numbers in the list S', for example, he invokes a certain imagined world and - inseparable from this invocation - he writes down an organized sequence of marks ending with the mark which is to be interpreted as the sum of S. These marks are signifiers of signs by virtue of their interpretation within this world. Within this world 'to add' might typically involve an infinite process, a procedure which requires that an infinity of actions be performed. This would be the case if, for example, S were the list of fractions 1, 1/2, 1/4, 1/8, etc. obtained by repeated halving. Clearly, in such a case, if 'add' is to be interpreted as an *action,* it has to be an imagined action, one performed not by the Mathematician – who can only manipulate very small finite sequences of written signs - but by an actor imagined by the Mathematician. Such an actor is not himself required to imagine anything. Unlike the Mathematician, the Agent is not reflective and has no intentions: he is never called upon to 'consider', 'define', or 'prove' anything, or indeed to attribute any significance or meaning to what he does; he is simply required to behave according to a prior pattern - do this then this then . . . -imagined for him by the Mathematician. The Agent, then, is a skeleton diagram of the Mathematician in two senses: he lacks the Mathematician's subjectivity in the face of signs; and he is free of the constraints of finitude and logical feasibility - he can perform infinite additions, make infinitely many choices, search through an infinite array, operate within nonexistent worlds -that accompany this subjectivity.

If the Agent is a truncated and idealized image of the Mathematician, then the latter is himself a reduced and abstracted version of the subject - let us call him the Person who operates with the signs of natural language and can answer to the agency named by the 'I' of ordinary nonmathematical discourse. An examination of the signs addressed to the Mathematician reveals that nowhere is there any mention of his being immersed in public historical or private durational time, or of occupying any geographical or bodily space. or of possessing any social or individualizing attributes. The Mathematician's psychology, in other words, is transcultural and disembodied. By writing its codes in a single tense of the constant present, within which addressees have no physical presence, mathematics dispenses entirely with the linguistic apparatus of dcixis: unlike the Person, for whom demonstrative and personal pronouns are available, the Mathematician is never called upon to interpret any sign or message whose meaning is inseparable from the physical circumstances - temporal, spatial, cultural – of its utterance. If the Mathematician's subjectivity is 'placed' in any sense, if he can be said to be physically selfsituated, his presence is located in and traced by the single point - the origin - which is required when any system of coordinates or process of counting is initiated: a replacement Hermann Wcyl once described as 'the necessary residue of' the extinction of the ego' (1949: 75).

I want now to being this trio of semiotic actors-Agent. Mathematician, and Person - together and to display them as agencies that operate in relation to each other on different levels of the same mathematical process: namely, the centrally important process of mathematical proof. In the extract quoted above, Peirce likened what he called 'reflective observation', in which a skeleton of the self takes part in a certain kind of thought experiment, to mathematical reasoning. For the Mathematician, reasoning is the process of giving and following proofs, of reading and writing certain highly specific and internally organized sequences of mathematical sentences - sequences intended to validate, test, prove, demonstrate, show that some particular assertion holds or is 'true' or is 'the case'. Proofs are tied to assertions, and the semiotic status of assertions, as we observed earlier, is inextricable from the nature of proof. So, if we want to give a semiotic picture of mathematical reasoning as a kind of Peircean thought experiment, an answer has to be given to the question we dodged before: how, as a business having to do with signs, are we to interpret the mathematical indicative?

The answer we propose runs as follows. A mathematical assertion is a *prediction,* a foretelling of the result of performing certain actions upon signs. In making an assertion the Mathematician is claiming to know what would happen if the sign activities detailed in the assertion *were* to he carried out. Since the actions in question are ones that fall within the Mathematician's own domain of activity, the Mathematician is in effect laying claim to knowledge of his own future signifying states. In Peirce's phrase, the sort of knowledge being claimed is 'what *would be true* of signs in all cases'. Thus, for example, the assertion '2 + 3 = 3 + 2' predicts that if the Mathematician concatenates 11 with 111, the result will be identical to hisconcatenating 111 with 11. And more generally, 'x + y = y + x' predicts that his concatenating any number of strokcs with any other number will turn out to he independent of the order in which these actions are performed. Or, to take a different kind of examplc, the assertion that the square root of 2 is irrational is the prediction that whatever particular integers x and y are taken to be, the result of calculating x SQUARED - 2ySQUARED will not be zero.

Obviously such claims to future knowledge need to he validated: the Mathematician has to persuade himself that if he performed the activities in question, the result would be as predicted. How is he to do this? In the physical sciences predictions are set against actualities: an experiment is carried out and, depending on the result, the prediction (or rather the theory which gave rise to it) is either repudiated or receives some degree of confirmation. It would seem to he the case that in certain very simple cases such a direct procedure will work in mathematics. Assertions like '2 + 3 = 3 + 2' or '101 is a prime number' appear to be directly validatable: the Mathematician ascertains whether they correctly predict what he would experience by carrying out an experiment - surveying strokes or examining particular numbers - which delivers to him precisely that experience. But the situation is not, as we shall see below, that straightforward; and in any case assertions of this palpable sort, though undoubtedly important in any discussion of the epistemological status of mathematical truths', are not the norm. (The Mathematician certainly cannot by *direct experiment* validate predictions like 'x + y = y + x' or 'the square root of 2 is irrational' unless he carries out infinitely many operations. lnstead, as we have observed, he must act indirectly and set up an imagined experience - a thought experiment -in which not he but his Agent, the skeleton diagram of himself, is required to perform the appropriate infinity of actions.

By observing his Agent performing in his stead, by 'reflective observation', the Mathematician becomes convinced -persuaded somehow by the thought experiment-that were *he* to perform these actions the result would be as predicted.

Now the Mathematician is involved in scribbling as well as thinking. The process of persuasion that a proof is supposed to achieve is an amalgam of fictive and logical aspects in dialectical relation to each other: Each layer of the thought experiment (that is, each stage of the journey undertaken by the Agent) corresponds to some written activity, some manipulation of written signs performed by the Mathematician: so that, for example, in reading/writing an inclusive imperative the Mathematician modifies or brings into being a suitable facet of the Agent, and in reading/writing an exclusive imperative he requires this Agent to carry out the action in question, observes the result, and then uses the outcome as the basis for a further bout of manipulating written signs. These manipulations form the steps of the proof in its guise as a logical argument: any given step is taken either as a premise, an outright assumption about which it is agreed no persuasion is necessary, or is taken because it is a conclusion logically implied by a previous step. The picture offered so far, then, is that a proof is a logically correct series of implications that the Mathematician is persuaded to accept by virtue of the interpretation given to these implications in the fictive world of a thought experiment.

Such a characterization of proof is correct but inadequate. Proofs are *arguments* and, as Peirce forcefully pointed out, every argument has an underlying idea - what he called a *leading principle* - which converts what would otherwise be merely an unexceptionable sequence of logical moves into an instrument of conviction. The leading principle, Peirce argued, is distinct from the premise and the conclusion of an argument, and if added to these would have the effect of requiring a new leading principle and so on, producing an infinite regress in place of a finitely presentable argument. Thus, though it operates through the logical sequence that embodies it, it is neither identical nor reducible to this sequence; indeed, it is only by virtue of it that the sequence is an argument and not an inert, formally correct string of implications.

The kading principle corresponds to a familiar phenomenon within mathematics. Presented with a new proof or argument, the first question the mathematician (but not, see below, the Mathematician) is likely to raise concerns 'motivation': he will in his attempt to understand the argument that is, follow and be convinced by it - seek the *idea behind the proof*. He will ask for the story that is belng told, the narrative through which the thought experiment or argument is organized. It is perfectly possible to follow a proof, in the more restricted, purely formal sense of giving assent to each logical step, without such an idea. If in addition an argument is based on accepted familiar patterns of inference, its leading principle will have been internalized to the extent of being no longer retrievable: it is read automatically as part of' the proof. Nonetheless a leading principle is always present, acknowledged or not - and attempts to read proofs in the absence of their underlying narratives are unlikely to result in the experience of felt necessity, persuasion, and conviction that proofs are intended to produce, and without which they fail to *be* proofs.

Now mathematicians - whether formalist, intuitionist, or platonist - when moved to comment on this aspect of their discourse might recognize the importance of such narratives to the process of persuasion and understanding, but they are inclined to dismiss them, along with any other 'motivational' or 'purely psychological' or merely 'esthetic' considerations, as ultimately irrelevant and epiphenomenal to the real business of doing mathematics. What are we to make of this?

It is certainly true, as observed, that the leading principle cannot be part of the proof itself: it is not, in other words, addressed to the subject who reads proofs that we have designated as the Mathematician. Indeed, the underlying narrative could not be so addressed, since it lies outside the linguistic resources mathematics makes available to the Mathematician. We might call the total of all these resources the mathematical Code, and mean by this the discursive sum of all legitimately defined signs and rigorously formulated sign practices that are permitted to figure in mathematical texts. At the same time let us designate by the *metaCode* the penumbra of informal, unrigorous locutions within natural language involved in talking about, referring to, and discussing the Code that mathematicians sanction. The Mathematician is the subject pertaining to the Code - the one who reads/writes its signs and interprets them by imagining experiments in which the actions inherent in them are performed by his Agent. We saw earlier that mathematicians prohibit the use of any deictic terms from their discourse: from which it follows that no description of himself is available to the Mathematician within the Code. Though he is able to imagine and observe the Agent as a skeleton diagram of himself, he cannot - within the vocabulary of the Code - articulate his relation to that Agent. He knows the Agent is a simulacrum of himself, but he cannot talk about his knowledge. And it is precisely in the articulation of this relation that the semiotic source of a proof's persuasion lies: the Mathematician cm be persuaded by a thought experiment designed to validate a prediction about his own actions only if he appreciates the resemblance - for the particular mathematical purpose at hand - between the Agent and himself. It is the business of the underlying narrative of a proof to articulate the nature of this resemblance. In short, the idea behind a proof is situated in the metaCode; it is not the Mathematician himself who can be persuaded by the idea behind a proof, but the Mathematician in the presence of the Person, the natural language subject of the metaCode for whom the Agent as a simulacrum of the Mathematician is an object of discourse.

What then, to return to the point above, is meant by the 'real business' of doing mathematics? In relation to the discussion so far, one can say this: if it is insisted that mathematical activity be described solely in terms of manipulations of signs within the Code, thereby restricting mathematical subjectivity to the Mathematician and dismissing the metaCode as an epiphenomenon, a domain of motivational and psychological affect, then one gives up any hope of a semiotic view of mathematical proof able to give a coherent account - in terms of sign use - of how proofs achieve conviction.

There are two further important reasons for refusing to assign to the metaCode the status of mere epiphenomenon. The first concerns the completion of the discussion of the indicative. As observed earlier, there are, besides the assertions within the Code that we have characterized as predictions, other assertions of undeniable importance that must be justified in thecourse of normal mathematical practice. When mathematicians make assertions like 'definition D is well-founded', or 'notation system N is coherent', they are plainly making statements that require some sort of justification. Equally plainly, such assertions cannot be interpreted as predictions about the Mathematician's future mathematical experience susceptible of proof' via a thought experiment. Indeed, indicatives such as 'assertion A is provable' or 'x is a counter-example to A'. where A is a predictive assertion within the Code, cannot themselves be proved mathematically without engendering an infinite regress of proofs.

It would seem that such metalingual indicatives-which of coursc belong to the metaCodc - admit 'proof' in the same way that the proof of the pudding is in the eating: one justifies the statement 'assertion A is provable' by exhibiting a proof of A. The second reason for treating the metaCode as important to a semiotic account of mathematics relates to the manner in which mathematical codes and sign usages come into being, since it can be argued (though I will not do so here) that Code and metaCode are mutually constitutive, and that a principal way in which new mathematics arises is through a process of catachresis - that is, through the sanctioning and appropriation of sign practices that occur in the first place as informal and unrigorous elements, in a merely descriptive, motivational, or intuitive guise, within the metaCodc.

The model I have sketched has required us to introduce three separate levels of mathematical activity corresponding to the sub-lingual imagined actions of an Agent, the lingual Coded manipulations of the Mathematician, and the metalingual activities of the Person, and then to describe how these agencies fit together. Normal mathematical discourse does not present itself in this way; it speaks only of a single unfractured agency, a 'mathematician', who simply 'does' mathematics. To justify the increase in complexity and artificiality of its characterization of mathematics, the model has to be useful; its picture of mathematics ought to illuminate and explain the attraction of the three principal ways of regarding mathematics we alluded to earlier.

Formalism, intuitionism, platonism

We extracted the idea of proof as a kind of thought experiment from Peirce's general remarks on reflective observation; we might have gotten a later and specifically mathematical version straight from Hilbert, from his formalist conception of metalogic as amounting to those 'considerations in the form of *thought-experiments* on objects, which can be regarded as concretely given' (Hilbert and Bernays 1934: 20). But this would have sidetracked us into a description of Hilbert's metamathematical program, which it was designed to serve,

The object of this progam was to show by means of mathematical reasoning that mathematical reasoning was consistent, that it was incapable of arriving at a contradiction. In order to reason about reasoning without incurring the obvious circularity inherent in such an enterprise, Hilbert made a separation between the reasoning that mathematicians use - which he characterized as formal manipulations, finite sequences of logically correct deductions performed on mathematical symbols - and the reasoning that the metamathematician would use, the metalogic, to show that this first kind of reasoning was consistent. The circularity would be avoided, he argued, if the metalogic was inherently safe and free from the sort of contradiction that threatened the object logic about which it reasoned. Since the potential source of contradiction in mathema tics was held to he the ocurrcnce of objects and processes that were interpreted by mathematicians is to be infinitary, the principal requirement of his metalogic was that it be finitary and that it avoid interpretations, that is. that it attribute no meanings to the subject matter about which it reasoned. Hilbert's approach to mathematics, then, was to ignore what mathematicians thought they meant or intended to mean, and instead to treat it as a formalism, as a system of meaningless written marks finitely manipulated by the mathematician according to explicitly stated formal rules. It was to this formalism that his rnetalogic, characterized as thought experiment on things, was intended to apply.

The first question to ask is to concern the 'things' that are supposed to figure in thought experiments: what are these concretely given entities about which thought takcs place? The formalist answer - objects concretely given as visible inscriptions, as definite but meaningless written marks - would have been open to the immediate objection that meaningless marks, while they can undoubtedly be manipulated and subjected to empirical (that *is,* visual) scrutiny, are difficult to equate to the sort of entities that figure in the finitary arithmetical assertions that form the basis for distinguishing experiments from *thought* experiments. Thus, in order to validate a unitary assertion like '2 + 3 = 3 + 2'. the formalist mathematician supposes that a *direct* experiment is all that is needed: the experimenter is convinced that concatenating '11' and '111 ' is the same as concatenating '111' and '11' through the purely empirical observation that in both cases the result is the assemblage of marks '11111'. But such an observation is a completely empirical validation a pure ad oculo demonstration free of any considerations of meaning - only the mathematical mark '1' is purely and simply an empirical mark, a mark all of whose significance lies in its visible appearance: and this, as philosophical critics of formalism from Frege onward have pointed out, is manifestly not the case. If it were, then arithmetical assertions would lack the generality universally ascribed to them; they would be about the physical perceptual characteristics of particular inscriptions—their exact shape. color, and size; their durability; the depth of their indentation on the page; the exact identity *of* one with another; and so on – and would need reformulating and revalidating every time one of these variables altered - a conclusion that no formalist of whatever persuasion would want to accept.

In fact, regardless of any considerations of intended meaning, the mathematical symbol 'I' cannot be identified with a mark at all. In Peirce's terminology 'I' is not a token (a concretely given visible object), but a *type,* an abstract pattern of writing, a general form of which any given material inscription is merely a perceptual instance. For us the mathematical symbol 'I' is an item of thinking/scribbling, a *sign*, and we can now flesh out one aspect of our semiotic model by elaborating what is to be meant by this. Specifically, we propose: the sign 'I' has for its signifier the type of the mark '1' used to notate it, and for its signified that relation in the Code - which we have yet to explicate - between thought and writing accorded to the symbol '1' by the Mathematician.

This characterization of 'I' has the immediate consequence that mathematical signifiers are themselves dependent on some prior signifying activity, since types as entities with meaningful attributes - abstract, general, unchanging, permanent, exact, and so on - can only come into being and operate through the semiotic separation between real and ideal marks. Now neither the creation nor recognition of this ideality is the business of what we have called the Mathematician: it takes place before his mathematical encounter with signs. Put differently, the Mathematician assumes this separation but is not, and cannot be, called upon to mention it in the course of interpreting signs within the Code; it forms no part of the meaning of these signs insofar as this meaning is accessible to him as the addressee of the Code. On the contrary, it is the Person, operating from a point exterior to the Mathematician, who is responsible for this ideality; for it is only in the metaCode, where mathematical symbols are discussed as signs, that any significance can be attributed to the difference between writing tokens and using types.

We can apply this view of signs to the question of mathematical 'experience' as it occurs within formalism's notion of a thought experiment. The pressure for inserting the presence of an Agent into our model of mathematical activity came from the fact that whatever unitary actions upon signs the Mathematician might *in fact* be able to carry out, such as concatenating '1I' and '111', these were the exception; for the most part the actions which he was called upon to perform (such as evaluating x + y for arbitrary integers x and y) could only be carried out *in principle*, and for these infinitary actions he had to invoke the activity of an Agent. I suggested earlier that for the former sort of actions there appeared to he no need for him to invoke an Agent and a *thought* experiment: that the Mathematician might experience for himself by direct experiment the validity of assertions such as '2 + 3 = 3 + 2'. This idea that (at least some) mathematical assertions are capable ol being directly 'experienced' is precisely what formalists - interpreting experience as meaning visual inspection of objects in space - claim. One needs therefore to be more specific about the meaning of. direct mathematical 'experience'. From what has been said so far, no purely physical manipulation of purely physical marks can, of itself, constitute mathematical persuasion. The Mathematician manipulates types: he can be persuaded that direct experiments with tokens constitute validations of assertions like '$2 + 3 = 3 + 2$' only by appealing to the relationship between tokens and types, to the way tokens stand in place of types. And this he, as opposed to the Person, cannot do. The upshot, from a semiotic point of view, is that in no case can thought experiments be supplanted by direct experirnentation; an Agent is always required. Validating finitary assertions is no different from validating assertions in general, where it is the Person who, by being able to articulate the relation between Mathematician and Agent within a thought experiment, is persuaded that a prediction about the Mathematician's future encounter with signs is to be accepted.

This way of seeing matters does not dissolve the difference between finitary and non-finitary assertions. Rather, it insists that the distinction - undoubtedly interesting, but for us limitedly so in comparison to its centrality within the formalist program - between mathematical actions executable in *fact* and *in principle* (in reified form: how big can a 'small' concretely surveyable collection of marks be before it becomes 'large' and unsurvcyable?) makes sense only in terms of what constitutes mathematical persuasion, which in turn can only be explicated by examining the possible relationships between Agent. Mathematician, and Person that mathematicians are prepared to countenance as legitimate.

Hilbert's program for proving the consistency of mathematics through a unitary metamathematics was, as is well known, brought to an effective halt by Gòdel's Theorem. But this refutation of what was always an overambitious project does not demolish formalism as a viewpoint; nor, without much technical discussion outside the scope of this essay, can it he made to shed much light on how formalism's inadequacies arise. Thus, from a *semiotic* point of view the problems experienced by formalism can he seen to rest on a chain of mis-identifications. By failing to distinguish between tokens and types, and thereby mistaking items possessing significance for pre-semiotic 'things', formalists simultaneously misdescribe mathematical reasoning as syntactical manipulation of meaningless marks and metamathematical reasoning as thought experiments that theoreticalize these manipulations–in the sense that the formalists 'experiment", of which the thought experiment is an imagined thoretical version, is an entirely empirical procss of checking the perceptual properties of visible marks.. As a consequence, the formalist account of mathematical agency–which distinguishes merely between a mathematician who manipulates mathematical symbols as if they were marks and a 'metamathematician' who reasons about the results of this manipulation - is doubly reductive of the picture offered by the present model, since at one point it shrinks the role of Mathematician to that of Agent and at another manages to absorb it into that of Person.

If formalism projects the mathematical amalgam of thinking/scribbling onto a plane of formal scribble robbed of meaning, intuitionism projects it onto a plane of thought devoid of any written trace. Each bases its truncation of the sign on the possibility of an irreducible mathematical 'experience' which is supposed to convey by its very unmediated directness what it takes to be essential to mathematical practice: formalism, positivist and suspicious in a behaviorist way about mental events, has to locate this experience in the tangible written product, surveyable and 'real'; intuitionism, entirely immersed in Kantian apriorism, identifies the experience as the process, the invisible unobservable construction in thought, whereby mathematics is created.

Brouwcr's intuitionism, like Hilbert's formalism, arose as a response to the paradoxes of the infinite that emerged at the turn of the century within the mathematics of infinite sets. Unlike Hilbert, who had no quarrel with the platonist conception of such sets and whose aim was to leave mathematics as it was by providing a post facto metamathernatical justification of its consistency in which all existing infinitary thought would be legitimated, Brouwer attacked the platonist notion of infinity itself and argued for a root and branch reconstruction from within in which large areas of classically secure mathematics - infinitary in character but having no explicit connection to any paradox - would have to he jettisoned as lacking any coherent basis in thought and therefore, meaningless. The problem, as Brouwer saw it, was the failure on the part of orthodox - platonist-inspired -mathematics to separate what for him are the proper objects of mathematics (namely constructions in the mind) from the secondary aspect of these objects, the linguistic apparatus that mathematicians might use to describe and communicate about in words the features and results of any particular such construction. Confusing the two allowed mathematicians to believe, Brouwer argued, that linguistic manipulation was an unexceptional route to the production of new mathematical entities. Since the classical logic governing such manipulation has its origins in the finite states of affairs described by natural language, the confusion results in a false mathematics of the infinite: verbiage that fails to correspond to any identifiable mental activity, since it allows forms of inference that make sense' only for finite situations, such as the law of excluded middle or the principle of double negation, to appear to validate what are in fact illegitimate assertions about infinite ones.

Thus the intuitionist approach to mathematics, like Hilbert's scheme for metamathematics, insists that a special and primary characteristic of logic lies in its appropriateness to finitary mathematical situations. And though they accord different functions to this logic-for Hilbert it has to validate unitary metamathematics in order to secure infinitary mathematics, while for Brouwer it is an after-the-fact formalization of the principles of correct mental constructions finitary and inflnitary - they each require it to be convincing: formalism grounding the persuasive force of its logic in the empirical certainty of *ad oculo* demonstrations, intuitionism being obliged to ground it, as we shall see, in the felt necessity and self-evidence of the mathematician's mental activity.

If mathematical assertions are construed platonistically, as unambiguous, exact, and precise statements of fact, propositions true or false about some objective state of allairs, Brouwer's rejection of the law of excluded middle (the principle of logic that declares that an assertion either is the case or is not the case) and his rejection of double negation (the rule that not being not the case is the same as being the case) seem puzzling and counterintuitive: about the particulars of mathematics conceived in this exact and determinate way there would appear to be no middle ground between truth and falsity, and no way of distinguishing between an assertion and the negation of the negation of that assertion. Clearly, truth and falsity of assertions will not mean for intuitionists if indeed they are to mean anything at all for them - what they do for orthodox mathematicians: and. since platonistic logic is founded on truth, intuitionists cannot be referring to the orthodox process of deduction when they talk about the validation of assertions.

For the intuitionist, an assertion is a claim that a certain mental construction has been carried out. To validate such claims the intuitionist must either exhibit the construction in question or, less directly, show that it *can* be carried out by providing an effective procedure, a finite recipe, for executing it. This effectivity is not an external requirement imposed on assertions after they have been presented, but is built into the intuitionist account of the logical connectives through which assertions are formulated. And it is from this internalized logic that principles such as the laws of excluded middle, double negation, and so on are excluded. Thus, in contrast to the platonist validation of an existential assertion (x exists if the assumption that it doesn't leads to a contradiction), for the intuitionist x can only be shown to exist by exhibiting it, or by showing effectively how to exhibit it. Again, to validate the negation of an assertion A, it is not enough - as it is for the platonist - to prove the existence of a contradiction issuing from the assumption A: the intuitionist must exhibit or show how one would exhibit the contradiction when presented with the construction that is claimed in A. Likewise for implication: to validate 'A implies B', the intuitionist must provide an effective procedure for converting the construction claimed to have been carried out in A into the construction being claimed in B.

Clearly, the intuitionist picture of mathematical assertions and proofs depends on the coherence and acceptability of what it means by an effective procedure and (inseparable from this) on the status of claims that mental constructions have been or can he carried out. Proofs, validations, and arguments, in order to 'show' or 'demonstrate' a claim, have before all else to be convincing; they need to persuade their addressees to accept 'what is claimed. Where then in the face of Brouwer's characterization that "intuitionist mathematics is an essentially languageless activity of the mind having its origin in the perception of a move in time" (1952), with its relegation of language - that is, all mathematical writing and speech - to an epiphenomenon of mathematical activity, a secondary and (because it is after the fact) theoretically unnecessary business of mere description, are we to locate the intuitionist version of persuasion? The problem is fundamental. Persuading, convincing, showing, and demonstrating are discursive activities whose business it is to achieve intersubjective agreement. But for Brouwer, the intersubjective collapses into the subjective: there is only a single cognizing subject privately carrying out constructions - sequences of temporally distinct moves - in the Kantian intuition of time. This means that, for the intuitionist, conviction and persuasion appear as the possibility of a replay, a purely mental reenactment within this one subjectivity: perform this construction in the inner intuition of time you share with me and you will - you must - experience what I claim to experience.

Validating assertions by appealing in this way to felt necessity, to what is supposed to be self-evident to the experiencing subject, goes back to Descartes' cogito, to which philosophers have raised a basic and (this side of solipsism) unanswerable objection: what is self-evidently the case for one may be not self-evident - or worse, may be self-evidently not the case - for another; so that, since there can be no basis other than subjective force for choosing between conflicting self-evidence, what is put forward as a process of rational validation intended to convince and persuade is ultimately no more than a refined reiteration o! the assertion it claims to he validating.

That intuitionism should be unable to give a coherent account of persuasion is what a semiotic approach which insists that mathematics is a business with and about signs, conceived as public, manifest amalgams of scribbling/thinking. would lead one to believe. The inability is the price intuitionism pays for believing it possible to first separate thought from writing and then demote writing to a description of this prior and languageless - presemiotic thought. Of course, this is not to assert that intuitionism's fixation on thinking to the exclusion of writing is not useful or productive; within mathematical practice, by providing an alternative to classical reasoning. it has been both. And indeed, insofar as a picture of mathematics-as-pure-thought is possible at all, intuitionism in some form or other could be said to provide it.

From a semiotic viewpoint, however, any such picture cannot avoid being a metonymic reduction, a *pars pro toto* which mistakes a part - the purely mental activities that seem undeniably to accompany all mathematical assertions and proofs - for the whole writing/thinking business of manipulating signs, and in so doing makes it impossible to recognize the distinctive role played by signifiers in the creation of mathematical meaning. Far from being the written traces of a language that merely *describes* prior mental constructions appearing as pre-semiotic events accessible only to private introspection, signifiers mark signs that are *interpreted* in terms of imagined actions which themselves have no being independent of their invocation in the presence of these very signifiers. And it is in this dialectic relation between scribbling and thinking, whereby each creates what is necessary for the other to come into being, that persuasion - as a tripartite activity involving Agent, Mathematician, and Person within a thought experiment - has to be located.

These remarks about formalism and intuitionism are intended to serve not as philosophical critiques of their claims about the nature of mathematics, but as means of throwing into relief the contrasting claims of our semiotic model. From he point of view of mainstream mathematical practice, moreover, the formalist and intuitionist descriptions of mathematics are of minor importance. True, formalism's attempt to characterize finitary reasoning is central to metamathematical investigations such as proof theory, and intuitionistic logic is at the front of any constructivist examination of mathematics; but neither exerts more than a marginal influence on how the overwhelming majority of mathematicians regard their subject matter. When they pursue their business mathematicians do so neither as formalist manipulators nor solitary mental constructors, but as scientific investigators engaged in publicly discovering objective truths. And they see these truths through platonistic eyes: eternal verities, objective irrefutably-the-case descriptions of some timeless, spaceless, subjectless reality of abstract 'objects'.

Though the question of' the nature of these platonic objects - what are numbers? - can be made as old as Western philosophy, the version of platonism that interests us (namely, the prevailing orthodoxy in mathematics) is a creation of nineteenth-century realism. And since our focus is semiotic and not philosophical, our primary interest is in the part played by a realist conception of language ir forming and legitimizing present-day mathematical platonism.

For the realist, language is an activity whose principàl function is that *of naming:* its character derives from the fact that its terms, locutions, constructions, and narratives are oriented outward, that they point to, refer to, denote some reality outside and prior to themselves. They do this not as a byproduct, consequentially on some more complex signifying activity, but essentially and genetically so in their formation: language, for the realist, arises and operates as a name for the pre-existing world. Such a view issues in a bifurcation of linguistic activity into a primary act of reference concerning what is 'real', given 'out there' within the prior world waiting to be labeled and denoted – and a subsidiary act of describing, commenting on, and communicating about the objects named. Frege, who never tired of arguing for the opposition between these two linguistic activities - what Mill's earlier realism distinguished as connotation/denotation and he called sense/reference - insisted that it was the latter that provided the ground on which mathematics was to be based. And if for technical reasons Frege's ground – the pre-existing world of pure logical objects - is no longer tenable and is now replaced by an abstract universe of sets, his insistence on the priority of reference over sense remains as the linguistic cornerstone of twentieth-century platonism.

What is wrong with it? Why should one not believe that mathematics is about some ideal timeless world populated by abstract unchanging objects; that these objects exist, in all their attributes, independent of any language used to describe them or human consciousness to apprehend them; and that what a theorem expresses is objectively the case, an eternally true description of a specific and determinate states of affairs about these objects?

One response might be to question immediately the semiotic coherence of a pre-linguistic referent:

Every attempt to establish what the referent of a sign *is* forces us to define the referent in terms of an abstract entity which moreover is only a cultural convention. (Eco 1976: 66)

If such is the case, then language–in the form of'cultural mediation - is inextricable from the process of referring. This will mean that the supposedly distinct and opposing categories of reference and sense interpenetrate each other, and that the object referred to can neither be separated from nor antedate the descriptions given of it. Such a referent will be a social historical construct; and, notwithstanding the fact that it might present itself as abstract, cognitively universal, pre-semiotic (as is the case for mathematical objects), it will he no more timeless, spaceless, or subjectless than any other social artifact. On this view mathematical platonism never gets off the ground, and Frege's claim that mathematical assertions are objectively true about eternal 'objects' dissolves into a psychologistic opposite that he would have abhorred: mathematics makes subjective assertions - dubitable and subject to revision - about entities that are time-hound and culturally loaded.

Another response to platonism's reliance on such abstract referents - one which is epistemological rather than purely semiotic, but which in the end leads to the same difficulty - lies in the questions 'How can one come to know anything about objects that exist outside space and time?' and 'What possible causal chain could there be linking such entities to temporally and spatially situated human knowers?' If knowledge is thought of as some form of justified belief, then the question repeats itself on the level of validation: what manner of conviction and persuasion is therc which will connect the platonist mathematician to this ideal and inaccessible realm of objects? Plato's answer - that the world of human knowers is a shadow of the eternal ideal world of pure form, so that by examining how what is perceivable partakes of and mimics the ideal, one arrives at knowledge of the eternal - succeeds only in recycling the question through the metaphysical obscurities of how concrete and palpable particulars arc supposed to partake of and be shadows of abstract universals. How does Frege manage to deal with the problem?

The short answer is that he doesn't. Consider the distinctions behind Frege's insistence that 'the thought we express by the Pythagorean theorem is surely timeless, eternal, unchangeable' (1967: 37). Sentences express thoughts. A thought is always the sense of some indicative sentence; it is 'something for which the question of truth arises' and so cannot be material, cannot belong to the 'outer world' of perceptible things which exists independently of truth. But neither do thoughts belong to the 'inner world', the world of sense impressions, creations of the imagination, sensations, feelings, and wishes. All these Frcgc calls 'ideas'. Ideas are experienced, they need an experient. a particular person to have them whom Frege calls their 'bearer'; as individual experiences every idea has one and only one bearer. It follows that if thoughts are neither inner ideas nor outer things, then

> A third realm must bc recognized. What belongs to this corresponds to ideas, in that it cannot be perceived by the senses, but with things, in that it needs no bearer to the contents of whose consciousness to belong. Thus the thought, for example, which we expressed in the Pythagorean theorem is timelessly true, true indcpendently of whether anyone takes it to be true. It needs no bearer. It is not true for the first time when it is discovered, but is like a planet which, already before anyonc has seen it, has been in interaction with other planets. (Frege 1967: 29)

The crucial question, however, remains: what is our relation to this non-inner, non-outer realm of planetary thoughts, and how is it realized? Frege suggests that we talk in terms of seeing things in the outer world, having ideas in the inner world, and thinking or *apprehending* thoughts in this third world; and that in apprehending a thought we do not create it but come to stand 'in a certain relation ... to what already existed before'. Now Frege admits that while 'apprehend' is a metaphor, unavoidable in the circumstances, it is not to be given any subjectivist reading, any interpretation which would reduce mathematical thought to a psychologism of ideas:

> The apprehension of a thought presupposes someone who apprehends it, who thinks it. He is the bearer of the thinking but not of the thought. Although the thought does not belong to the thinker's consciousness yet something in his consciousness must be aimed at that thought. But this should not be confused with the thought itself. (Frege 1967: 35)

Frcge gives no idea, explanation, or even hint as to what this 'something' might be which allows the subjective, temporally located bearer to 'aim' at an objective, changeless thought. Certainly he sees that there is a difficulty in connecting the eternity of the third realm to the time-bound presence of bearers: he exclaims. 'And yet: What value could there be for us in the eternally unchangeable which could neither undergo effects nor have effect on us?' His concern, however, is not an epistemological one about human knowing (how we can know thoughts), but a reverse worry about 'value' conceived in utilitarian terms (how can thoughts be useful to us who apprehend them). The means by which we manage to apprehend them are left in total mystery.

It does look as if platonism, if it is going to insist on timeless truth, is incapable of giving a coherent account of knowing, and *a fortiori* of how mathematical practice comes to create mathematical knowledge.

But we could set aside platonism's purely philosophical difficulties about knowledge and its aspirations to eternal truths (though such is the principal attraction to its adherents), and think semiotically: we could ask whether what Frege wants to understand by thoughts might not be interpreted in terms of the amalgamations al' thinking/scribbling have called signs. Thus, can one not replace Frcge's double exclusion (thoughts are neither inner subjectivities nor outer materialities) with a double inclusion (signs are both materially based signifiers and mentally structured signifieds), and in this way salvage a certain kind of semiotic sense from his picture of mathematics as a science of objective truths? Of course, such 'truths' would have to relate to the activities of a sign-interpreting agency; they would not be descriptions of some non-temporal extra-human realm of objects but laws - freedoms and limitations - of the mathematical subject. A version of such an anthropological science seems to have occurred to Frege as a way of recognizing the 'subject' without at the same time compromising his obsessive rejection of any form of psychologism:

> Nothing would be a greater misunderstanding of mathematics than its subordination to psychology. Neither logic nor mathematics has the task of investigating minds and the contents of consciousness whose bearer is a single person. Perhaps their task could be represented rather as the investigation of the mind, of the mind not minds, (Frege 1967: 35)

But in the absence of any willingness to understand that both minds and mind are but different aspects of a single process of semiosis, that both are inseparable from the social and cultural creation of meaning by sign-using subjects, Frege's opposition of mind/minds degenerates into an unexamined Kantianism that explains little (less than Brouwer's intuitionism, for example) about how thoughts - that is, in this suggested reading of him, the signifieds of assertion signs - come to inhabit and be 'apprehended' as objective by individual subjective minds.

In fact, any attempt to rescue platonism from its incoherent attachment to 'eternal' objects can only succeed by destroying what is being rescued: the incoherence, as we said earlier, lies not in the supposed eternality of its referents but in the less explicit assumption, imposed by its realist conception of language, that they are pre-linguistic, pre-semiotic, precultural. Only by being so could objects - existing, already out there, in advance of language that comes after them - possess 'objective' attributes untainted by 'subjective' human interference. Frege's anti-psychologism and his obsession with eternal truth correspond to his complete acceptance of the two poles of the subjective/objective opposition - an opposition which is the *sine qua non* of nineteenth-century linguistic realism. If this opposition and the idea of a 'subject' it promotes is an illusion, then so too is any recognizable form of platonism.

That the opposition is an illusion becomes apparent once one recognizes that mathematical signs play a *creative* rather than merely descrip-tive function in mathematical practice. Those things which are 'described'– thoughts, signifieds, notions - and the means by which they are described - scribbles - are mutually constitutive: each causes the presence of the other; so that mathematicians at the same time think their scribbles and scribble their thoughts. Within such a scheme the attribution of 'truth' to mathematical assertions becomes questionable and problematic, and with it the platonist idea that mathematical reasoning and conviction consists of giving assent to deductive strings of truth-preserving inferences.

On the contrary, as the model that we have constructed demonstrates, the structure of mathematical reasoning is more complicated and interesting than any realist interpretation of mathematical language and mathematical 'subjectivity' can articulate. Persuasion and the dialectic of thinking/scribbling which embodies it is a tripartite activity: the Person constructs a narrative, the leading principle of an argument, in the metaCode; this argument or proof takes the Form of a thought experiment in the Code; in following the proof the Mathematician imagines his Agent to perform certain actions and observes the results; on the basis of these results, and in the light of the narrative, the Person is persuaded that the assertion being proved - which is a prediction about the Mathematician's sign activities - is to be believed.

By reducing the function of mathematical signs to the naming of presemiotic objects, platonism leaves a hole at precisely the place where the thinking/scribbling dialectic occurs. Put simply, platonism occludes the Mathematician by flattening the trichotomy into a crude opposition: Frege's bearer-subjective, changeablc, immersed in language, mortal - is the Person, and the Agent - idealized, infinitary—is the source (though he could not say so) of objective eternal 'thoughts'. And, as has become clear, it is precisely about the middle term, which provides the epistemological link between the two, that platonism is silent.

If platonic realism is an illusion, a myth clothed in the language of some supposed scientific 'objectivity', that issues from the metaphysical desire for absolute eternal truth rather than from any non-theistic wish to characterize mathematical activity, it does nonetheless - as the widespread belief in it indicates - answer to some practical need. To the ordinary mathematician, unconcerned about the nature of mathematical signs, the ultimate status of mathematical objects, or the semiotic basis of mathematical persuasion, it provides a simple working philosophy: it lets him get on with the normal scientific business of research by legitimizing the feeling that mathematical language describes entities and their properties that are 'out there', waiting independently of mathematicians, to be neither invented nor constructed nor somehow brought into being by human cognition, but rather *discovered* as planets and their orbits are discovered.

It is perfectly possible, however, to accomodate the force of this feeling without being drawn into any elaborate metaphysical apparatus of eternal referents and the like. All that is needed is the very general recognition -familiar since Hegel -that human products frequently appear to their producers as strange, unfamiliar, and surprising: that what is created need bear no obvious or transparent markers of its human (social, cultural, historical, psychological) agency, but on the contrary can, and for the most part does, present itself as alien and prior to its creator.

Marx, who was interested in the case where the creative activity was economic and the product was a commodity, saw in this masking of agency a fundamental source of social alienation, whereby the commodity appeared as a magical object, a fetish, separated from and mysterious to its creator; and he understood that in order to be bought and sold commmodities *had* to be fetishized, that it was a condition of their existence and exchangeability within capitalism. Capitalism and mathematics are intimately related: mathematics functions as the grammar of techno-scientific discourse which every form of capitalism has relied upon and initiated. So it would be feasible to read the widespread acceptance of mathematical platonism in terms of the effects of this intimacy, to relate the exchange of meaning within mathematical languages to the exchange of commodities, to see in the notion of a 'timeless, eternal, unchangeable' object the presence of a pure fetishized meaning, and so on: feasible, in other words, to see in the realist account of mathematics an ideological formation serving certain (techno-scientific) ends within twentieth-century capitalism.

But it is unnecessary to pursue this reading. Whether one sees realism as a mathematical adjunct of capitalism or as a theistic wish for eternity, the semiotic point is the same: what present-day mathematicians think they are doing - using mathematical language as a transparent medium for describing a world of pre-semiotic reality - is semiotically alienated from what they are, according to the present account, doing – namely, creating that reality through the very lanuage which claims to 'describe' it.

What Is mathematics 'about'?

To claim, as I have done, that mathematical thought and scribbling enter into each other, that mathematical language creates as well as talks about its worlds of objects, is to urge a thesis antagonistic not only to the present-day version of mathematical platonism, but to any interpretation of mathematical signs, however sanctioned and natural, that insists on the separateness of objects from their descriptions.

Nothing is nearer to mathematical nature than the integers, the progression of those things mathematicians allow to be called the 'natural' numbers. And no opposition is more sanctioned and acknowledged as obvious than that between these *numbers* and their names, the *numerals,* which denote and label them. The accepted interpretation of this opposition runs as follows: first (and the priority is vital) there are numbers, abstract entities of some sort whose ultimate nature, mysterious though it might be, is irrelevant for the distinction in hand; then there arc numerals - notations, names such as 1111111111, X, 10, and so on - which are attached to them. According to this interpretation, the idea that numerals might precede numbers, that the order of creation might be reversed or neutralized, would be dismissed as absurd: for did not the integers named by Roman X or Hindu 10 exist before the Romans took up arithmetic or Hindu mathematicians invented the place notation with zero? and does not the normal recognition that X, 10, 'ten', 'dix', etc., name the same number require one to accept the priority of that number as the common referent of these names?

Insisting in this way on the prior status of the integers, and with it the posterior status of numerals, is by no means a peculiarity of Fregean realism. Hilbert's formalism, for all its programmatic abolition of meaningful entities, had in practice to accept that the whole numbers are in some sense given at the outset - as indeed does constructivism, either in the sense of Brouwcr's intuitionism, where they are *a priori* constructions in the intuition of time, or in the version urged by Kronecker according to which 'God made the integers, all the rest being the work of man'.

In the face of such a universal and overwhelming conviction that the integers - whether conceived as eternal platonic entities, pre-formal givens, prior intuitions, or divine creations - are *before* us, that they have always been there, that they are not social, cultural, historical artifacts but *natural* objects, it is necessary to he more specific about the semiotic answer to the fundamental question of what (in terms of sign activity) the whole numbers are or might be.

However possible it is for them to be individually instantiated, exemplified, ostensively indicated in particular, physically present. pluralities such as piles of stones, collections of marks, fingers, and so on, numbers do not arise, nor can they be characterized, as single entities in isolation from each other: they form an ordered sequence, a *progression.* And it seems impossible to imagine what it means for 'things' to be the elements of this progression except in terms of their production through the process of *counting,* And since counting rests on the repetition of an identical act, any semiotic explanation of the numbers has to start by invoking the familiar pattern of figures

1. 11, 111, 1111, 11111, 111111, 1111111, etc.

created by iterating the operation of writing down some fixed but arbitrarily agreed upon symbol type. Such a pattern achieves mathematical meaning as soon as the type '1' is interpreted as the signifier of a mathematical sign and the 'etc.' symbol as a command, an imperative addressed to the mathematician, which instructs him to enact the rule: copy previous inscription then add to it another type. Numbers, then, appear as soon as there is a subject who counts. As Lorenzen - from an operationalist viewpoint having much in common with the outlook of the present project - puts it: 'Anybody who has the capacity of producing such figures can at any time speak of numbers' (1955). With the semiotic model that we have proposed, the subject to whom the imperative is addressed is the Mathematician, while the one who enacts the instruction, the one who is capable of this unlimited written repetition, is his Agent. Between them they create the possibility of a progression of numbers, which is exactly the ordered sequence of signs whose signifiers are '1', '11', and so on.

Seen in this way, numbers are things in *potentia,* theoretical availabilities of sign production, the elementary and irreducible signifying acts that the Mathematician, the one-who-counts, can imagine his Agent to perform via a sequence of iterated ideal marks whose paradigm is the pattern 1, 11, 111, etc. The meaning that numbers have -what in relation to this pattern they are capable of signifying within assertions -lies in the imperatives and thought experiments that mathematics can devise to prove assert ions; that is, can devise to persuade the mathematician that the predictions being asserted about his future encounters with number signs are to be believed.

Thus, the numbers are objects that result - that is, are capable of resulting - from an amalgam of two activities, thinking (imagining actions) and scribbling (making ideal marks), which are inseparable: mathematicians think about marks they themselves have imagined into potential existence. In no sense can numbers be understood to precede the signifiers which bear them; nor can the signifiers occur in advance of the signs (the numbers) whose signifiers they are. Neither has meaning without the other: they are coterminous, co-creative, and co-significant.

What then, in such a scheme, is the status of numerals? Just this: since it seems possible to imagine pluralities or collections or sets or concatenates of marks only in the presence of notations which 'describe' these supposedly prior pluralities, it follows that every system of numerals gives rise to its own progression *of'* numbers. But this seems absurd and counter-intuitive. For is it not so that the 'numbers' studied by Babylonian, Greek, Roman, and present-day mathematicians, though each of these mathematical cultures presented them through a radically different numeral system, are the *same* numbers? If they arc not, then (so the objection would go) how can we even understand, let alone include within current mathematics, theorems about numbers produced by, say, Greek mathcmatics? The answer is that we do so through a backward appropriation: mathematics is historically cumulative not because both we and Greek mathematicians are talking about the same timeless 'number' - which is essentially the numerals-name-numbers view - but because we refuse to mean anything by 'number' which does not square with what we take them to have meant by it. Thus, Euclid's theorem 'given any prime number one can exhibit a larger one' is not the same as the modern theorem 'there exist infinitely many prime numbers' since, apart from any other considerations, the nature of Greek numerals makes it highly unlikely that Greek mathematicians thought in terms of an infinite progression of numbers. That the modern form subsumes the Greek version is the result not of the timelessness of mathematical objects, but of a historically imposed continuity - an imposition that is by no means explicitly acknowl-edged, on the contrary presenting itself as the obvious 'fact' that mathematics is timeless.

In relation to this issue one can make a more specific claim, one which I have elaborated elsewhere (Rotman 1987), that the entire modern conception of integers in mathematics,was made possible by the system of signifiers provided by the Hindu numerals based on zero; so that it was the introduction of the sign zero - unknown to either classical Greek or Roman mathematicians - into Renaissance mathematics that *created* the present-day infinity of numbers.

Insofar, then, as the subject matter of mathematics is the whole numbers, we can say that its objects - the things which it countenances as existing and which it is said to be 'about' – are unactualizcd possibles, the potential sign productions of a counting subject who operates in the presence of a notational system of signifiers. Such a thesis, though. is by no means restricted to the integers. Once *it* is accepted that the integers can be characterized in this way, essentially the same sort of analysis is available for numbers in general. The real numbers, for example, exist and are created as signs in the presence of the familiar extension of Hindu numerals - the infinite decimals - which act as their signifiers, Of course, there are complications involved in the idea of signifiers being infinitely long, but from a semiotic point of view the problem they present is no different from that presented by *arhitrarily* long finite signifiers. And moreover, what is true of numbers is in fact true of the entire totality of mathematical objects: they are all signs – thought/scribbles - which arise as the potential activity of a mathematical subject.

Thus mathematics, characterized here as a discourse whose assertions are predictions about the future activities of its participants, is 'about' -insofar as this locution makes sense - itself. The entire discourse refers to, is 'true' about, nothing other than its own signs. And since mathematics is entirely a human artifact, the truths it establishes - if such is what they are - are attributes of the mathematical subject: the tripartite agent' of Agent/Mathematician/Person who reads and writes mathematical signs and suffers its persuasions.

But in the end, 'truth' seems to be no more than an unhelpful relic of the platonist obsession with a changeless eternal heaven. The question of whether a mathematical assertion, a prediction, can be said to be 'true' (or accurate or correct) collapses into a problem about the tense of the verb. A prediction - about some determinate world for which true and false make sense-might in the future be seen to be true, but only after what it foretold has come to pass; for only then, and not before, can what was pre-dicted be dicted. Short of' fulfillment, as is the condition of all but trivial mathematical cases, predictions can only be believed to be true. Mathematicians believe because they are persuaded to believe; so that what is salient about mathematical assertions is not their supposed truth about some world that precedes them, but the inconceivability of persuasively creating a world ii which they are denied. Thus, instead of a picture of logic as a form of truth-preserving inference, a semiotics of mathematics would see it as an inconceivability-preserving mode of persuasion - with no mention of 'truth' anywhere.

References

Berry. Margaret (1975). *Introduction to Systematic Linguistics* London: T. J. Batsford, Ltd.

Brouwer, L. E.J. ***(1952). Hiistorical back***ground. principles and methods *of* intuitionism ***South African Journal of Science 49, 139-146***

Buchler, Justus ***(1940). The Philosophy of Peirce: Selected Writings***. London: Routledgc.

Eco, Umberto ***(1976).*** *A Theory of Semiotics London*: Macmillan.

Frcgc, Gottlob ***(1967)***. The thought: ***A logical*** enquiry. In *Philosophical Logic. P. F* Strawson (ed.). Oxford: Oxford University Press.

Hilbert, *D.* and Bernays, ***P. (1934)***. *Grunddlagen der* ***Mathmatik, vol. 1***. Berlin: Springer Verlag.

Lorenzen, P. ***(1955)***. *Einfuhrung in die operative Logik und Mathematik,* Berlin; Springer Verlag.

Rotman, Brian ***(1987)***. *Signifying Nothing: TheSemiotics of* ***Zero***. London: Macmillan.

\Vcyl. H. **(1949)**. *Philosophy of Mathematics and Natural Science*. Princeton: Princeton University Press.

7

Computers and the Sociology of Mathematical Proof

DONALD MACKENZIE

1. Introduction

In this paper, I shall explore the relationship between mathematical proof and the digital computer. This relationship is at the heart of the six scientific and technological activities shown in Figure 1.

Proofs are conducted *about* computers in at least three areas: those systems upon which human lives depend; key aspects of some microprocessors; and those systems upon which national security depends. These proofs about computers are themselves normally conducted *using* computer programs that prove theorems: automated theorem provers. But mathematicians themselves have also turned to the computer for assistance in proofs of great complication, and automated theorem provers are of considerable interest and importance within artificial intelligence. They raise, for example, the question of whether a computer can be an "artificial mathematician."

I have explored in other papers aspects of the evolution of these six areas since the 1950s: for an outline chronology, see appendix[1]. Here, I shall explore the questions how knowledge *about* computers, and knowledge produced *using*

[1] The research reported on here was funded by the UK Economic and Social Research Council under the Program in Information and Communication Technologies (A35250006) and research grants R000234031 and R00029008; also by the Engineering and Physical Sciences Research Council under grants GR/J58619, GR/H74452, GR/L37953 and GR/N13999. More detailed aspects are discussed in (MacKenzie 1991, 1994, 1995, 1999, 2000); (MacKenzie and Pottinger 1997), and (MacKenzie and Tierney 1996); an integrated discussion of this work will be found in (MacKenzie 2001). This paper is a development of three earlier papers attempting to summarize the main results of this work: (MacKenzie 1993b, 1993a, 1996).

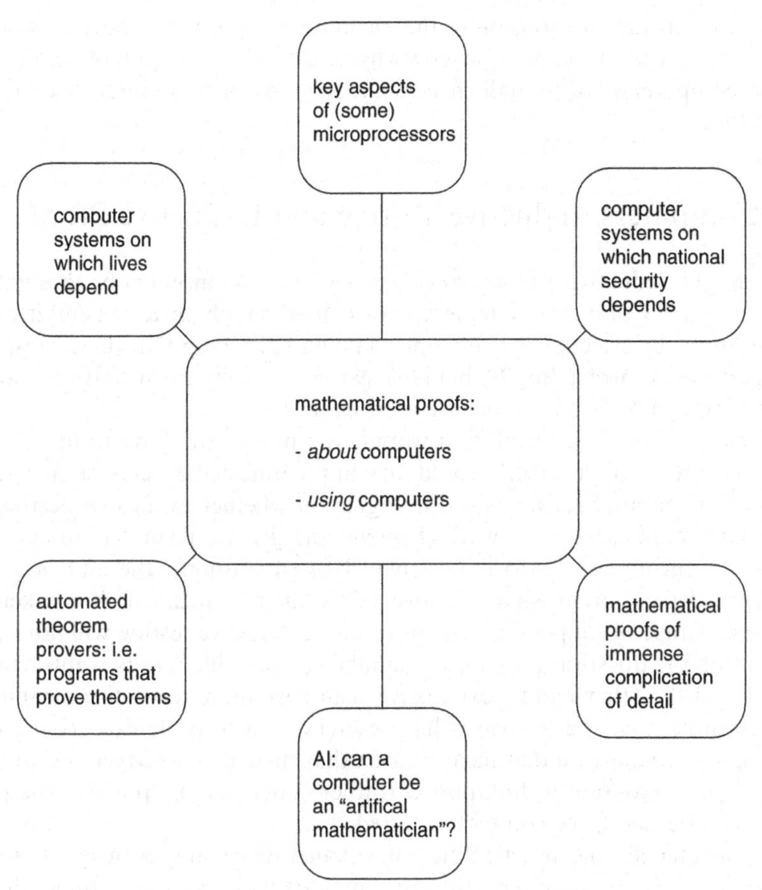

FIGURE 1.

computers, become seen as trustworthy? In particular, I shall address the question of variation: the way in which demonstrations that are convincing to some actors are unconvincing to others; the way in which "mathematical proof" can mean different things to different people. In Galison's terminology,[2] we are dealing here with a "trading zone": an arena within which different communities, with different traditions of demonstration, achieve a measure of practical coordination of their activities. I have focused upon differences in usage of "proof," but in a sense this is an arbitrary choice: it would also be possible to examine how they have developed what Galison calls a "pidgin," which allows fruitful communication between, for example, the National Security Agency, mathematicians, philosophers, and computer hardware designers.

[2] (Galison 1997).

Before turning to variation in the meaning of "proof," however, it is first necessary for me to explain briefly why proof *about* computers, and proof *using* computers, are thought necessary by many in these interacting communities.

2. Computers, Inductive Testing and Deductive Proof

Why might deductive, mathematical, proofs *about* computers be thought to be necessary? Computer systems can, of course, simply be tested empirically, for example by selecting sets of input data and checking that the corresponding output is correct. Surely that is a clear-cut, inductive way to demonstrate that computer systems will behave as intended?

For testing to yield certainty it would, at a minimum,[3] have to be exhaustive: otherwise, errors clearly could lurk in the untested aspects of computer systems. Computer scientists do not agree on whether exhaustive testing of computer hardware or software (I intend the generic term "computer systems" to encompass both) is possible.[4] Not surprisingly, the answer given tends to depend upon what "exhaustive" is taken to mean. If it is taken to mean coverage of all possible sets of inputs, exhaustive testing will normally be out of the question because the number of possible inputs is intractably large.[5] On the other hand, "exhaustive" can have more restrictive meanings: for example, that the test cases have exercised all possible execution paths through a program. In that meaning of exhaustive, then as Myers's standard textbook of program testing puts it (rather equivocally), "possibly the program can be said to be completely tested."[6]

Unfortunately, the practicalities of exhaustive testing, even in the more restrictive meanings, are typically daunting. Myers's text, for example, shows that the number of possible execution paths through even simple programs - ten to twenty lines long -can be enormously large, of the order of as much as 10^{14}, or 100 trillion.[7] It takes only modest complexity in the system being

---

[3] There are other issues here too, for example concerning (a) what it means for output to be correct and (b) the possibility of physical failure of a computer system or its components.

[4] In interviews and other discussions, proponents of testing have told me emphatically that exhaustive testing is possible, while proponents of formal verification frequently deny the possibility of exhaustive empirical testing.

[5] There may be only a finite number of valid inputs (e.g. integers may be restricted to a certain size), but it can be argued that "to be sure of finding all ... errors, one has to test using not only all valid inputs, but all possible inputs" ((Myers 1979, p. 8). In other situations, the number even of valid inputs will be, effectively, infinite. The valid inputs to a compiler, for example, are all the possible executable programs written in the corresponding language.

[6] (Myers 1979, p. 10).

[7] (Myers 1979, p. 10.)

tested to render exhaustive testing infeasible in practice, even with the fastest and most highly automated testing techniques currently available.

There are two broad responses to the practical limitations of exhaustive testing. On the one hand, proponents of testing point out that it can nevertheless be a very powerful way to discover errors, if enough resources are devoted to it, if the systems in question are designed to facilitate it, if it is systematic, and if it is planned and conducted by suitably skilled and suitably motivated people (especially people not directly involved in the development of the system in question). The world of commercial computing - of mainframe computers, microprocessors, packaged software and the like - has almost always accepted that (in combination, perhaps, with review of designs and programs by persons other than their authors) testing, however imperfect, is the best practical means of gaining knowledge of the properties of computer systems in advance of their use. In an environment where priorities are shaped by the need to control development costs and to move products to the market quickly, the general feeling has been that there is no alternative but to live with the limitations of testing. Intriguingly, there is evidence that the practical efficacy of the testing of deployed systems is considerably greater than its abstract statistical analysis might suggest.[8] One possible explanation is that programmers develop a good understanding of where mistakes are likely to lie, and choose test cases with this in mind.[9]

Since the late 1960s, however, a significant strand of thought - originating with academic computer scientists, but influential also in government (especially, in the U.S., in the national security community) and in limited sectors of industry - has begun to take a different approach.[10] Practitioners of this approach seek to verify programs and hardware designs mathematically, rather than just by empirical testing. The key practical appeal of this approach is that a deductive, mathematical analysis can claim to cover *all* cases, not merely the finite number that can be subject to empirical testing. A mathematician who wants to demonstrate the truth of a theorem about, for example, triangles does not embark on the endless task of drawing and checking all possible triangles. He or she looks for a compelling argument -a mathematical proof-that the result must be true for every triangle. The proponents of "formal verification" in computer science have argued since the 1960s that the way to get round the limitations of empirical testing is to proceed as the geometer does: to prove theorems about computer systems.

To verify mathematically the properties of a computer system requires the drawing up of an exact specification of its intended behaviour. Because of their many ambiguities, natural languages such as English are not considered adequate: the specification has to be expressed in a mathematical formalism.

[8] See (Hoare 1994).

[9] It is also possible, however, that there is an element of selection bias here: that systems that come into widespread use are better than those that are not brought to market, or are little used; and widely-used systems are, of course, "debugged" by their many users as well as by their developers.

[10] For examples, see (MacKenzie 2001).

From this specification one can attempt to derive deductively a detailed design that would implement it. More commonly, a program is written, or a hardware design produced, by more traditional methods, and then a mathematical proof is constructed to show that the program or design is a correct implementation of the specification.

3. Proofs Using Computers

In the first academic work in the 1960s on formal, mathematical, program verification, computer scientists used the traditional mathematician's tools of pencil and paper to prove that some simple examples of computer programs corresponded to their specifications. Mainstream opinion in this field, however, quickly came to the conclusion that hand-proof is of limited use: the reasoning involved is too tedious, intricate and error-prone, especially for real-world systems rather than "toy" examples. Since the 1970s, therefore, the field has turned to specially-written computer programs for assistance in proof construction: either proof checkers ("book-keeping" programs, intended to store proofs and to detect errors in them) or automated theorem provers (which are programs with a limited capacity to find proofs for themselves).[11]

Computer scientists interested in formal verification were, however, not the only people to use computer systems to conduct proofs. The earliest automated theorem provers were developed by those who wanted to construct "thinking machines" to demonstrate the feasibility of artificial intelligence: by common consent, the first AI program was Newell, Shaw, and Simon's "Logic Theory Machine," designed to prove simple theorems in the propositional calculus. However, early work on automated theorem proof was also done by philosophers and logicians such as Martin Davis, Hao Wang, and Hilary Putnam. Wang, in particular, seemed to see the artificial intelligence work as incompetent intrusion into the terrain of the logician, and there certainly was tension between the two approaches; tension replicated within artificial intelligence as the latter field bifurcated in the 1970s into contending camps of logicist "neats" and proceduralist "scruffies."[12]

Mathematicians, too, also turned to the computer for proof, though not, in general, to automated theorem-provers: their use of the computer was rather for ad hoc assistance in particular parts of complicated proofs. The best-known, and most debated, instance is the use of computer analysis by Kenneth Appel and Wolfgang Haken of the University of Illinois in their 1976 proof of the four-colour conjecture (that four colours suffice to colour in any map drawn upon a plane in such a way that countries which share a border are given different colours). First put forward in 1852, the conjecture had

[11] For an outline history of the latter, see (MacKenzie 1995).

[12] For these debates, see (MacKenzie 1995). For "neats" and "scruffies" see (Hayes 1987) and (Crevier 1993).

become perhaps the most famous unsolved problem in mathematics, resisting a multitude of efforts at proof for over a century. Appel and Haken's demonstration rested upon computerized analysis, occupying 1,200 hours of computer time, of over 1,400 graphs. The analysis of even one of those graphs typically went beyond what an unaided human being could plausibly do: the ensemble of their demonstration certainly could not be checked in detail by human beings. In consequence, whether that demonstration constituted "proof" was deeply controversial.[13] It will be of great interest to discover whether the computerized proof of the Kepler conjecture[14] announced in August 1998 by Thomas Hales of the University of Michigan will be equally controversial, since that is the first instance since the four-colour theorem of computer proof of a "famous" conjecture.

4. A Prediction and its (Near) Confirmation

That there might be variation in the kinds of argument that are taken to constitute mathematical proof should come as no surprise to either the sociologist of scientific knowledge or the historian of mathematics. Although nearly all subsequent empirical work in the sociology of scientific knowledge has concerned the natural sciences, rather than mathematics or logic, David Bloor clearly intended his "strong programme" to be applied to those latter fields as well.[15] In the work of Bloor, and also, for example, that of Eric Livingston, at least schematic arguments can be found to suggest that a sociology of mathematical proof should be possible.[16]

The history of mathematics is even more unequivocal in revealing variation in the kinds of argument that have been taken as constituting mathematical proof.[17] Eighteenth-century work in calculus, for example, often relied upon manipulating infinitesimally small quantities or infinite series in ways that became unacceptable in the nineteenth century. Early twentieth-century mathematics was riven by dispute over the acceptability, in proofs involving infinite sets, of the law of the excluded middle (the principle that if a proposition is meaningful, either it or its negation must be true).

These are clues that mathematical proof is a less straightforward, absolute matter than it is often taken to be. In 1987, colleagues and I drew upon this evidence to make a prediction about the effort (by then widely pursued) to apply

[13] For a history, see (MacKenzie 1999).
[14] The Kepler conjecture is that there is no tighter way of packing identical spheres in infinite space than the "face-centered cubic" packing, in which infinite triangular layers of spheres are placed one on top of the other, the higher layer being positioned so that its spheres sit in the depressions between the spheres of the lower layer, in a way similar to that in which grocers often stack oranges.
[15] (Bloor 1973, 1976).
[16] See (Bloor 1983, 1994, 1997) and (Livingston 1986), (Barnes et al. 1996, chapter 7).
[17] There is a useful survey in (Kleiner 1991).

mathematical proof to computer systems. We noted that this effort involved moving mathematical proof into a commercial and regulatory arena. We speculated that the pressures of that arena would force potential variation in the meaning of proof out into the open, but that disputes about proof would no long simply be academic controversies. We suggested that it might not be long before a court of law had to rule on what a mathematical proof is.[18]

That prediction was nearly borne out in 1991, when litigation broke out in Britain over the application of mathematical proof to a microprocessor chip called VIPER (Verificable Integrated Processor for Enhanced Reliability), which had been developed by computer scientists working for the Ministry of Defence's Royal Signals and Radar Establishment. At stake was whether the chain of mathematical reasoning connecting the detailed design of VIPER to its specification was strong enough and complete enough to be deemed a proof. Some members of the computer-system verification community denied that it was,[19] and (largely for unconnected reasons) sales of VIPER were disappointing. Charter Technologies Ltd., a firm which had licensed aspects of VIPER technology from the Ministry of Defence, took legal action against the Ministry, alleging, amongst other things, that VIPER's design had not been proven to be a correct implementation of its specification.

No "bug" had been found in the VIPER chips; indeed, their design had been subjected to an unprecedented amount of testing, simulation, checking and mathematical analysis. At issue was whether or not this process, as it stood immediately prior to the litigation,[20] amounted to a mathematical proof. Matters of fact about what had or had not been done were not central; the key questions that had been raised by critics were about the status, adequacy and completeness, from the viewpoint of mathematical proof, of particular kinds of argument. With the Ministry of Defence vigorously contesting Charter's allegations, the case failed to come to court only because Charter became bankrupt before the High Court heard it. Had it come to court, it is hard to see how the issue of what, in this context, mathematical proof consists in could have been avoided.

The VIPER controversy has been reported elsewhere,[21] and a single episode has inevitable idiosyncrasies. Let me turn, therefore, to wider issues about mathematical proof raised by the attempt to apply it to computer systems.

5. Mathematical Proof, Intentions and the Material World

The first issue is not contentious amongst specialists in the mathematical verification of computer systems, but it is important to highlight it nonetheless

[18] (Peláez et al. 1987, p. 5).
[19] (Brock and Hunt 1990); see also (Cohn 1989).
[20] There has been considerable further proof work since then on VIPER.
[21] (MacKenzie 1991, 1993b).

because it reveals a potential gap between what these specialists mean by "proof" and what the layperson might take them as meaning. Specialists agree that human intentions, as psychological entities, and computer systems, as physical artefacts, cannot be the objects of mathematical proof. For example, one specialist interviewed for the research upon which this paper is based observed that:

> When the President of the United States says, "Shoot down all incoming Russky missiles," that's not quite in predicate logic. ... To translate the requirement into a formal specification introduces a gulf which is not verifiable in any formal system.

Specialists in the field believe that "proof" applied to a computer program or computer hardware can only be proof of correspondence between two mathematical models: a model (the formal specification) of the intended behaviour of the program or hardware; and a model of the program or of the detailed design of the hardware.

To specialists, therefore, "proof," or "formal verification," refers to the correspondence between two mathematical models, not to the relationship between these models and either psychological or physical reality. This immediately opens up a potential discrepancy between specialist and lay views of the assurance conveyed by terms such as "proven" or "verified." To the specialist, they do not mean that systems are correctly designed, or cannot fail (as the lay person might infer), merely that a mathematical model of their detailed design corresponds to their formal specification.[22]

The potential discrepancy matters because lay people's views of the trustworthiness of computer systems can be of considerable importance. The safety of many systems depends on the interaction between computers and human users (who will normally be lay people in terms of their knowledge of computer system verification). Computer-related fatal accidents are caused more commonly by problems in this interaction than by failures of the computer system itself, and users' undue trust in computer systems plays a significant role in these interaction problems.[23]

6. Formal Proof and Rigorous Argument

Substantively important though it is, the potential discrepancy between lay and specialist views of "proof" is less interesting intellectually than variation amongst specialist views. This variation is caused less by the fact that proof is applied *to* computer systems than by the fact that proof is conducted *using* computer systems; but it also goes beyond the latter.

If we examine proofs within the disciplines (mathematics, logic and computer science) where deductive proof is central, we find a large variety of

[22] For a trenchant statement of this by a specialist, see (Cohn 1989). The celebrated critique of program verification by (Fetzer 1988) amounts largely to the same point.
[23] (MacKenzie 1994).

types of argument. Much of this variety is, of course, simply the result of variation in the subject-matter being reasoned about. However, it is also possible to class proofs into two more over-arching categories:

1. *Formal proof.* A formal proof is a finite sequence of formulae, in which each formula is either an axiom or derived from previous formulae by application of rules of logical inference such as modus ponens.[24] The steps in a formal proof are "mechanical" or syntactic applications of inference rules; their correctness can therefore be checked without understanding the meaning of the formulae involved.
2. *Rigorous argument.* This broader category includes all those arguments that are accepted by mathematicians (or other relevant specialists) as constituting mathematical proofs, but that are not formal proofs in the above sense. The proofs of ordinary Euclidean geometry, for example, are rigorous arguments, not formal proofs: even if they involve deducing a theorem from axioms (and some involve reasoning that it is not, at least directly, of this form), the steps in the deduction are typically not simply applications of rules of logical inference. This is not simply a reflection of the antiquity of Euclidean geometry: articles in modern mathematics journals, whatever their subjects, almost never contain formal proofs.[25]

The distinction between formal proof and rigorous argument bears a close - though not one-to-one -relation to the computerization of proof. Formal proof has been relatively easy to automate. The application of rules of inference to formulae, considered simply as strings of symbols, can be implemented on a digital computer using syntactic pattern matching.[26] The automation of rigorous argument, on the other hand, has been a far more difficult problem. There are some parts of specific rigorous arguments that can be reduced to calculation or algorithmic checking, and there the potential for automation is high. Furthermore, there are now widely-used commercial programs that automate

---

[24] Let p and q be any propositions. If p and p implies q are either axioms or previous formulae in the sequence, then (according to modus ponens) we can derive q. The "meanings" of p and q do not enter into the derivation. On the more general conception of formal proof used here, see, for example, (Boyer and Moore 1984, p. 181).

[25] Although those in this field would all recognise something akin to the distinction drawn here, there is no entirely standard terminology for capturing it. Thus many mathematicians would call deductions from axiomatic set theory "formal proofs," even if these deductions are not simply applications of rules of logical inference. What I am calling "rigorous argument" is sometimes called "informal proof", but I avoid that phrase because it carries the connotation of inferiority vis-à-vis formal proof. I draw the term "rigorous argument" from (Ministry of Defence 1991, part 2, p. 28): see (MacKenzie 1996).

[26] Many of the theorem-prover bugs discussed below were caused by subtle mistakes in these algorithms. Participants recognize these as "errors," rather than matters of dispute, indicating a degree of consensus about the basic process.

symbol manipulation in fields like algebra and calculus.[27] There are, however, as yet no "artificial mathematicians," in the sense of automated systems capable of handling the full spectrum of rigorous arguments used in different fields of mathematics. The proof checkers and theorem provers used in computer system verification automate formal proof, not - in general - rigorous argument.

Formal proof and rigorous argument are not necessarily *inherently* opposed. It is widely held, for example, that rigorous arguments are "sketches" or "outlines" of formal proofs: arguments with gaps that could, in principle, usually be filled by sequences of applications of rules of logical inference.[28] Yet formal proof and rigorous argument remain *available* to be counterposed, for there is a sense in which they have complementary virtues. Because the steps in formal proof do not depend, at least directly, upon the meaning of the formulae being manipulated, they avoid appeals, often implicit, to intuitions of meaning - appeals that can contain subtle, deep pitfalls. That formal proofs can be checked mechanically is a great advantage in a field, like computer system verification, where proofs are typically not "deep" (in the mathematicians' sense of involving profound concepts) but are large and intricate, and where it is particularly desirable to have a corrective to human wishful thinking.

Rigorous arguments, on the other hand, typically have the virtue of surveyability. They are nearly always very much shorter than corresponding formal proofs, and are thus easier for human beings to read and to understand. By virtue of their appeal to the meaning of formulae, rigorous-argument proofs can produce a psychological sense of conviction ("it *must* be so") that is hard to achieve with formal proof, and they have a certain robustness. Typographical mistakes in them, for example, are commonplace, but what carries conviction is the overall structure of the argument, which is more than an aggregate of individual steps. Because this overall structure can be grasped mentally, it can be checked by others, used by them in other contexts, and rejected or improved upon as necessary.

At its most extreme, defenders of one variety of proof can deny the label "proof" to the other. That, for example, was the basis of a famous attack on the formal verification of computer programs mounted in the late 1970s by the American computer scientists Richard DeMillo, Richard Lipton and Alan Perlis. Program verifications, they argued, were mere formal manipulations, incapable of being read, understood and assessed like "real" mathematical proofs. A proof, they said "is not ... [an] abstract object with an independent existence," independent of these "social processes." Since program verifications could not, they claimed, be subject to these social processes, they should

[27] For a critical discussion of their soundness, see (Stoutemyer 1991).

[28] There is, however, general acceptance of Kurt Gödel's celebrated conclusion (the "incompleteness theorem") that any finite formal system rich enough to encompass arithmetic must contain a true statement for which there is no formal proof within the system.

not be seen as proofs.[29] A similar, albeit less sweeping, contrast between formal proof and rigorous argument also underpinned the most explicit defence of the VIPER proof, which was mounted by a leading figure in the U.K. software industry, Martyn Thomas, head of the software house Praxis. "We must beware," he wrote, "of having the term 'proof' restricted to one, extremely formal, approach to verification. If proof can only mean axiomatic verification with theorem provers, most of mathematics is unproven and unprovable. The 'social' processes of proof are good enough for engineers and other disciplines, good enough for mathematicians, and good enough for me.... If we reserve the word 'proof' for the activities of the followers of Hilbert [David Hilbert, 1862–1943, leader of 'formalism' within mathematics], we waste a useful word, and we are in danger of overselling the results of their activities."[30]

7. Proof and Disciplinary Authority

Over and above particular contexts in which formal proof and rigorous argument are counterposed is something of a disciplinary divide. Although logic and mathematics are often though of as similar enterprises, and there is in a general sense a considerable overlap between their subject-matter, they are, to a degree, socially distinct. Logic's origins as a discipline are closer to philosophy than to mathematics, and even now logicians distinguish between mathematical logic and philosophical logic. Even the former is sometimes not seen by university mathematics departments as a genuine part of their province, and mathematics undergraduates often learn no more than the most elementary formal logic. "I was going to be a mathematician, so I didn't learn any logic," said one interviewee.

The two general notions of proof - formal proof and rigorous argument - are, to an extent, underpinned by this social divide. Logic has provided the notation and conceptual apparatus that make formal proof possible, and provides a viewpoint from which proof as conducted by mathematicians can be seen as unsatisfactory. One logician, Peter Nidditch, wrote, in 1957, that "in the whole literature of mathematics there is not a single valid proof in the logical sense."[31] An interviewee reported that even as an undergraduate:

> I knew, from the point of view of formal logic, what a proof was ... [and was] already annoyed by the vagueness of what constitutes a mathematics proof.

Mathematics, on the other hand, provides the exemplars of proof as rigorous argument. The main resource for criticism of formal proof is, as we have seen, appeal to the practice of mathematics.

---

[29] (DeMillo et al. 1979, esp. p. 273); see (MacKenzie 2001, chapter 6)

[30] (Thomas 1991), quoted in (MacKenzie 1993b).

[31] (Nidditch 1957, p. v.)

Note, however, that attitudes to proof are not in any simple sense determined by disciplinary background. Disciplines are not homogeneous, biographies are complex, and other factors are at work. Many of those who work in the formal aspects of computer science have training in both mathematics and logic, and so have three disciplinary identities potentially open to them. When logicians themselves do proofs about formal logical systems, these proofs are typically rigorous arguments, not formal proofs. Those who have wished to computerize proof have had little alternative but to automate formal proof, whatever their disciplinary background. Rather, the connection between discipline and proof is more a matter of the disciplinary authorities of mathematics and, in a weaker sense, logic being *available* to those who, whatever their backgrounds (Martyn Thomas, for example, was trained as a biochemist before entering the computer industry), defend rigorous argument against enthusiasts for formal proof, or vice versa.

8. Logics, Bugs and Certainty

Those who perform computer system verifications are often, in practice, adherents of formal proof,[32] and the regulatory standard in this area that most sharply distinguishes formal proof from rigorous argument, the U.K. Ministry of Defence's interim standard governing safety-critical software, came down in favour of the former, at least for those situations where the highest level of assurance is desirable.[33] (One reason this area is fascinating is that a defence procurement standard is forced onto philosophical terrain such as this!) There is, therefore, perhaps potential consensus - at the heart of this "trading zone," within the field of computer system verification, not more widely - that "proof" should mean formal, mechanized proof. Would such consensus lead to complete agreement on "proof," and remove the possibility for future dispute and litigation?

Agreement on the formal notion of proof still leaves open the question of the precise nature of the logical system to be used to manipulate formulae. There are different formal logics. The early twentieth century dispute over the law of the excluded middle has echoes in a continuing divide between classical logic (which allows one to prove that a mathematical object exists by showing that its non-existence would imply a contradiction) and constructive logic (which requires demonstration of how to construct of the object in question). Computer science has, furthermore, been an important spur to work on "non-standard" logics, such as modal and temporal logics. "Relevance logic" - long thought to be a philosophers' plaything - is being automated with a view to

[32] The main exception to this generalization concerns decision procedures (see below). There is also variation in attitude to the importance of the surveyability of formal proofs.

[33] (Ministry of Defence 1991; see (Tierney 1992).

practical applications in inference from databases.[34] There is even a vogue, and several practical applications, for "fuzzy logic," which permits degrees of simultaneous adherence to both a proposition and its negation, so violating what some have felt to be a cultural invariant.[35]

This zoo of diverse, sometimes exotic, formal logics is currently a less potent source of dispute than might be imagined. Not all of them are regarded as suitable for mathematical reasoning or computer system verification. Furthermore, there has been a subtle shift in attitudes to formal logic. Nowadays, few specialists take a unitary view of logic in which there is only one true logic and all else is error. It is much more common (especially in computer science) to find a pluralist, pragmatic attitude, in which different logics are seen as technical tools appropriate in different circumstance. Direct clashes between the proponents of different logical systems are far less common than might be expected by extrapolation from, for example, the bitter earlier disputes between classical logic and constructivism. One interviewee, for example, told me that the decision about whether or not to include the law of excluded middle in his theorem prover, a modern version of the issue that had caused great philosophical angst early in the twentieth century, had been simply "a marketing decision"!

Nevertheless, the possibility remains that a pluralist, pragmatic attitude to different logics is a product of the early, exploratory, academic phase of the application of formal logic to computer systems.[36] It cannot be guaranteed that it would remain intact in a situation where there are major financial or political interests in the validity or invalidity of a particular chain of formal reasoning. Specialists who are called upon to *justify* their formal logic, under courtroom cross-examination by a well-briefed lawyer, may face difficulties, for that process of justification is fraught with philosophical problems.[37]

At least equal in importance to the diversity of formal logics as a source of potential dispute is the simple fact that the tools of mechanized formal proof (proof checkers and automated theorem provers) are themselves computer programs. As such, they - especially automated theorem provers, which are quite complicated programs - may contain design faults. In interviews for this

---

[34] (Thistlewaite et al. 1988). Relevance logic excludes the elementary theorem of standard logic that a contradiction implies any proposition whatsoever (see (Bloor 1983, 123–32)). The database issue is the possible need to make sensible inferences in the presence of contadictory items of data.

[35] See (Archer 1987). A major issue about fuzzy logic is whether it is indeed a logic, or a version of probability theory. Claude Rosenthal of the École des Mines has recently completed a PhD thesis dealing with the development of fuzzy logic.

[36] (MacKenzie 1993b).

[37] Best known is Kurt Gödel's theorem that a finite formal system rich enough to express arithmetic cannot be proven consistent without use of a more powerful system, the consistency of which in turn would have to be proven, hence beginning an endless regress. On more general problems of circularity and regress in the justification of deduction, see (Haack 1976).

research, designers of automated theorem provers often reported experience of bugs in their systems that would have allowed "theorems" that they knew to be false nevertheless to be proven. Such bugs were not large in number, they were corrected whenever they were discovered, and no interviewee reported such a bug causing a false result whose falsity had not been detected readily. But no designer seemed able to give an unequivocal guarantee that no such bugs remained.

No automated theorem prover has, to date, itself been subject in its entirety to formal verification.[38] Even if one had been verified - or even if the results of an unverified prover were checked by an independent proof checking program (something several interviewees advocated, but which is rare in practice) - many specialists in the field would still not regard this as guaranteeing complete certainty. The reasons given by them ranged from "Gödelian" concerns about the consistency of formal systems[39] to the possibility of coincident errors in different automated theorem provers or proof checkers.

Reactions to the difficulty of achieving certainty with automated theorem provers varied amongst those interviewed. One interviewee suggested that it indicates that the overall enterprise of formal verification is flawed, perhaps because of what he felt to be the impoverishment of its notion of proof:

> You've got to prove the theorem-proving program correct. You're in a regression, aren't you? ... That's what people don't seem to realise when they get into verification. They have a hairy great thing they're in doubt about, so they produce another hairy great thing which is the proof that this one's OK. Now what about this one which you've just [used to perform the proof]? ... I say that serves them jolly well right.

That is not the response of program and hardware verification "insiders." While paying considerable attention to soundness, they feel that theorem-prover bugs are not important practical worries compared to ensuring that the specification of a system expresses what, intuitively, is intended:

> If you ... ask where the risks are, and what are the magnitudes of the risks, the soundness of the logic is a tiny bit, a really tiny bit, and the correctness of the proof tool implementing the logic is slightly larger [but] actually ... quite a small risk.

Insiders almost all share the perception that the risk of a serious mistake in computer system verification being caused by a bug in an automated prover is small, but they are also wary of claiming that the risk can ever be shown to be *zero*. Their judgments differ as to what measures are necessary to allow claims of proof safely to be made. Some would not be happy without a full formal proof checked by an independent proof checking program; others feel that this would be a waste of effort, compared to the need for attention to more likely dangers, notably deficient specifications.

---

[38] The Boyer-Moore prover (a leading U.S. system) is being re-engineered with a view to using it to verify its own correctness.
[39] See note 37 above.

One potentially contentious issue is the use of decision procedures: procedures that can decide, in a wholly deterministic, algorithmic way, whether or not mathematical statements in particular domains are true. Typically, decision procedures return simply the judgement "true" or "false," not a formal proof of truth or falsity. To some interviewees, decision procedures are a necessary and harmless part of the "real world" of computer system verification. They offer the great practical advantage that their use does not require the time-consuming skill that a generic theorem prover typically demands. To other practitioners, however, decision procedures must themselves be verified formally (which is currently rare). Otherwise, using them is "like selling your soul to the Devil - you get this enormous power, but what have you lost? You've lost proof, in some sense."

9. Conclusion

That there is variation in what proof is taken as consisting in, and that insiders do not believe absolute certainty is achievable, are not arguments against seeking to apply mathematical proof to computer systems. Even in its current state, the effort to do so has shown its practical worth in finding errors not found by conventional methods.[40] Furthermore, the field is not static, and some of the issues I have discussed (such as the verification of decision procedures) are active research topics. Even the overall divide between formal proof and rigorous argument is not necessarily unbridgeable, with many insiders believing it to be possible to construct proofs that are formal in their detailed steps but that still have a humanly surveyable overall structure. It is true that the end of the Cold War, and subsequent reductions in defence budgets, have caused a precipitous decline in the traditional chief funding source for formal verification: computer security. Nevertheless, a well-established tradition of academic research, and the emergence of new spheres of application for formal methods (especially to hardware), give grounds for at least modest confidence that the field will survive its short-term difficulties.

Historians and sociologists of science certainly have grounds for hoping that it does so! The automation of proof, largely fuelled as a practical activity by the desire to do proofs about computer systems, is a fascinating intellectual experiment. It highlights the fact that our culture contains not one, but two, ideals of mathematical proof: rigorous argument and formal proof. The distinction, of course, predates the digital computer. But when David Hilbert, for example, put forward the notion of formal proof in the 1920s, he did not intend that mathematicians should actually do their proofs in this way. Instead, formal proof was a tool of his "metamathematics," a way of

[40] (Glanz 1995).

making mathematical proof the subject of mathematical analysis by providing a simple, syntactic, model of it. Nidditch was still able, as we have seen, to claim in 1957 that the "whole literature of mathematics" contained not a single formal proof.

From the point of view of the history, philosophy, and sociology of science, the significance of the automation of proof is that it has turned formal proof from a tool of metamathematical inquiry into a practical possibility. To some extent at least, a *choice* between formal proof and rigorous argument has been opened up. The mainstream mathematical community remains firmly wedded to rigorous argument and sceptical of "computer proof," as was demonstrated by reactions to the Appel and Haken proof of the four colour theorem, or, more recently, by the indifference that greeted "QED," an ambitious proposal to create a giant computerized encyclopaedia of formally, mechanically proven mathematics. Nevertheless, formal, mechanized proof is now well entrenched in a variety of niches outside of pure mathematics. Although, to be sure, there are many areas of mathematics where such proof is not, or at least is not yet, a viable alternative to conventional forms of demonstration, its very existence indicates something of the contingency of the latter.

Furthermore, the automation of proof throws interesting light upon formal proof itself. Just as a choice between formal proof and rigorous argument has opened up, so those committed to formal, mechanized proof face decisions. Which logic shall they adopt? How shall they ensure that a putative formal proof is indeed a proof? Is an independent mechanized proof checker a necessity? Should a theorem prover itself be subject to formal verification? Going beyond the current practical state of the art, further questions can be anticipated: must the compiler for the prover's language be verified? What about the machine upon which it runs?

Questions such as these are not merely of academic interest. Formal verification is, as I have emphasized, a practical activity, often conducted in a commercial and regulatory arena. As the VIPER controversy showed, epistemological questions can, in that context, take on a sharp "real-world" significance, as practitioners have to decide what forms of demonstration constitute mathematical proof. The VIPER episode is as yet *sui generis*: no other dispute about formal verification or automated proof has led to litigation. If, however, formal verification prospers, it will be of great interest to discover whether (in Galison's terminology) a stable creole, with a stable meaning of "proof," will emerge in this trading zone, or whether the VIPER episode becomes the first of many disputes over what proof consists in.

10. Appendix: An Outline Chronology

1955/56: Newell, Shaw, Simon: Logic Theory Machine
1959: Herbert Gelernter: Geometry Machine
1960: Hao Wang champions algorithmic against heuristic theorem provers

1962: John McCarthy: "Towards a Mathematical Science of Computation"
1963: Alan Robinson: Resolution, a machine-oriented inference rule
1966: Peter Naur: "Proof of Algorithms by General Snapshots"
1967: R. W. Floyd: "Assigning Meanings to Programs"
1967: First public discussion of computer security vulnerabilities of US defence systems
1968: NATO conference on Software Engineering at Garmisch: diagnosis of "software crisis"
1969: C. A. R. Hoare: "An Axiomatic Basis for Computer Programming"
1971: Robert Boyer and J Strother Moore begin collaboration that leads to their theorem prover
1973-80: SRI PSOS project. Provably Secure Operating System
1976: Appel and Haken prove four-colour conjecture, using 1200 hours computer time
1978-87: SRI proof work on SIFT fault-tolerant avionics computer
1979: DeMillo, Lipton and Perlis attack program proof in *Communications of the ACM*
1983: "Orange Book." Design proof required for A1 secure systems
1983: SIFT Peer Review
1984-85: VIPER proof effort begins at Royal Signals and Radar Establishment
1985: Honeywell Secure Communications Processor: first system to achieve A1 rating.
1986-90: Disputes over Trip Computer Software delay licensing of Darlington, Ontario, nuclear power station
1986: First clear-cut software error deaths: overdoses from Therac-25 radiation therapy machines
1987: Inmos T800 transputer: floating-point unit microcode formally verified
1987-88: Commercialisation of VIPER begins
1988-89: Entry into service of Airbus A320, first fly-by-wire airliner
1988: Crash of A320 at Habsheim air show triggers fierce controversy in France
1988: James Fetzer attacks program proof in *Communications of the ACM*
1989: Formally verified SACEM train protection software enters service on RER, Paris
1991: UK Interim Defence Standard 00-55: proof required for systems most critical to safety
1991: VIPER law suit
1994: Hackers seize control of computer network at USAF Rome (N.Y.) Lab
1994: Pentium divide bug
1995: Entry into service of Boeing 777, most computer-intensive airliner to date
1996: New computer-assisted proof of four-colour theorem by Robertson et al.
1998: Thomas Hales announces computer-assisted proof of Keplers conjecture

References

Archer, M. S. (1987). Resisting the Revival of Relativism, *International Sociology* **2**: 235–50.

Barnes, B., Bloor, D. and Henry, J. (1996). *Scientific Knowledge: A Sociological Analysis*, Athlone, London.

Bloor, D. (1973). Wittgenstein and Mannheim on the Sociology of Mathematics, *Studies in the History and Philosophy of Science* **4**: 173–91.

Bloor, D. (1976). *Knowledge and Social Imagery*, Routledge, London.

Bloor, D. (1983). *Wittgenstein: A Social Theory of Knowledge*, Macmillan, London.

Bloor, D. (1994). "What can the Sociologist of Knowledge say about 2 + 2 = 4", in: *Mathematics, Education and Philosophy* (P. Ernest, ed., London, Falmer), pp. 21–32.

Bloor, D. (1997). *Wittgenstein, Rule and Institutions*, Routledge, London.

Boyer, R. S. and Moore, J. S. (1984). Proof Checking the RSA Public Key Encryption Algorithm, *American Mathematical Monthly* **91**: 181–89.

Brock, B. and Hunt, W. A. (1990). *Report on the Formal Specification and Partial Verification of the VIPER Microprocessor*, Computational Logic, Inc, Austin, Texas.

Cohn, A. (1989). The Notion of Proof in Hardware Verification, *Journal of Automated Reasoning* **5**: 127–39.

Crevier, D. (1993). *AI: The Tumultuous History of the Search for Artificial Intelligence*, Basic Books, New York.

DeMillo, R., Lipton, R. and Perlis, A. (1979). Social Processes and Proofs of Theorems and Programs, *Communications of the ACM* **22**: 271–80.

Fetzer, J. H. (1988). Program Verification: The Very Idea, *Communications of the ACM* **31**: 1048–63.

Galison, P. (1997). *Image and Logic: A Material Culture of Microphysics*, University of Chicago Press, Chicago.

Glanz, J. (1995). Mathematical Logic Flushes out the Bugs in Chip Designs, *Science (20 January)* **267**: 332–33.

Haack, S. (1976). The Justification of Deduction, *Mind* **80**: 112–19.

Hayes, P. J. (1987). A Critique of Pure Treason, *Computational Intelligence* **3**: 179–185.

Hoare, C. A. R. (1994). "How Did Software Get So Reliable Without Proof?" Talk to the Awareness Club in Computer Assisted Formal Reasoning, Heriot Watt University, Edinburgh, 21 March 1994.

Kleiner, I. (1991). Rigor and Proof in Mathematics: A Historical Perspective, *Mathematics Magazine* **64**: 291–314.

Livingston, E. (1986). *The Ethnomethodological Foundations of Mathematics*, Routledge, London.

MacKenzie, D. (1991). The Fangs of the VIPER, *Nature* **352**: 467–68.

MacKenzie, D. (1993a). The Social Negotation of Proof: An Analysis and a further Prediction, in: *Formal Methods in Systems Engineering* (Peter Ryan and Chris Sennett, eds., London, Springer), 23–31.

MacKenzie, D. (1993b). Negotiating Arithmetic, Constructing Proof: The Sociology of Mathematics and Information Technology, *Social Studies of Science* **23**: 37–65.

MacKenzie, D. (1994). Computer-Related Accidental Death: An Empirical Exploration, *Science and Public Policy* **21**: 233–48.

MacKenzie, D. (1995). The Automation of Proof: An Historical and Sociological Exploration, *IEEE Annals of the History of Computing* **17**(3): 7–29.

MacKenzie, D. (1996). Proof and the Computer: Some Issues Raised by the Formal Verification of Computer Systems, *Science and Public Policy* **23**: 45–53.

MacKenzie, D. (1999). Slaying the Kraken: The Sociohistory of a Mathematical Proof, *Social Studies of Science* **29**: 7–60.

MacKenzie, D. (2000). A Worm in the Bud? Computers, Systems, and the Safety-Case Problem, in *Systems, Experts, and Computation: The Systems Approach in Management and Engineering, World War II and After* (Hughes, A. C., and Hughes, T. P., eds., Cambridge, Mass., MIT Press), 161–190.

MacKenzie, D. (2001). *Mechanizing Proof: Computing, Risk, and Trust*, MIT Press, Cambridge, Mass.

MacKenzie, D. and Pottinger, G. (1997). Mathematics, Technology, and Trust: Formal Verification, Computer Security, and the U.S. Military, *IEEE Annals of the History of Computing* **19**(3): 41–59.

MacKenzie, D. and Tierney, M. (1996). Safety-Critical and Security-Critical Computing in Britain: An Exploration, *Technology Analysis and Strategic Management* **9**: 355–79.

Ministry of Defence (1991). *Interim Defence Standard 00-55: The Procurement of Safety Critical Software in Defence Equipment*, Ministry of Defence, Directorate of Standardization, Glasgow.

Myers, G. J. (1979). *The Art of Software Testing*, Wiley, New York.

Nidditch, P. H. (1957). *Introductory Formal Logic of Mathematics*, University Tutorial Press, London.

Peláez, E., Fleck, J. and MacKenzie, D. (1987). "Social Research on Software", Paper presented to workshop of the Economic and Social Research Council, Programme on Information and Communication Technologies, Manchester, December 1987.

Stoutemyer, D. R. (1991). Crimes and Misdemeanors in the Computer Algebra Trade, *Notices of the American Mathematical Society* **38**: 778–85.

Thistlewaite, P., McRobbie, M. A. and Meyer, R. K. (1988). *Automated theorem proving in non-classical logics*, Pitman, London.

Thomas, M. (1991). "VIPER Lawsuit withdrawn." Electronic mail communication, 5 June (1991).

Tierney, M. (1992). Software Enginering Standards: The 'Formal Methods Debate' in the UK, *Technology Analysis and Strategic Management* **4**: 245–78.

8

From G.H.H. and Littlewood to XML and Maple: Changing Needs and Expectations in Mathematical Knowledge Management

TERRY STANWAY

Abstract: This paper concerns changing needs and expectations in the way mathematics is practiced and communicated. The time frame is mainly the early twentieth century to the present and the scope is all activity that can be considered to fall under the purview of the mathematical community. Unavoidably, the idea of a mathematical community is confronted; what it means to claim ownership in this community and how knowledge management practices affect the community. Finally, a description of an extendable mathematical text-based database which can be used to manage user defined forms of mathematical knowledge is presented.

Hardy and Littlewood: A Study in Collaboration

The anecdote is related by C.P. Snow in his introduction to *A Mathematician's Apology*. It is a pleasant May evening some time in the 1930's and Hardy is in his fifties. He and Snow are walking at Fenner's cricket ground when the 6 o'clock chimes ring out from the nearby Catholic chapel. "It is rather unfortunate", Hardy remarks, "that some of the happiest hours of my life should have been spent within sound of a Roman Catholic church".

One of the preeminent mathematicians of his day and indeed, of the century, it is perhaps not entirely surprising that, living as he did, in the intellectual communities of Cambridge and Oxford, Gottfried Harold Hardy was an atheist. While our immediate concern is not his views on religion, it is germane only that Hardy was, in this fundamental domain, a non-conformist in an age that put a great deal of stock in conformity. Our concern *is* with his mathematics and, in particular, with his lively and productive collaboration with John Edensor Littlewood. Commencing in 1912, spanning some 36 years, and resulting in enough papers to fill several thick volumes of

Hardy's seven volume *Collected Papers* the Hardy-Littlewood collaboration stands as one of the most celebrated and productive academic partnerships of the twentieth century. It also provides an excellent starting point for an examination of mathematical knowledge management; but first, a little background is in order.

Hardy was already established in his career when at 33 years old he first met the 25 year-old John Edensor Littlewood at Cambridge in 1910. Prior to their meeting, each man had distinguished himself as a first rate mathematician. Falling into the broader definitions of the fields of Number Theory and Analysis, the work that they did together contributed immensely to the reputation of each and helped to bring about a renewal in English mathematics which had, for some time, been overshadowed by work from schools on the continent. There are several reasons why the Hardy-Littlewood collaboration provides a good backdrop to a discussion about mathematical knowledge management. One of the most important is the very fact that it was a collaboration between peers and thus entailed sharing of mathematical knowledge. The fact that their efforts were prolific and well-documented means that resources are relatively easy to obtain. Equally important is that both men were very much part of the mathematical community of their day and they conducted themselves, at least in the way they practised mathematics, according to community standards. But there is another reason that their collaboration bears examination in the context of mathematical knowledge management and that is that they lived and worked in the proverbial *interesting times* and it is to the intellectual climate of those times that we first turn our attention.

Towards Foundational Pluralism...

> Hardy 'asked 'What's your father doing these days. How about that esthetic measure of his?' I replied that my father's book was out. He said, 'Good, now he can get back to real mathematics'. Garrett Birkhoff

The date is 1910 and the location is Cambridge; the time and place of the first encounter between Hardy and Littlewood. The first world war was still four years away but tremors were already being felt in old orders both inside and outside the socio-political domain. In art and literature the modernist perspective had informed such works as Picasso's early cubist piece, *Les Desmoiselles d'Avignon* and Santayana's *The Life of Reason*. In physics, Einstein had published *The Special Theory of Relativity*, challenging the determinism of Newtonian mechanics. Mathematics did not emerge unscathed. In his book, *What Is Mathematics, Really?*, Reuben Hersh describes the fractures that arose in the philosophy of mathematics after the widely accepted idea that all of mathematics could be ultimately derived from the principles of Euclidean geometry fell victim first to logically consistent non-euclidean geometries and second to geometrically counter-intuitive concepts such as space filling curves and such unavoidable consequences of analysis as continuous everywhere but nowhere differentiable curves. In Hersh's words:

> The situation was intolerable. Geometry served from the time of Plato as proof that certainty is possible in human knowledge - including religious certainty. Descartes and Spinoza followed the geometrical style in in establishing the existence of God. Loss of certainty in geometry threatened loss of all certainty.

The response of mathematicians concerned with the philosophy of their subject was an attempt to replace geometry at the foundation of mathematical knowledge with arithmetic and set theory; thus giving rise to the field of Mathematical Logic. It is reasonable to state that the culmination of these efforts was the enunciation by David Hilbert of what came to be known as *Hilbert's Program*. In an address entitled *The Foundations of Mathematics* given in July of 1927 at the Hamburg Mathematical Seminar, he stated:

> ...I pursue a significant goal, for I should like to eliminate once and for all the questions regarding the foundations of mathematics, in the form that they are now posed, by turning every mathematical proposition into a formula that can be concretely exhibited and strictly derived, thus recasting mathematical definitions and inferences in such a way that they are unshakable and yet provide an adequate picture of the whole science.

Hilbert's objective and the objectives of others, such as Brouwer, von Neumann, and Weyl, whose goals were similar, albeit while starting from slightly different sets of assumptions, were famously shown to be impossible to achieve by the 1931 incompleteness result of the Austrian mathematician, Kurt Gödel. The result is that by the mid 1930's, all hope of a unified perspective regarding the foundations of mathematics is lost. In the main, at least initially, the competing perspectives on the foundations of mathematics are Hilbert's *formalism* and Brouwer's *intuitionism*. Later, *constructivism* articulated by among others, the American mathematician Erret Bishop, would emerge as a radical extension of some of the ideas of intuitionism. From the point of view of mathematical knowledge management, the specifics of these ideologies are not important; what is important is that, in the end, there is a plurality of perspectives regarding the foundations of mathematics. Which brings the discussion back to Hardy and Littlewood and the question of how all of the turmoil in the foundation of their science affected the lives of the "working mathematician".

If we could have asked Hardy and Littlewood about how the shifting philosophical ground affected the way they think, talk, and write about mathematics as individual mathematicians and, in particular, how the philosophical upheaval affected the nature of their collaboration, it is quite likely that their reply would have been that the philosophical debate was of little or no consequence. Like most practicing mathematicians of the time and today, when pressed, they seemed to adhere to a Neo-Platonist perspective of mathematical investigation. Here's Hardy from the *Apology*:

> I believe that mathematical reality lies outside us, that our function is to discover or observe it, and that the theorems which we prove, and which we describe grandiloquently as our "creations", are simply the notes of our observations.

This was all the philosophy that any practicing mathematician needed and all the philosophy that any mathematician continues to need. It explains mathematics as a process of discerning truths.

So, what about these two whose intent it was to simply get on with the business of doing mathematics? What influenced their day to day experience as mathematicians and what influenced the nature of their collaboration? The answer lies in the structures of the broader mathematical community in which they existed and in which, to be certain, both individually and in collaboration, they played prominent roles.

The Mathematical Community

> A man is necessarily talking error unless his words can claim membership in a collective body of thought. Kenneth Burke

The notion of community is loosely defined and can be used to refer to a lot of quite different types of social groupings; it is important to spend some time with the definition of "community". Intentionally, we will broadly define the mathematics community to include those involved with advancing the understanding of mathematics; either at its frontiers, the primary occupation of researchers, or within the existing body of mathematical knowledge such as teachers and students. The boundary is a porous one and relatively few would claim full time membership. Many others are interlopers, jumping in and out as the need arises or circumstances dictate. This idea of community will need to be passed through a prism, allowing us to consider separately four inter-related factors that help to bind the community: the *language* of the community, the *purposes* of the community, the *methods* of the community, and the *meeting places* of the community.

The Language of the Community

> A precisian professor had the habit of saying: '...quartic polynomial $ax^4+bx^3+cx^2+dx+e$, where e need not be the base of natural logarithms.' J.E. Littlewood

It is tempting, but tautological, to state that the language of the community is the language of mathematics and it only extends the tautomerism to state that anyone who claims membership in the community knows what this statement means. In reality, it may be argued that the paradigm for mathematical discourse is the language of the published research paper. All other discourse approximates the paradigm by degree according to what level of rigor is appropriate to the situation and audience.

The special symbols of mathematics present a particular challenge to expressing mathematics in mechanically type set or digital forms. An individual claiming membership in the mathematical community can generally be assumed to have some understanding of how to overcome those challenges.

The Purposes of the Community

> If intellectual curiosity, professional pride, and ambition are the dominant incentives to research, then assuredly, no one has a fairer chance of gratifying them than a mathematician. G. H. Hardy

In the sense it is used here, 'purpose' does not refer to the overriding *raison d'être* of the community; that has already been defined to be an interest in the advancement of mathematics. Rather, purpose here refers to what motivates an individual to seek membership in the community; and there are many. There is a professional motive which expresses itself by the simple statement that "I am involved with mathematics because this is how I earn my living". There is an egotistical motive which is expressed in the statement that "I am involved with mathematics because I take pleasure from proving to myself and others that I can overcome the challenges that the field affords". There is a social motive which is evident in the statement that "I am involved with mathematics because I benefit from the company of others who are involved with mathematics". And, finally, there is an aesthetic motive that is reflected in statements like "I am involved with mathematics because I wish to help unlock the beauty of mathematics".

With respect to mathematical knowledge management, an individual's reasons for being involved with mathematics strongly affects the individual's role in the community. This sense of purpose in the community in turn helps to determine the individual's information management needs.

The Methods of the Community

Next, we look at the *methods of the community*. Under this rubric, we examine the question of how do mathematicians do what they do and what tools do they use. Traditionally, and certainly through the period encompassed by the Hardy-Littlewood collaboration, mathematics has always been one of the most purely cerebral of the sciences, depending, for its practice, on little more than pencil and paper. This austerity is tightly associated with underlying philosophical assumptions about the nature of mathematics. The foundational shifts of the last century and developments in computer technology paved the way to the situation we find at present, with mathematicians lining up with theoretical physicists, molecular biologists, and others to claim time on the world's most powerful super computers. An important consideration regarding the question of how mathematicians do mathematics is the question of how and to whom do mathematicians *express* their mathematics. This warrants separate treatment.

The Meeting Places of the Community

J.J. Sylvester sent a paper to the London Mathematical Society. His covering letter explained, as usual, that this was the most important result in the subject for 20 years. The secretary replied that he agreed entirely with Sylvester's opinion of the paper; but Sylvester had actually published the results in the L.M.S. five years before. J. E. Littlewood

Tightly associated with the methods of the community, is the notion of the *meeting places of the community*. These are the venues in which mathematics is presented and discussed. Not only the offices, classrooms, seminar rooms, labs, and conference halls, but also the notes, postcards, letters, journals, and, in our electronic age, their digital equivalents.

The factors that bind the mathematical community help to explain how Hardy and Littlewood, and, indeed, all mathematicians, could continue their particular practice of mathematics despite the state of disarray in the underlying ideas which attempted to bind their subject on an intellectual level. Their relationship to the community and their role in the community prevented questions about the intellectual foundations of mathematics from getting in the way of *their* mathematics. On a formal level, these roles and relationships were defined by the unwritten rules of community membership. In Hardy and Littlewood's case both were, after all, professors at very established universities and as such, were expected to mix teaching responsibilities with research and to submit papers in the accepted form to acceptable journals. On an informal level however, they were completely free, as were all community members, to define their own rules of engagement. And they did. In his foreword to *A Mathematician's Miscellany*, Béla Bollobás quotes a letter in which Harald Bohr describes Hardy and Littlewood's four "axioms" for successful collaboration:

> The first [axiom] said that when one wrote to the other (they often preferred to exchange thoughts in writing instead of orally), it was completely indifferent whether what they said was right or wrong. As Hardy put it, otherwise they could not write completely as they pleased, but would have to feel a certain responsibility thereby. The second axiom was to the effect that, when one received a letter from the other, he was under no obligation whatsoever to read it, let alone answer it, because, as they said, it might be that the recipient of the letter would prefer not to work at that particular time, or perhaps that he was just then interested in other problems....The third axiom was to the effect that, although it did not really matter if they both thought about the same detail, still, it was preferable that they should not do so. And, finally, the fourth, and perhaps most important axiom, stated that it was quite indifferent if one of them had not contributed the least bit to the contents of a paper under their common name; otherwise there would constantly arise quarrels and difficulties in that now one, and now the other, would oppose being named co-author.

This example of informal mathematical knowledge management on the micro scale provides an excellent point of departure for an accelerated trip through the remainder of the twentieth century with the goal of examining the impact of digital technology on the exchange and management of knowledge within the mathematical community. However, before traveling forwards in time, it will be useful to travel backwards; back to the interface between scribal and typographic culture in the fifteenth and sixteenth centuries.

From Scribal Culture to Typographic Culture

> The difference between the man of print and the man of scribal culture is nearly as great as between the non-literate and the literate. The components of Gutenberg technology were not new. But when brought together in the fifteenth century there was an acceleration of social and personal action tantamount to "take off" in the sense that W.W. Rostow develops this concept in *The Stages of Economic Growth* "that decisive interval in the history of a society in which growth becomes its normal condition." Marshall McLuhan

In *The Gutenberg Galaxy*, Marshall McLuhan describes the change in cultural orientations, expectations, and assumptions that occurred with the wide spread adoption of typography in the fifteenth and early sixteenth centuries and is occurring today with the adoption of electronic media. Typical of McLuhan's style, his main ideas are developed from a number of different perspectives in a non-sequential fashion. Examining the impact of the printing press, McLuhan argues that, while pre-typographic culture was characterized by localized production and limited distribution of production - most abbeys would have at least one scribe but a single scribe can only produce so many manuscripts - typographic culture would come to be characterized by centralized production and mass distribution; a limited number of publishing houses producing and distributing many copies of individual texts. McLuhan suggests that, with printing, came notions of authority, authorship, and intellectual property that were completely unknown in scribal culture. He cites E.P. Goldschmidt, a scholar in medieval studies:

> One thing is immediately obvious: before 1500 or thereabouts, people did not attach the same importance to ascertaining the precise identity of the author of a book they were reading or quoting as we do now. We very rarely find them discussing these points...Not only were users of manuscripts, writes Goldschmidt, mostly indifferent to the chronology of authorship and to the "identity and personality of the author of the book he was reading, or in the exact period at which this particular piece of information was written down, equally little, did he expect his future readers to be interested in himself."

Despite the fact that Samuel Morse had brought in the age of electronic communication with his "What hath God wrought?" transmission of 1844, electronic technology had made little impact in the early part of the twentieth century. Phones were few and far between and calls were expensive. The telegraph had found its niche in the long distance communication of simple messages. With advancements in the technology of typography, the typographic age was at its apogee. It can reasonably be argued that in so far as their reputation was earned primarily through the reception by the mathematical community of their published works, the *public identities* of Hardy and Littlewood were creations of typographic culture. The question that confronts us today, in this new era of "take off", is what is the effect of electronic media on the factors that bind the mathematical community and, in particular, what is the effect of electronic media on the organization of mathematical knowledge.

From Typographic Culture to Electronic Culture

> Today, with the arrival of automation, the ultimate extension of the electromagnetic form to the organization of production, we are trying to cope with such new organic production as if it were mechanical mass production. Marshall McLuhan

While the transition from scribal culture to typographic culture represented a shift from loose notions of authorship with distributed loci of publication and limited distribution to firm notions of authorship with centralized loci of publication and mass distribution, the transition to electronic culture turns the equation inside out, presenting the possibility of distributed authorship via mass collaboration and multiple nodes of production with various forms of near instantaneous mass publication. The transformation that occured in the foundations of mathematics, from a unified perspective to a plurality of perspectives, finds resonance in the media environment in which the mathematical community exists and has the potential to affect the language, purposes, methods, and meeting places of the community. In a speech entitled *The Medieval Future of Intellectual Culture: Scholars and Librarians in the Age of the Electron*, professor Stanley Chodorow states:

> In the not-so-distant future, our own intellectual culture will begin to look something like the medieval one. Our scholarly and information environment will have territories dominated by content, rather than by distinct individual contributions. The current geography of information was the product of the seventeenth-century doctrine of copyright. We are all worrying about how the electronic medium is undermining that doctrine. In the long run, the problem of authorship in the new medium will be at least as important as the problem of ownership of information. ...
>
> Works of scholarship produced in and through the electronic medium will have the same fluidity - the same seamless growth and alteration and the same de-emphasis of authorship - as medieval works had. The harbingers of this form of scholarship are the listservs and bulletin boards of the current electronic environment. In these forums, scholarly exchange is becoming instantaneous and acquiring a vigor that

even the great scholarly battlers of old - the legendary footnote fulminators - would admire. Scholars don't just work side by side in the vineyard; they work together on common projects

Applied to mathematics, Chodorow's ideas suggest the possibility that the community's elites, long having been composed of those individuals who demonstrate a particular "*individual* vision and brilliance", may undergo a process of reconstruction, resulting in elites whose members are those who have learned how to start with good ideas and develop them by using the internet to harness the intellectual power of the community. In an age of massively parallel mathematical computation, the potential exists for massively parallel mathematical collaboration. Perhaps the best idea of what a fully digital mathematical scholarship and teaching environment *might* look like can be gleaned from the "hacker culture" of the open source programming community. The meeting places of this community are primarily email, threaded bulletin boards, and implementations of the Concurrent Version System. Those who identify themselves as members, speak of the community's "gift culture" which rewards the most talented and generous of members with status in the community meritocracy. In the opening section of *The Cathedral and the Bazaar*, Eric S. Raymond describes hacker culture:

Many people (especially those who politically distrust free markets) would expect a culture of self-directed egoists to be fragmented, territorial, wasteful, secretive, and hostile. But this expectation is clearly falsified by (to give just one example) the stunning variety, quality and depth of Linux documentation. It is a hallowed given that programmers hate documenting; how is it, then, that Linux hackers generate so much of it? Evidently Linux's free market in egoboo [coined by the author for 'ego boost'] works better to produce virtuous, other-directed behavior than the massively-funded documentation shops of commercial software producers.

He goes on to invoke the idea of a "community of interest":

I think the future of open-source software will increasingly belong to people who know how to play Linus's game, people who leave behind the cathedral and embrace the bazaar. This is not to say that individual vision and brilliance will no longer matter; rather, I think that the cutting edge of open-source software will belong to people who start from individual vision and brilliance, then amplify it through the effective construction of voluntary communities of interest.

If, indeed, *doing mathematics* in the digital age were to develop in a similar fashion to the way that *doing software development* has in the open source community, then the mathematics community must prepare itself for the loss of fixed notions of authorship and ownership and the accountability and economic models that those notions sustain. There are good reasons to believe, however, that despite changes in patterns of collaboration, doing mathematics in the twenty-first century, will not be too unlike doing mathematics in the twentieth century.

Implications and a Proposal

> Mathematics books and journals do not look as beautiful as they used to. It is not that their mathematical content is unsatisfactory, rather that the old and well-developed traditions of type-setting have become too expensive. Fortunately, it now appears that mathematics itself can be used to solve this problem.
> Donald Knuth

In *Digital Typography*, Donald Knuth describes his efforts to capture the traditions of mathematical type-setting in a digital publishing environment; efforts which ultimately led to the development of TEX and METAFONT. It is noteworthy that, faced with a new technology for communicating and presenting mathematics, one of the first major projects related to mathematical publishing was a truly awesome effort to preserve the most valued qualities of traditional typography. The enormous success of Knuth's enterprise stands as a testament to the high value that is placed upon traditional methods of representing mathematical knowledge. It is an interesting aside that the TEX and METAFONT projects were among the first open source programming efforts, functioning without the support of the web, the code being released to collaborators via email.

Apart from the value that members of the mathematical community evidently place on good quality digital typography, there are a number of other reasons to believe that traditional methods of knowledge representation such as the refereed journal and the bound textbook together with the ideas of copyright and accountability that they encapsulate, will not completely fall victim to the distributed modes of digital technology. The educational and research institutions that support traditional forms of knowledge representation are well established and, as evidenced by the success of firms which offer support for open source software, there is every reason to believe that there will continue to be a market for mathematical content that comes with some form of explicit or implicit guarantee and accountablity.

This is not to imply that digital modes of expression can or should be left to develop unscrutinized. The complexity of modern mathematics and the volume of work produced has led to classification schema that are being adapted and extended to digital publication. The use of metatags to describe digital documents has an interesting antecedent in medieval scholarship. In his book *Medieval Theory of Authorship*, A.J. Minnis describes the use of formal prologues found at the beginning of manuscripts. Here, he describes the so-called 'type C' prologue:

> In the systematisation of knowledge which is characteristic of the twelfth century, the 'type C' prologue appeared at the beginning of commentaries on textbooks of all disciplines: the arts, medicine, Roman law, canon law, and theology. Its standard headings, refined by generations of scholars and to some extent modified through the influence of other types of prologue, may be outlined as follows: *Titulus*, the title of the work...*Nomen auctoris*, the name of the author...*Intentio Auctoris*, the intention of the author...*Materia Libri*, the subject matter of the work...*Modus agendi*,

> the method of didactic procedure...Ordo libri, the order of the book...Utilitas, utility...Cui parti philosophiae supponitur, the branch of learning to which the work belonged.

Cast as medieval metadata, these prologues indicate some effort on the part of medieval scholars to protect the integrity of scholarly works in a distributed publishing environment. The question arises as to whether or not the unique modes of digital publishing that members of the mathematical community may create to express mathematical thought might be supceptible to some form of classification via metadata. For example, what might Hardy and Littlewood's correspondence look like in a digital environment? The default answer is that it would *look* much the same as long as they were to access their work on devices that presented a viewing area that resembled the paper and postcards that they used. The qualified answer is that it would look much the same unless they were reading the information using a display with quite different geometry from note paper or postcards. If, for example, they were to attempt to read the data using a cell phone, then ideally the logic of the the digital environment would make the appropriate adaptations and do the best job possible to make the data readable. The correct answer is that the correspondence wouldn't look like anything at all because it would never be in digital form; Hardy would refuse to have anything to do with it. By all accounts, he was a true Luddite who mistrusted the telephone and would refuse even to use ball point pens. Hardy's likely reluctance aside, however, it is possible to imagine how the appropriate digital environment might have been tremendously useful to the Hardy-Littlewood collaboration. If their correspondence had been instantly stored in a database, then a minimal effort invested in specifying metadata would permit them efficient searches and the other digital data manipulations that we now take for granted such as cutting and pasting. Depending on what permissions they chose to grant to the data, a wider audience could be included as observers, or partial or full collaborators.

Emkara, the Extensible Mathematical Knowledge Archiving and Retrieval Agent, is the working title of a project at Simon Fraser University's *Centre for Experimental and Constructive Mathematics* that is designed to investigate how user defined mathematical knowledge construction might conform to metadata driven information management. Emkara is, in effect, a database management system which affords qualified users the ability to create their own data structures, while encouraging thoughtful use of metadata to describe what they are creating. The act of creating a data structure is a table creation operation on the database. All user-defined metadata is stored internally as well-formed XML and, at present, mathematical content fields store data as mathML embedded in xhtml. When a qualified user creates a new mathematical object, emkara generates a default 'edit mode' script, a default 'view mode' script, and the corresponding style sheet for each. These scripts and style sheets are accessible to the object creator for modification or replacement. The object creation interface requests certain metadata by default. These include:

- basic elements of the *Math-Net* metadata set
- some elements of *eduML* mark-up
- information concerning the *copyright* of the object data.

The research incentive of this project is two-fold; to gain insight into what forms of metadata "work" in electronic mathematics knowledge management and to develop a framework for describing the characteristics of electronic mathematics interfaces that meet their objectives.

Sequel...

In the theory of software development processes, *Conway's Law* is cited as a caveat regarding the tendency of a software project's logical design to take on the characteristics of the organizational structure of the work groups that create it. The full statement of the law describes a set of complementary forces in which software architecture informs organizational structure and vice versa. Ultimately, however, the two become aligned and it is therefore important in the early stages of a project to build as much flexibility as possible into both architecture and organizational structure. While these statements about the software development process were never intended to apply to an undertaking such as the development of systems for mathematical knowledge management, the relationship between organizational structure and system design is worth considering.

Earlier, a broad definition of *mathematical community* was adopted encompassing all "those involved with advancing the understanding of mathematics; either at its frontiers, the primary occupation of researchers, or within the existing body of mathematical knowledge such as teachers and students". This definition is at odds with the experience of most who might claim either full or part time membership in the community. If there is truth to the idea that our individual experience of "community" is formed by the group of people with whom we exchange ideas, then it can be argued that any undertaking that attempts to define systems of mathematical knowledge management is, perforce, also defining the structure of the mathematical community.

It is possible to argue that many of the factors that have determined the current divisions within the broad mathematics community, such as domain specialization in research and age group specialization in education, have their origin in the perspectives of typographic culture. It is not necessary here that these arguments be completed. Rather, it is important to point out that there is the possibility of defining management systems that can support the exchange of ideas between smaller communities within the broad mathematics community. These systems would need to create "meeting places" that bring the broader community together. The schism that presently exists between school level and university level mathematics provides a good example. There is no reason that those interested in exchanging ideas regarding a topic in a high school mathematics curriculum could not visit the same electronic mathematical community centre as members of a particular research community. While members of one group may never use nor even look at the resources that are designed for members of the other group, the fact that they pass through the same digital front doors and may even linger, looking at the notices posted in the digital entrance hall, offers hope that each group, even if only accidentally, might gain a better understanding of the priorities and concerns of the other group.

Conclusion

Men despise religion; they hate it, and they fear it is true.
Pascal, from *Pensées*, 1670.

The idea of organized religion representing a fixed, centralized view of the world has come up several times in this paper starting with the anecdote concerning Hardy's atheism and most recently with the citations from the metaphorically titled *The Cathedral and the Bazaar*. With some exceptions, it is probably inaccurate to imply that modern churches are unyielding and inflexible in their outlook and community structure. It would be even more inaccurate to suggest that the broad mathematical community has much in common with organized religion. If this were true, then in his day, Hardy would certainly have been one of the high priests; an idea that he surely would have found either very amusing or very annoying or both. What is true however is that over the course of this century and particularly with the accelerated spread of computer technology that has occured in the last fifteen years, the mathematical community has been faced with the challenge of adapting its language, methods, and meeting places to new technology. The affect that digital technology had on the methods of the community was clearly reflected in the types of knowledge considered valid by the community. For example, it is today not unusual to find mathematical papers with blocks of Maple code. Networking technology is having a more complicated effect on the forms of knowledge that are accepted by the community. Bearing in mind that, in 1969, the Culler-Fried Interactive Mathematics Center at the University of California at Santa Barbara became the third node on the arpanet, it is fair to say that mathematicians have seen the potential of the network and encountered its problems from its genesis. At this still early stage in the development of network technologies for mathematical knowledge management, it is important to consider the effect that those technologies can have on the meeting places of the community and, among other things, who is invited to those meeting places.

9

Do *Real* Numbers Really Move? Language, Thought, and Gesture: The Embodied Cognitive Foundations of Mathematics

RAFAEL NÚÑEZ

Abstract: Robotics, artificial intelligence and, in general, any activity involving computer simulation and engineering relies, in a fundamental way, on mathematics. These fields constitute excellent examples of how mathematics can be applied to some area of investigation with enormous success. This, of course, includes embodied oriented approaches in these fields, such as Embodied Artificial Intelligence and Cognitive Robotics. In this chapter, while fully endorsing an embodied oriented approach to cognition, I will address the question of the nature of mathematics itself, that is, mathematics not as an application to some area of investigation, but as a human conceptual system with a precise inferential organization that can be investigated in detail in cognitive science. The main goal of this piece is to show, using techniques in cognitive science such as cognitive semantics and gestures studies, that concepts and human abstraction in general (as it is exemplified in a sublime form by mathematics) is ultimately embodied in nature.

1. A challenge to embodiment: The nature of Mathematics

Mathematics is a highly technical domain, developed over several millennia, and characterized by the fact that the very entities that constitute what Mathematics is are idealized mental abstractions. These entities cannot be perceived directly through the senses. Even, say, a point, which is the simplest entity in Euclidean geometry, can't be actually *perceived*. A point, as defined by Euclid is a dimensionless entity, an entity that has only location but no extension. No super-microscope will ever be able to allow us to actually perceive a point. A point, after all, with its precision and clear identity, is an idealized abstract entity. The imaginary nature of mathematics becomes more evident when the entities in question are related to *infinity* where, because of the finite nature of our bodies and brains, no direct experience can exist with the infinite itself. Yet, infinity in mathematics is essential. It lies at the very core of many fundamental concepts such as limits, least upper bounds, topology, mathematical induction, infinite sets, points at infinity in projective geometry, to mention only a few. When studying the very nature of mathematics, the challenging and intriguing question that comes to mind is the following: if mathematics is the product of human ideas, how can we explain the nature of mathematics with its unique features such as precision, objectivity, rigor, generalizability, stability, and, of course, applicability to the real world? Such a question doesn't represent a real problem for approaches inspired in platonic philosophies, which rely on the existence of transcendental worlds of ideas beyond human existence. But this view doesn't have any support based on scientific findings and doesn't provide any link to current empirical work on human ideas and conceptual systems (it may be supported, however, as a matter of faith, not of science, by many Platonist scientists and mathematicians). The question doesn't pose major problems to purely formalist philosophies either, because in that worldview mathematics is seen as a manipulation of meaningless symbols. The question of the origin of the meaning of mathematical ideas doesn't even emerge in the formalist arena. For those studying the human mind scientifically, however (e.g., cognitive scientists), the question of the nature of mathematics is indeed a real challenge, especially for those who endorse an *embodied* oriented approach to cognition. How can an embodied view of the mind give an account of an abstract, idealized, precise, sophisticated and powerful domain of ideas if direct bodily experience with the subject matter is not possible?

In *Where Mathematics Comes From*, Lakoff and Núñez (2000) give some preliminary answers to the question of the cognitive origin of mathematical ideas. Building on findings in mathematical cognition, and using mainly methods from Cognitive Linguistics, a branch of Cognitive Science, they suggest that most of the idealized abstract technical entities in Mathematics are created via human cognitive mechanisms that extend the structure of bodily experience (thermic, spatial, chromatic, etc.) while preserving the inferential organization of these domains of bodily experience. For example, linguistic expressions such as "send her my *warm* helloes" and "the teacher was very *cold* to me" are statements that refer to the somewhat abstract domain of Affection. From a purely literal point of view, however, the language used belongs to the domain of Thermic experience, not Affection. The meaning of these statements and the inferences one is able to draw from them is structured by precise mappings from the Thermic domain to the domain of Affection: Warmth is mapped onto presence of affection, Cold is mapped onto lack of affection, X is warmer than Y is mapped onto X is more affectionate than Y, and so on. The ensemble of inferences is modeled by one conceptual metaphorical mapping, which in this case is called AFFECTION IS

WARMTH[1]. Research in Cognitive Linguistics has shown that these phenomena are not simply about "language," but rather they are about thought. In cognitive science the complexities of such abstract and non/literal phenomena have been studied through mechanisms such as conceptual metaphors (Lakoff & Johnson, 1980; Sweetser, 1990; Lakoff, 1993; Lakoff & Núñez, 1997; Núñez & Lakoff, in press; Núñez, 1999, 2000), conceptual blends (Fauconnier & Turner, 1998, 2002; Núñez, in press), conceptual metonymy (Lakoff & Johnson, 1980), fictive motion and dynamic schemas (Talmy, 1988, 2003), and aspectual schemas (Narayanan, 1997). Based on these findings Lakoff and Núñez (2000) analyzed many areas in mathematics, from set theory to infinitesimal calculus, to transfinite arithmetic, and showed how, via everyday human embodied mechanisms such as conceptual metaphor and conceptual blending, the inferential patterns drawn from direct bodily experience in the real world get extended in very specific and precise ways to give rise to a new emergent inferential organization in purely imaginary domains[2]. For the remainder of this chapter we will be building on these results as well as on the corresponding empirical evidence provided by the study of human speech-gesture coordination. Let us now consider a few mathematical examples.

2. Limits, curves, and continuity

Through the careful analysis of technical books and articles in mathematics, we can learn a good deal about what structural organization of human everyday ideas have been used to create mathematical concepts. For example, let us consider a few statements regarding limits in infinite series, equations of curves in the Cartesian plane, and continuity of functions, taken from mathematics books such as the now classic *What is Mathematics?* by R. Courant & H. Robbins (1978).

a) *Limits of infinite series*

In characterizing limits of infinite series, Courant & Robbins write:

"We describe the behavior of s_n by saying that the sum s_n *approaches* the limit 1 as *n tends* to infinity, and by writing

$$1 = 1/2 + 1/2^2 + 1/2^3 + 1/2^4 + \ldots$$" (p. 64, our emphasis)

---

[1] Following a convention in Cognitive Linguistics, the name of a conceptual metaphorical mapping is capitalized.

[2] The details of how conceptual metaphor and conceptual blending work go beyond the scope of this piece. For a general introduction to these concepts see Lakoff & Núñez (2000, chapters 1-3), and the references given therein.

Strictly speaking, this statement refers to a sequence of discrete and motionless partial sums of s_n (real numbers), corresponding to increasing discrete and motionless values taken by n in the expression $1/2^n$ where n is a natural number. If we examine this statement closely we can see that it describes some facts about numbers and about the result of discrete operations with numbers, but that there is *no motion* whatsoever involved. No entity is actually *approaching* or *tending* to anything. So, why then did Courant and Robbins (or mathematicians in general, for that matter) use dynamic language to express static properties of static entities? And what does it mean to say that the "sum s_n approaches," when in fact a sum is simply a fixed number, a result of an operation of addition?

b) *Equations of lines and curves in the Cartesian Plane*

Regarding the study of conic sections and their treatment in analytic geometry, Courant & Robbins' book says:

> "The hyperbola *approaches* more and more nearly the two straight lines $qx \pm py = 0$ as we *go out farther and farther* from the origin, but it never actually *reaches* these lines. They are called the asymptotes of the hyperbola." (p. 76, our emphasis).

And then the authors define hyperbola as "the locus of all points P the *difference* of whose distances to the two points $F(\sqrt{(p^2 + q^2)}, 0)$ and $F'(-\sqrt{(p^2 + q^2)}, 0)$ is $2p$." (p. 76, original emphasis).

Strictly speaking, the definition only specifies a "*locus* of all points P" satisfying certain properties based exclusively on arithmetic differences and *distances*. Again, no entities are actually moving or approaching anything. There are only statements about static differences and static distances. Besides, as Figure 1 shows, the authors provide a graph of the hyperbola in the Cartesian Plane (bottom right), which in itself is a static illustration that doesn't have the slightest insinuation of motion (like symbols for arrows, for example). The figure illustrates the idea of *locus* very clearly, but it says nothing about motion. Moreover the hyperbola has *two* distinct and separate *loci*. Exactly which one of the two is then "the" moving agent (3rd person singular) in the authors' statement "*the* hyperbola *approaches* more and more nearly the two straight lines $qx \pm py = 0$ as we *go out farther and farther* from the origin"?

c) *Continuity*

Later in the book, the authors analyze cases of continuity and discontinuity of trigonometric functions in the real plane. Referring to the function $f(x) = \sin 1/x$ (whose graph is shown in Figure 2) they say: "... since the denominators of these fractions increase without limit, the values of x for which the function $\sin(1/x)$ has the values 1, −1, 0, will cluster nearer and nearer to the point $x = 0$. Between any such point and the origin there will be still an infinite number of *oscillations* of the function" (p. 283, our emphasis).

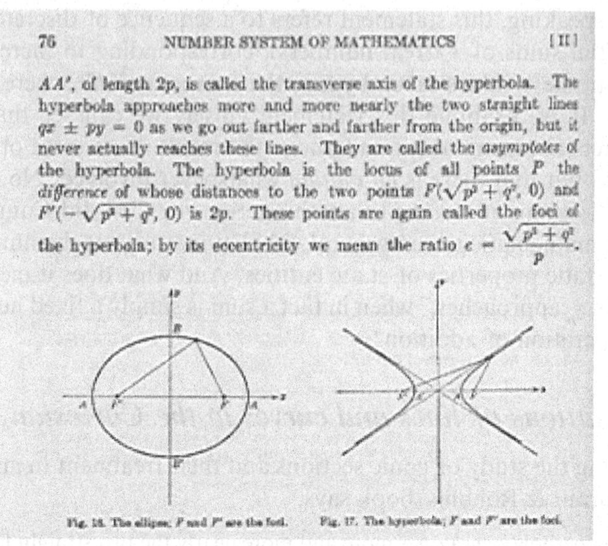

76 NUMBER SYSTEM OF MATHEMATICS [II]

AA', of length $2p$, is called the transverse axis of the hyperbola. The hyperbola approaches more and more nearly the two straight lines $qx \pm py = 0$ as we go out farther and farther from the origin, but it never actually reaches these lines. They are called the *asymptotes* of the hyperbola. The hyperbola is the locus of all points P the *difference* of whose distances to the two points $F(\sqrt{p^2+q^2}, 0)$ and $F'(-\sqrt{p^2+q^2}, 0)$ is $2p$. These points are again called the foci of the hyperbola; by its eccentricity we mean the ratio $e = \frac{\sqrt{p^2+q^2}}{p}$.

Fig. 16. The ellipse; F and F' are the foci. Fig. 17. The hyperbola; F and F' are the foci.

FIGURE 1. Original text analyzing the hyperbola as published in the now classic book *What is Mathematics?* by R. Courant & H. Robbins (1978).

Once again, if, strictly speaking, a function is a mapping between elements of a set (coordinate values on the x-axis) with one and only one of the elements of another set (coordinate values on the y-axis), all that we have is a static correspondence between points on the x-axis with points on the y-axis. How then can the authors (or mathematicians in general) speak of "*oscillations* of the function," let alone an infinite number of them?

These three examples show how ideas and concepts are described, defined, illustrated, and analyzed in mathematics books. You can pick your favorite

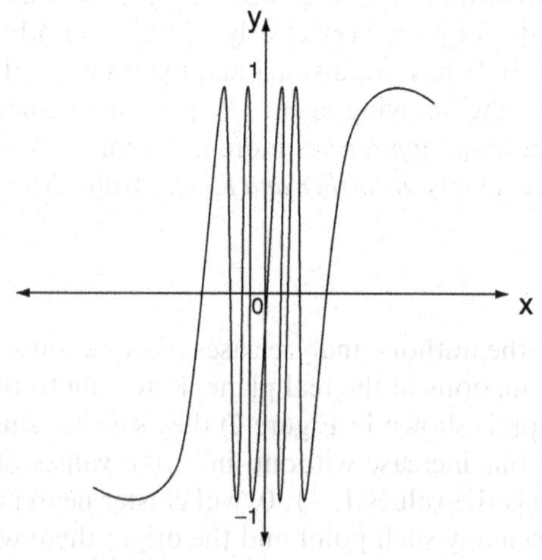

FIGURE 2. The graph of the function $f(x) = \sin 1/x$.

mathematics books and you will see similar patterns. You will see them in topology, fractal geometry, space-filling curves, chaos theory, and so on. Here, in all three examples, static numerical structures are involved, such as partial sums, geometrical loci, and mappings between coordinates on one axis with coordinates on another. Strictly speaking, absolutely no dynamic entities are involved in the formal definitions of these terms. So, if no entities are really moving, why do authors speak of "approaching," "tending to," "going farther and father," and "oscillating"? Where is this motion coming from? What does dynamism mean in these cases? What role is it playing (if any) in these statements about mathematics facts?

We will first look at pure mathematics to see whether we can find answers to these questions. Then, in order to get some deeper insight into them, we will turn to human language and real-time speech-gesture coordination.

3. Looking at pure Mathematics

Among the most fundamental entities and properties the above examples deal with are the notions of real number and continuity. Let us look at how pure mathematics defines and provides the inferential organization of these concepts.

In pure mathematics, entities are brought to existence via formal definitions, formal proofs (theorems) or by axiomatic methods (i.e., by declaring the existence of some entity without the need of proof. For example, in set theory the axiom of infinity assures the existence of infinite sets. Without that axiom, there are no infinite sets). In the case of real numbers, ten axioms taken together, fully characterize this number system and its inferential organization (i.e., theorems about real numbers). The following are the axioms of the real numbers.

1. Commutative laws for addition and multiplication.
2. Associative laws for addition and multiplication.
3. The distributive law.
4. The existence of identity elements for both addition and multiplication.
5. The existence of additive inverses (i.e., negatives).
6. The existence of multiplicative inverses (i.e., reciprocals).
7. Total ordering.
8. If x and y are positive, so is $x + y$.
9. If x and y are positive, so is $x \bullet y$.
10. The Least Upper Bound axiom.

The first 6 axioms provide the structure of what is called a *field* for a set of numbers and two binary operations. Axioms 7 through 9, assure ordering constraints. The first nine axioms fully characterize ordered fields, such as the rational numbers with the operations of addition and multiplication. Up to here we have already a lot of structure and complexity. For instance we can characterize and prove theorems about all possible numbers that can be expressed as the division of two whole numbers (i.e., rational numbers). With the rational numbers we can describe with any given (finite) degree of precision the proportion given by the perimeter of a circle and its diameter (e.g., 3.14; 3.1415; etc.). We can also locate along a line (according to their magnitude) any two different rational numbers and be sure (via proof) that there will always be infinitely many more rational numbers between them (a property referred to as density). With the rational numbers, however, we can't "complete" the points on this line, and we can't express with infinite exactitude the magnitude of the proportion mentioned above ($\pi = 3.14159 \ldots$). For this we need the full extension of the real numbers. In axiomatic terms, this is accomplished by the tenth axiom: the Least Upper Bound axiom. All ten axioms characterize a complete ordered field.

In what concerns our original question of where is motion coming from in the above mathematical statements about infinite series and continuity, we don't find any answer in the first nine axioms of real numbers. All nine axioms simply specify the existence of static properties regarding binary operations and their results, and properties regarding ordering. There is no explicit or implicit reference to motion in these axioms. Since what makes a real number a real number (with its infinite precision) is the Least Upper Bound axiom, it is perhaps this very axiom that hides the secret of motion we are looking for. Let's see what this axiom says:

10. Least Upper Bound axiom: every nonempty set that has an upper bound has a least upper bound.

And what exactly is an upper bound and a least upper bound? This is what pure mathematics says:

Upper Bound

b is *an upper bound* for S if

$x \leq b$, for every x in S.

Least Upper Bound

b_0 is a *least upper bound* for S if

- b_0 is an upper bound for S, and
- $b_0 \leq b$ for every upper bound b of S.

Once again, all that we find are statements about motionless entities such as universal quantifiers (e.g., for every x; for every upper bound b of S), membership relations (e.g., for every x in S), greater than relationships (e.g., $x \leq b$; $b_0 \leq b$), and so on. In other words, there is absolutely no indication of motion in the Least Upper Bound axiom, or in any of the other nine axioms. In short, the axioms of real numbers, which are supposed to completely characterize the "truths" (i.e., theorems) of real numbers don't tell us anything about a sum "approaching" a number, or a number "tending to" infinity (whatever that may mean!).

Let's try continuity. What does pure mathematics say about it?

Mathematics textbooks define continuity for functions as follows:

- A function f is continuous at a number a if the following three conditions are satisfied:
  1. f is defined on an open interval containing a,
  2. $\lim_{x \to a} f(x)$ exists, and
  3. $\lim_{x \to a} f(x) = f(a)$.

Where by $\lim_{x \to a} f(x)$ what is meant is the following:

Let a function f be defined on an open interval containing a, except possibly at a itself, and let L be a real number. The statement

$$\lim_{x \to a} f(x) = L$$

means that $\forall \varepsilon > 0, \exists \delta > 0,$

such that if $0 < |x - a| < \delta,$

then $|f(x) - L| < \varepsilon.$

As we can see, pure formal mathematics defines continuity in terms of limits, and limits in terms of

- static universal and existential quantifiers predicating on static numbers (e.g., $\forall \varepsilon > 0, \exists \delta > 0,$), and
- on the satisfaction of certain conditions which are described in terms of motionless arithmetic difference (e.g., $|f(x) - L|$) and static smaller than relations (e.g., $0 < |x - a| < \delta$).

That's it. Once again, these formal definitions don't tell us anything about a sum "approaching" a number, or a number "tending to" infinity, or about a function "oscillating" between values (let alone doing it infinitely many times, as in the function $f(x) = \sin 1/x$).

But this shouldn't be a surprise. Lakoff & Núñez (2000), using techniques from cognitive linguistics showed what well-known contemporary mathematicians had already pointed out in more general terms (Hersh, 1997; Henderson, 2001):

- The structure of human mathematical ideas, and its inferential organization, is richer and more detailed than the inferential structure provided by formal definitions and axiomatic methods. Formal definitions and axioms neither fully formalize nor generalize human concepts.

We can see this with a relatively simple example taken from Lakoff & Núñez (2000). Consider the function $f(x) = x \sin 1/x$ whose graph is depicted in Figure 3.

$$f(x) = \begin{cases} x \sin 1/x & \text{for } x \neq 0 \\ 0 & \text{for } x = 0 \end{cases}$$

According to the ε -δ definition of continuity given above, this function is continuous at every point. For all x, it will always be possible to find the specified ε's and δ's necessaries to satisfy the conditions for preservation of closeness. However, according to the everyday notion of continuity—*natural*

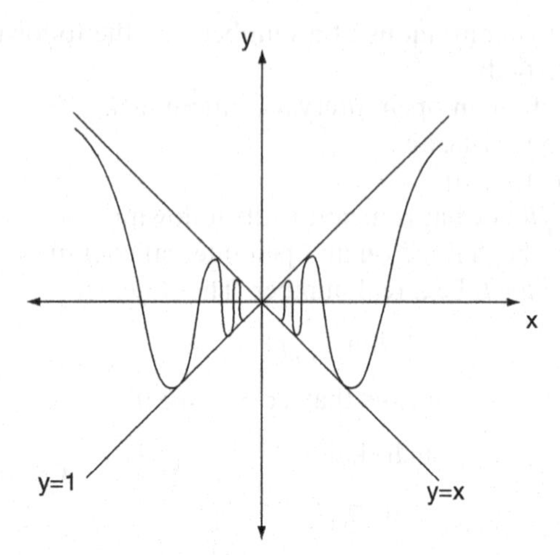

FIGURE 3. The graph of the function $f(x) = x \sin 1/x$.

continuity (Núñez & Lakoff, 1998)—as it was used by great mathematicians such as Kepler, Euler, and Newton and Leibniz, the inventors of infinitesimal calculus in the 17th Century, this function is *not* continuous. According to the inferential organization of natural continuity, certain conditions have to be met. For instance, in a naturally continuous line we are supposed to be able to tell how long the line is between two points. We are also supposed to be able to describe essential components of the motion of a point along that line. With this function we can't do that. Since the function "oscillates" infinitely many times as it "approaches" the point (0, 0) we can't really tell how long the line is between two points located on the left and right sides of the plane. Moreover, as the function approaches the origin (0, 0) we can't tell, say, whether it will "cross" from the right plane to the left plane "going down" or "going up." This function violates these basic properties of natural continuity and therefore it is not continuous. The function $f(x) = x \sin 1/x$ is thus ε-δ continuous but it is not naturally continuous. The point is that the formal ε-δ definition of continuity doesn't capture the inferential organization of the human everyday notion of continuity, and it doesn't generalize the notion of continuity either.

The moral here is that what is characterized formally in mathematics leaves out a huge amount of inferential organization of the human ideas that constitute mathematics. As we will see, this is precisely what happens with the dynamic aspects of the expressions we saw before, such as "approaching," "tending to," "going farther and farther," "oscillating," and so on. Motion, in those examples, is a genuine and constitutive manifestation of the nature of mathematical ideas. In pure mathematics, however, motion is not captured by formalisms and axiomatic systems.

4. Embodied Cognition

It is now time to look, from the perspective of embodied cognition, at the questions we asked earlier regarding the origin of motion in the above mathematical ideas. In the case of limits of infinite series, motion in "the sum s_n *approaches* the limit 1 as *n tends* to infinity" emerges *metaphorically* from the successive values taken by *n* in the sequence as a whole. It is beyond the scope of this chapter to go into the details of the mappings involved in the various underlying conceptual metaphors that provide the required dynamic inferential organization (for details see Lakoff & Núñez, 2000). But we can at least point out some of the many conceptual metaphors and metonymies[3] involved.

- There are conceptual metonymies in cases such as a partial sum *standing for* the entire infinite sum;
- there are conceptual metaphors in cases where we conceptualize the sequence of these metonymical sums as a *unique trajector*[4] moving in space (as it is indicated by the use of the 3rd person singular in *the* sum s_n approache*s*);
- there are conceptual metaphors for conceiving infinity as a single location in space such that a metonymical *n* (standing for the entire sequence of values) can "tend to;"
- there are conceptual metaphors for conceiving 1 (not as a mere natural number but as an infinitely precise *real* number) as the result of the infinite sum; and so on.

Notice that none of these expressions can be *literal*. The facts described in these sentences don't exist in any real perceivable world. They are metaphorical in nature. It is important to understand that these conceptual metaphors and metonymies are not simply "noise" added on top of pre-defined formalisms. They are in fact *constitutive* of the very embodied ideas that make mathematical ideas possible. It is the inferential organization provided by our embodied understanding of "approaching" and "tending to" that is at the core of these mathematical ideas.

In the case of the hyperbola, the moving agent is one holistic object, the hyperbola in the Real plane. This object, which has two distinct separate parts, is conceptualized as one single trajector metaphorically moving away

---

[3] A conceptual metonymy is a cognitive mechanism that allow us to conceive a part of a whole standing for the whole, as when we say *Washington and Paris have quite different views on these issues*, meaning the governments of two entire nations, namely, United States and France.

[4] In cognitive linguistics, "trajector" is a technical term used to refer to the distinct entity that performs the motion traced by a trajectory. The trajector moves against a background called "landscape."

from the origin. Via conceptual metonymies and metaphors similar to the ones we saw for the case of infinite series, the hyperbola is conceived as a trajector tracing the line, which describes the geometrical locus of the hyperbola itself. In this case, of course, because we are dealing with real numbers, the construction is done on non-countable infinite ($>\aleph_0$) discrete real values for x, which are progressively bigger in absolute terms. The direction of motion is stated as moving away from the origin of the Cartesian coordinates, and it takes place in both directions of the path schemas defined by the two branches of the hyperbola, *simultaneously*. The hyperbola not "reaching" the asymptotes is the cognitive way of characterizing the mathematically formalized fact that there are no values for x and y that satisfy equations

$$qx \pm py = 0 \text{ and } (x^2/p^2) - (y^2/q^2) = 1$$

Notice that characterizing the hyperbola as "not reaching" the asymptotes provides the same *extensionality* (i.e., it gives the same resulting cases) as saying that there is an "absence of values" satisfying the above equations. The inferential organization of these two cases, however, is cognitively very different[5].

Finally, in what concerns our "oscillating" function example, the moving object is again one holistic object, the trigonometric function in the Real plane, constructed metaphorically from non-countable infinite ($>\aleph_0$) discrete real values for x, which are progressively smaller in absolute terms. In this case motion takes place in a specific manner, towards the origin from two opposite sides (i.e., for negative and positive values of x) and always between the values $y = 1$ and $y = -1$. As we saw, a variation of this function, $f(x) = x \sin(x)$, reveals deep cognitive incompatibilities between the dynamic notion of continuity implicit in the example above and the static ε-δ definition of continuity coined by Weierstrass in the second half of the 19th century (based on quantifiers and discrete Real numbers) and which has been adopted ever since as "the" definition of what Continuity really is (Núñez & Lakoff, 1998; Lakoff & Núñez, 2000). These deep cognitive incompatibilities between dynamic-wholistic entities and static-discrete ones may explain important aspects underlying the difficulties encountered by students all over the world when learning the modern technical version of the notions of limits and continuity (Núñez, Edwards, and Matos, 1999).

---

[5] In order to clarify this point, consider the following two questions: (a) What Alpine European country does not belong to the European Union?, and (b) What is the country whose currency is the Swiss Franc? The extensionality provided by the answers to both questions is the same, namely, the country called "Switzerland." This, however, doesn't mean that we have to engage in the same cognitive activity in order to correctly answer these questions.

5. Fictive Motion

Now that we are aware of the metaphorical (and metonymical nature) of the mathematical ideas mentioned above, I would like to analyze more in detail the dynamic component of these ideas. From where do these ideas get motion? What cognitive mechanism is allowing us to conceive static entities in dynamic terms? The answer is *fictive motion.*

Fictive motion is a fundamental embodied cognitive mechanism through which we unconsciously (and effortlessly) conceptualize static entities in dynamic terms, as when we say *the road goes along the coast*. The road itself doesn't actually move anywhere. It is simply standing still. But we may conceive it as moving "along the coast." Fictive motion was first studied by Len Talmy (1996), via the analysis of linguistic expressions taken from everyday language in which static scenes are described in dynamic terms. The following are linguistic examples of fictive motion:

- The Equator *passes through* many countries
- The border between Switzerland and Germany *runs along* the Rhine.
- The California coast *goes all the way down* to San Diego
- After Corvisart, line 6 *reaches* Place d'Italie.
- Right after *crossing* the Seine, line 4 *comes to* Chatêlet.
- The fence *stops* right after the tree.
- Unlike Tokyo, in Paris there is no metro line that *goes around* the city.

Motion, in all these cases, is fictive, imaginary, not real in any literal sense. Not only do these expressions use verbs of action, but they also provide precise descriptions of the quality, manner, and form of motion. In all cases of fictive motion there is a *trajector* (the moving agent) and a *landscape* (the background space in which the trajector moves). Sometimes the trajector may be a real object (e.g., the *road* goes; the *fence* stops), and sometimes it is an imaginary entity (e.g., the *Equator* passes through; the *border* runs). In fictive motion, real world trajectors don't move but they have the potential to move or the potential to enact movement (e.g., a car moving along that road). In Mathematics proper, however, the trajector has always a metaphorical component. That is, the trajector as such can't be literally capable or incapable of enacting movement, because the very nature of the trajectory is imagined via metaphor (Núñez, 2003). For example, a point in the Cartesian Plane is an entity that has location (determined by its coordinates) but has no extension. So when we say "point P *moves* from A to B" we are ascribing motion to a metaphorical entity that only has location. First, as we saw earlier, entities which have only location (i.e., points) don't exist in the real world, so, as such, they don't have the potential to move or not to move in any literal sense. They simply don't exist in the real world. They are metaphorical entities. Second, literally speaking, point A and point B are distinct *locations*, and no point can change location while preserving its identity. That is, the trajector (point P, uniquely determined by its coordinates) can't preserve its identity throughout the process of motion from A to B, since that would mean that it is changing the very properties that are defining it, namely, its coordinates.

We now have a basic understanding of how conceptual metaphor and fictive motion work, so we are in a position to see the embodied cognitive mechanisms underlying the mathematical expressions like the ones we saw earlier. Here we have similar expressions:

- sin $1/x$ *oscillates* more and more as x *approaches* zero
- $g(x)$ never *goes beyond* 1
- If there exists a number L with the property that $f(x)$ *gets closer and closer to* L as x gets larger and larger; $\lim_{x\to\infty} f(x) = L$.

In these examples Fictive Motion operates on a network of precise *conceptual metaphors*, such as NUMBERS ARE LOCATIONS IN SPACE (which allows us to conceive numbers in terms of spatial positions), to provide the inferential structure required to conceive mathematical functions as having motion and directionality. Conceptual metaphor generates a purely imaginary entity in a metaphorical space, and fictive motion makes it a moving trajector in this metaphorical space. Thus, the progressively smaller numerical values taken by x which determine numerical values of sin $1/x$, are via the conceptual metaphor NUMBERS ARE LOCATIONS IN SPACE conceptualized as spatial locations. The now metaphorical spatial locus of the function (i.e., the "line" drawn on the plane) now becomes available for fictive motion to act upon. The progressively smaller numerical values taken by x (now metaphorically conceptualized as locations progressively closer to the origin) determine corresponding metaphorical locations in space for sin $1/x$. In this imaginary space, via conceptual metaphor and fictive motion now sin $1/x$ can "oscillate" more and more as x "approaches" zero.

In a similar way the infinite precision of real numbers themselves can be conceived as limits of sequences of rational numbers, or limits of sequences of nested intervals. Because, as we saw, limits have conceptual metaphor and fictive motion built in, we can now see the fundamental role that these embodied mechanisms play in the constitution of the very nature of the real numbers themselves.

5. Dead Metaphors?

Up to now, we have analyzed some mathematical ideas through methods in cognitive linguistics, such as conceptual metaphor, conceptual metonymy, and fictive motion. We have studied the inferential organization modeling *linguistic expressions*. But so far not much has been said of actual people speaking, writing, explaining, learning, or gesturing in real-time when involved in mathematical activities. The analysis so far has been almost exclusively at the level of written and oral linguistic expressions. We must know whether there is any psychological (and presumably neurological) reality underlying these linguistics expressions. The remaining task now is to show that all these cases are not, as some scholars have suggested, mere instances of so-called *dead metaphors*, that is, expressions that once in the past had a metaphorical dimension but that now, after centuries of usage, have lost their metaphorical component becoming "dead." Dead metaphorical expressions are those that have lost their psychological (and cognitive semantic) original reality, becoming simply new "lexical items." Perhaps in the cases we have seen in mathematics, what once was a metaphorical expression has now become a *literal* expression whose meaningful origin speakers of English don't know anymore (very much like so many English words whose Latin or Greek etymology may have been known by speakers at a certain point in history, but whose original meaning is no longer evoked by speakers today). Is this what is happening to cases such as "approaching" limits, "oscillating" functions, or hyperbolae not "reaching" the asymptotes? Maybe, after all, all that we have in the mathematical expressions we have examined, is simply a story of dead metaphors, with no psychological (or neurological) reality whatsoever. As we will see, however, the study of human *gesture* provides embodied convergent evidence showing that this is not the case at all. Gesture studies, via a detailed investigation of real-time cognitive and linguistic production, bodily motion (mainly hands and arms), and voice inflection, show that the conceptual metaphors and fictive motion involved in the mathematical ideas analyzed above, far from being dead, do have a very embodied psychological (and presumably neurological) reality.

6. Gesture as Cognition

Human beings from all cultures around the world gesture when they speak. The philosophical and scientific study of human language and thought has largely ignored this simple but fundamental fact. Human gesture constitutes the forgotten dimension of thought and language. Chomskian linguistics, for instance, overemphasizing syntax, saw language mainly in terms of abstract grammar, formalisms, and combinatorics, you could study by looking at written statements. In such a view there was simply no room for meaningful (semantic) "bodily production" such as gesture. In mainstream experimental psychology gestures were left out, among others, because being produced in a spontaneous manner, it was very difficult to operationalize them, making rigorous experimental observation on them extremely difficult. In mainstream cognitive science, which in its origins was heavily influenced by classic artificial intelligence, there was simply no room for gestures either. Cognitive science and artificial intelligence were heavily influenced by the information-processing paradigm and what was taken to be essential in any cognitive activity was a set of body-less abstract rules and the manipulation of physical symbols governing the processing of information. In all these cases, gestures were completely ignored and left out of the picture that defined what constituted genuine subject matters for the study of the mind. At best, gestures were considered as a kind of epiphenomenon, secondary to other more important and better-defined phenomena.

But in the last decade or so, this scenario has changed in a radical way with the pioneering work of A. Kendon (1980), D, McNeill (1992), S. Goldin-Meadow & C. Mylander (1984), and many others. Research in a large variety of areas, from child development, to neuropsychology, to linguistics, and to anthropology, has shown the intimate link between oral and gestural production. Finding after finding has shown, for instance, that gestures are produced in astonishing synchronicity with speech, that in children they develop in close relation with speech, and that brain injuries affecting speech production also affect gesture production. The following is a (very summarized) list of nine excellent sources of evidence supporting (1) the view that speech and gesture ae in reality two facets of the same cognitive linguistic reality, and (2) an embodied approach for understanding language, conceptual systems, and high-level cognition:

1) *Speech accompanying gesture is universal.* This phenomenon is manifested in all cultures around the world. Gestures then provide a remarkable "back door" to linguistic cognition (McNeill, 1992; Iverson & Thelen 1999; Núñez & Sweetser, 2001).
2) *Gestures are less monitored than speech*, and they are, to a great extent, unconscious. Speakers are often unaware that they are gesturing at all (McNeill, 1992)
3) *Gestures show an astonishing synchronicity with speech.* They are manifested in a millisecond-precise synchronicity, in patterns which are specific to a given language (McNeill, 1992).
4) *Gestures can be produced without the presence of interlocutors.* Studies of people gesturing while talking on the telephone, or in monologues, and studies of conversations among congenitally blind subjects have shown that there is no need of visible interlocutors for people to gesture (Iverson & Goldin-Meadow, 1998).
5) *Gestures are co-processed with speech.* Studies show that stutterers stutter in gesture too, and that impeding hand gestures interrupts speech production (Mayberry and Jaques, 2000).
6) *Hand signs are affected by the same neurological damage as speech.* Studies in neurobiology of sign language show that left hemisphere damaged signers manifest similar phonological and morphological errors as those observed in speech aphasia (Hickok, Bellugi, and Klima, 1998).
7) *Gesture and speech develop closely linked.* Studies in language acquisition and child development show that speech and gesture develop in parallel (Iverson & Thelen 1999; Bates & Dick, 2002).
8) *Gesture provides complementary content to speech content.* Studies show that speakers synthesize and subsequently cannot distinguish information taken from the two channels (Kendon, 2000).

9) *Gestures are co-produced with abstract metaphorical thinking.* Linguistic metaphorical mappings are paralleled systematically in gesture (McNeill, 1992; Cienki, 1998; Sweetser, 1998; Núñez & Sweetser, 2001).

In all these studies, a careful analysis of important parameters of gestures such as handshapes, hand and arm positions, palm orientation, type of movements, trajectories, manner, and speed, as well as a careful examination of timing, indexing, preservation of semantics, and the coupling with environmental features, give deep insight into human thought[6]. An important feature of gestures is that they have three well-defined phases called preparation, stroke, and retraction (McNeill, 1992). The stroke is in general the fastest part of the gesture's motion, and it tends to be highly synchronized with speech accentuation and semantic content. The preparation phase is the motion that precedes the stroke (usually slower), and the retraction phase is the motion observed after the stroke has been produced (usually slower as well), when the hand goes back to a resting position or to whatever activity it was engaged in.

With these tools from gesture studies and cognition, we can now analyze mathematical expressions like the ones we saw before, but this time focusing on the gesture production of the speaker. For the purposes of this chapter, an important distinction we need to make concerns the gestures that refer to real objects in the real world, and gestures that refer to some abstract idea that in itself doesn't exist in the real world.

An example of the first group is shown in Figure 4, which shows renowned physicist Professor Richard Feynman giving a lecture on physics of particles at Cornell University many years ago. In this sequence he is talking about particles moving in all directions at very high speeds (Figure 4, a through e), and a few milliseconds later he completes his utterance by saying "once in a while hit" (Figure 4f). The action shown in the first five pictures correspond to the gesture characterizing the random movements of particles at high speeds. The precise finger pointing shown in figure 4f occurs when he says "once in a while hit" (the stroke of the gesture). The particle being indexed by the gesture is quite abstract and idealized, in the sense that it doesn't preserve some properties of the real referent, such as the extremely high speed at which particles move, for instance. But the point here is that although Prof. Feynman's talk was about a very abstract domain (i.e., particle physics), it is still the case that with his finger he is *indexing* a "particle," an object with location, extension, and mass, which does exist in the real world. The trajector in this dynamic scene is, an extremely small and fast object, but nonetheless a real entity in the real world.

[6] An analysis of the various dimensions and methodological issues regarding the scientific study of gestures studies is beyond the scope of this chapter. For details see references mentioned above.

FIGURE 4. Professor Richard Feynman giving a lecture on physics. He is talking about particles moving in all directions at very high speeds (a through e), which "once in a while hit" (f).

Now, the gestures we are about to analyze below are similar in many respects, but they are even more abstract. In these cases the entities that are indexed with the various handshapes are purely imaginary entities, like points and numbers in mathematics. Figure 5, for instance shows a professor of mathematics lecturing on convergent sequences in a university level class. In this particular situation, he is talking about a case in which the real values of an infinite sequence do not get closer and closer to a single real value as n increases, but "oscillate" between two fixed values. His right hand, with the palm towards his left, has a handshape called *baby O* in American Sign Language and in gesture studies, where the index finger and the thumb are touching and are slightly bent while the other three fingers are fully bent. In this gesture the touching tip of the index and the thumb are metaphorically indexing a metonymical value standing for the values in the sequence as n increases (it is almost as if the subject is carefully holding a very tiny object with those two fingers). Holding that fixed handshape, he moves his right arm horizontally back and forth while he says "oscillating."

Hands and arms are essential body parts involved in gesturing. But often it is also the entire body that participates in enacting the inferential structure of an idea. In the following example (Figure 6) a professor of mathematics is lecturing on some important notions of calculus at a university level course. In this scene he is talking about a particular theorem regarding monotone sequences.

As he is talking about an unbounded monotone sequence, he is referring to the important property of "going in one direction." As he says this he is producing frontwards iterative unfolding circles with his right hand, and at the

FIGURE 5. A professor of mathematics lecturing on convergent sequences in a university level class. Here he is referring to a case in which the real values of a sequence "oscillate" (horizontally).

same time he is walking frontally, accelerating at each step (Figure 6a through 6e). His right hand, with the palm toward his chest, displays a shape called *tapered O* (Thumb relatively extended and touching the upper part of his extended index finger bent in right angle, like the other fingers), which he keeps in a relatively fixed position while doing the iterative circular movement. A few milliseconds later he completes the sentence by saying "it takes off to infinity" at the very moment when his right arm is fully extended and his hand shape has shifted to an extended shape called *B spread* with a fully (almost over) extension, and the tips of the fingers pointing frontwards at eye-level.

It is important to notice that in both cases the blackboard is full of mathematical expressions containing formalisms like the ones we saw earlier (e.g., existential and universal quantifiers \exists and \forall): formalisms, which have no indication of, or reference to, motion. The gestures (and the linguistic expressions used), however, tell us a very different conceptual story. In both cases, these mathematicians are referring to fundamental dynamic aspects of the mathematical ideas they are talking about. In the first example, the oscillating gesture matches, and it is produced synchronically with, the linguistic expressions used. In the second example, the iterative frontally-unfolding circular gesture matches the inferential structure of the description of the iteration involved in the increasing monotone sequence, where even the entire body moves forwards as the sequence unfolds. Since the sequence is unbounded, it "takes off to infinity," idea which is precisely characterized in a synchronous way with the full frontal extension of the arm and the hand.

The moral we can get from these gesture examples is two-fold.

- First, gestures provide converging evidence for the psychological and embodied reality of the linguistic expressions analyzed with classic techniques in cognitive linguistics, such as metaphor and blending analysis. In these cases gesture analyses show that the metaphorical expressions we saw

FIGURE 6. A professor of mathematics at a university level class talking about an unbounded monotone sequence "going in one direction" (a through e), which "takes off to infinity" (f).

earlier are not cases of dead metaphors. The above gestures show, in real time, that the dynamism involved in these ideas have full psychological and cognitive reality.

- Second, these gestures show that the fundamental dynamic contents involving infinite sequences, limits, continuity, and so on, are in fact *constitutive* of the inferential organization of these ideas. Formal language in mathematics, however, is not as rich as everyday language and cannot capture the full complexity of the inferential organization of mathematical ideas. It is the job of embodied cognitive science to characterize the full richness of mathematical ideas.

7. Conclusion

We can now go back to the original question asked in the title of this chapter: Do real numbers really move? Since fictive motion is a real cognitive mechanism, constitutive of the very notion of a real number, the answer is yes. Real numbers are metaphorical entities (with a very sophisticated inferential organization), and they do move, metaphorically. But, of course, this was not the main point of this chapter. The main point was to show that even the most abstract conceptual system we can think of, mathematics(!), is ultimately embodied in the nature of our bodies, language, and cognition. It follows from this that if mathematics is embodied in nature, then *any* abstract conceptual system is embodied.

Conceptual metaphor and fictive motion, being a manifestation of extremely fast, highly efficient, and effortless cognitive mechanisms that preserve inferences, play a fundamental role in bringing many mathematical concepts into being. We analyzed several cases involving dynamic language in mathematics, in domains in which, according to formal definitions and axioms in mathematics, no motion was supposed to exist at all. Via the study of gestures, we were able to see that the metaphors involved in the linguistic metaphorical expressions were not simply cases of "dead" linguistic expressions. Gesture studies provide real-time convergent evidence supporting the psychological and cognitive reality of the embodiment of mathematical ideas, and their inferential organization. Building on gestures studies we were able to tell that the above mathematics professors, not only were using metaphorical linguistic expressions, but that they were in fact, in real time, thinking dynamically!

For many, mathematics is a timeless set of truths about the universe, transcending our human existence. For others, mathematics *is* what is characterized by formal definitions and axiomatic systems. From the perspective of our work in the cognitive science *of* mathematics (itself), however, a very different view emerges: Mathematics doesn't exist outside of human cognition. Formal definitions and axioms in mathematics are themselves created by human ideas (although they constitute a very small and specific fraction of human cognition), and they only capture very limited aspects of the richness of mathematical ideas. Moreover, definitions and axioms often neither formalize nor generalize human everyday concepts. A clear example is provided by the modern definitions of limits and continuity, which were coined after the work by Cauchy, Weierstrass, Dedekind, and others in the 19th century. These definitions are at odds with the inferential organization of natural continuity provided by cognitive mechanisms such as fictive and metaphorical motion. Anyone who has taught calculus to new students can tell how counter-intuitive and hard to understand the epsilon-delta definitions of limits and continuity are (and this is an extremely well-documented fact in the mathematics education literature). The reason is (cognitively) simple. Static epsilon-delta formalisms neither formalize nor generalize the rich human dynamic concepts underlying continuity and the "approaching" of locations.

By finding out that real numbers "really move," we can see that even the most abstract, precise, and useful concepts human beings have ever created are ultimately *embodied*.

References

1. Bates, E. & F. Dick. (2002). Language, Gesture, and the Developing Brain. *Developmental Psychobiology*, 40(3), 293–310.
2. Cienki, A. (1998). Metaphoric gestures and some of their relations to verbal metaphoric expressions. In J-P Koenig (ed.) *Discourse and Cognition*, pp. 189–204. Stanford CA: CSLI Publications.

3. Courant, R. & Robbins, H. (1978). *What is Mathematics?* New York: Oxford.
4. Fauconnier, G. & M. Turner. (1998). Conceptual Integration Networks. *Cognitive Science* 22:2, 133–187.
5. Fauconnier, G. & M. Turner. (2002). *The Way We Think: Conceptual Blending and the Mind's Hidden Complexities.* New York: Basic Books.
6. Goldin-Meadow, S. & C. Mylander. (1984). Gestural communication if deaf children: The effects and non-effects of parental input on early language development. *Monographs of the Society for Research in Child Development*, 49(3), no, 207.
7. Henderson, D. (2001). *Experiencing geometry*. Upper SaddleRiver, NJ: Prentice Hall.
8. Hersh, R. (1997). *What is mathematics, really*? New York: Oxford Univ. Press.
9. Hickok, G., Bellugi, U., and Klima, E. (1998). The neural organization of language: Evidence from sign language aphasia. *Trends in Cognitive Sciences*, 2(4), 129–136.
10. Iverson, J. & S. Goldin-Meadow. (1998). Why people gesture when they speak. *Nature* 396, Nov. 19, 1998. p. 228.
11. Iverson, J. & E. Thelen, E. (1999). In R. Núñez & W. Freeman (Eds.), *Reclaiming cognition: The primacy of action, intention, and emotion*, pp. 19–40. Thorverton, UK: Imprint Academic.
12. Kendon, A. (1980). Gesticulation and Speech: Two aspects of the process of utterance. In M. R. Key (ed.), *The relation between verbal and nonverbal communication*, pp. 207–227. The Hague: Mouton.
13. Kendon, A. (2000). Language and gesture: unity or duality? In D. McNeill (Ed.), *Language and gesture* (pp. 47–63). Cambridge: Cambridge University Press.
14. Lakoff, G. (1993). The contemporary theory of metaphor. In A. Ortony (Ed.), *Metaphor and Thought* (2nd ed.), pp. 202–251. Cambridge: Cambridge University Press.
15. Lakoff, G. & M. Johnson. (1980). *Metaphors we live by*. Chicago: University of Chicago Press.
16. Lakoff, G., & R. Núñez (1997). The metaphorical structure of mathematics: Sketching out cognitive foundations for a mind-based mathematics. In L. English (ed.), *Mathematical Reasoning: Analogies, Metaphors, and Images.* Mahwah, N.J.: Erlbaum.
17. Lakoff, G. and Núñez, R. (2000). *Where Mathematics Comes From: How the Embodied Mind Brings Mathematics into Being*. New York: Basic Books.
18. McNeill, D. (1992). *Hand and Mind: What Gestures Reveal About Thought*. Chicago: Chicago University Press.
19. Mayberry, R. & Jaques, J. (2000) Gesture production during stuttered speech: insights into the nature of gesture-speech integration. In D. McNeill (ed.) *Language and Gesture*. Cambridge, UK: Cambridge University Press.
20. Narayanan, S. (1997). Embodiment in Language Understanding: Sensory-Motor Representations for Metaphoric Reasoning about Event Descriptions. Ph.D. dissertation, Department of Computer Science, University of California at Berkeley.
21. Núñez, R. (1999). Could the Future Taste Purple? In R. Núñez and W. Freeman, *Reclaiming cognition: The primacy of action, intention, and emotion*, pp. 41–60. Thorverton, UK: Imprint Academic.
22. Núñez, R. (2000). Mathematical idea analysis: What embodied cognitive science can say about the human nature of mathematics. Opening plenary address in *Pro-*

ceedings of the 24th International Conference for the Psychology of Mathematics Education, 1:3–22. Hiroshima, Japan.
23. Núñez, R. (2003). Fictive and metaphorical motion in technically idealized domains. *Proceedings of the 8th International Cognitive Linguistics Conference*, Logroño, Spain, July 20–25, p. 215.
24. Núñez, R.(in press). Creating Mathematical Infinities: The Beauty of Transfinite Cardinals. *Journal of Pragmatics.*
25. Núñez, R., Edwards, L., Matos, J.F. (1999). Embodied Cognition as grounding for situatedness and context in mathematics education. *Educational Studies in Mathematics*, *39*(1–3): 45–65.
26. Núñez, R. & G. Lakoff. (1998). What did Weierstrass really define? The cognitive structure of natural and ε-δ continuity. *Mathematical Cognition*, *4*(2): 85–101.
27. Núñez, R. & Lakoff, G. (in press). The Cognitive Foundations of Mathematics: The Role of Conceptual Metaphor. In J. Campbell (ed.) *Handbook of Mathematical Cognition*. New York: Psychology Press.
28. Núñez, R. & E. Sweetser, (2001). *Proceedings of the 7th International Cognitive Linguistics Conference*, Santa Barbara, USA, July 22–27, p. 249–250.
29. Sweetser, E. (1990). *From Etymology to Pragmatics: Metaphorical and Cultural Aspects of Semantic Structure*. New York: Cambridge University Press.
30. Sweetser, E. (1998). Regular metaphoricity in gesture: bodily-based models of speech interaction. In *Actes du 16e Congrès International des Linguistes*. Elsevier.
31. Talmy, L. (1996). Fictive motion in language and "ception." In P. Bloom, M. Peterson, L. Nadel, & M. Garrett (eds.), *Language and Space.* Cambridge: MIT Press.
32. Talmy, L. (1988). Force dynamics in language and cognition. *Cognitive Science,* 12: 49–100.
33. Talmy, L. (2003). *Toward a Cognitive Semantics. Volume 1: Concept Structuring Systems*. Cambridge: MIT Press.

10

Does Mathematics Need a Philosophy?

WILLIAM TIMOTHY GOWERS

Introduction

There is a philosophical doctrine known as bialetheism, with, apparently, many adherents in Australia, who take the view that there can be true contradictions. I do not wish to defend this view, but nevertheless in the next 45 minutes I will be arguing, sincerely, that mathematics both does and does not need a philosophy. Of course, the apparent contradiction can in my case be resolved in a boringly conventional way: the statement "Mathematics needs philosophy" has at least two reasonable interpretations, and my contention is that one of them is false and another is true. But before I tell you what these interpretations are, I would like to say just a bit about what I hope to achieve in this talk, since as the first speaker in the series I have no precedents to draw on and have therefore had to decide for myself what kind of talk would be appropriate.

First, as you can see, I am reading from a script, something I would never do when giving a mathematics lecture. This, I fondly imagine, is how philosophers do things, at least some of the time, and it is how I prefer to operate when I need to choose my words carefully, as I do today.

Secondly, I would like to stress that it is not my purpose to say anything original. I have read enough philosophy to know that it is as hard to be an original philosopher, at least if one wishes to be sensible at the same time, as it is to be an original mathematician. When I come across a philosophical problem, I usually know pretty soon what my opinion is, and how I would begin to defend it, but if I delve into the literature, I discover that many people have had similar instincts and have worked out the defence in detail, and that many others have been unconvinced, and that whatever I think has already been labelled as some "-ism" or other. Even if my tone of voice occasionally makes it sound as though I think that I am the first to make some point, I don't. I have done a bit of remedial reading and re-reading over the last couple of weeks, but there are many large gaps and I will often not give credit where it is due, and for that I apologize to the philosophers here.

Thirdly, I am very conscious that I am talking to an audience with widely varying experience of mathematics and philosophy, so I hope you will be patient if quite a lot of what I say about your own discipline seems rather elementary and old hat.

Finally, I shall try to make the talk somewhat introductory. Two weeks have not been enough to develop a fully worked-out position on anything, so, rather than looking at one small issue in the philosophy of mathematics, I shall discuss many of the big questions in the subject, but none of them in much depth. I hope this will give us plenty to talk about when the discussion starts - it certainly ought to as many of these questions have been debated for years.

If you ask a philosopher what the main problems are in the philosophy of mathematics, then the following two are likely to come up: what is the status of mathematical truth, and what is the nature of mathematical objects? That is, what gives mathematical statements their aura of infallibility, and what on earth are these statements *about*?

Let me very briefly describe three main (overlapping) schools of thought that have developed in response to these questions: Platonism, logicism and formalism.

The basic Platonist position is rather simple. Mathematical concepts have an objective existence independent of us, and a statement such as "2+2=4" is true because two plus two really does equal four. In other words, for a Platonist mathematical statements are pretty similar to statements such as "that cup is on the table" even if mathematical objects are less tangible than physical ones.

Logicism is an attempt to justify our extreme confidence in mathematical statements. It is the view that all of mathematics can be deduced from a few simple and undeniably true axioms using simple and undeniably valid logical steps. Usually these axioms come from set theory, and they are supposed to form the secure foundation on which the entire edifice of modern mathematics rests. Notice that one can be a Platonist and a logicist at the same time.

Formalism is more or less the antithesis of Platonism. One can caricature it by saying that the formalist believes that mathematics is nothing but a few rules for replacing one system of meaningless symbols with another. If we start by writing down some axioms and deduce from them a theorem, then what we have done is correctly apply our replacement rules to the strings of symbols that represent the axioms and ended up with a string of symbols that represents the theorem. At the end of this process, what we know is not that the theorem is "true" or that some actually existing mathematical objects have a property of which we were previously unaware, but merely that a certain statement can be obtained from certain other statements by means of certain processes of manipulation.

There is another important philosophical attitude to mathematics, known as intuitionism, but since very few working mathematicians are intuitionists, I shall not discuss it today. Let me just say to those mathematicians who know a little about intuitionism that certain aspects of it that seem very off-putting, such as the rejection of the law of the excluded middle or the idea that a mathematical statement can "become true" when a proof is found, should not be dismissed as ridiculous: perhaps on a future occasion we should have a debate here about whether classical logic is the only logic worth considering. For what it is worth, I myself am prepared to countenance other ones.

For the rest of this talk, I shall discuss some fairly specific questions, in what I judge to be ascending order of complexity, with the idea that they will give us a convenient focus for discussion of more general issues of the kind I have been describing. But before I do so, let me give you a very quick idea of where my own philosophical sympathies lie.

I take the view, which I learnt recently goes under the name of *naturalism*, that a proper philosophical account of mathematics should be grounded in the actual practice of mathematicians. In fact, I should confess that I am a fan of the later Wittgenstein, and I broadly agree with his statement that "the meaning of a word is its use in the language". [*Philosophical Investigations* Part I section 43 —actually Wittgenstein qualifies it by saying that it is true "for a large class of cases".] So my general approach to a philosophical question in mathematics is to ask myself how a typical mathematician would react to it, and why. I do not mean by this that whatever an average mathematician thinks about the philosophy of mathematics is automatically correct, but rather than try to make precise what I do mean, let me illustrate it by my treatment of the questions that follow.

1. What is 2+2?

The first question I would like to ask is this: could it make sense to doubt whether 2+2=4? Let me do what I promised, and imagine the reaction of a typical mathematician to somebody who did. The conversation might run as follows.

Mathematician: Do you agree that 2 means the number after 1 and that 4 means the number after the number after 2?
Sceptic: Yes.

M: Do you agree that 2=1+1?

S: Yes.

M: Then you are forced to admit that 2+2=2+(1+1).

S: Yes.

M: Do you agree that addition is associative?

S: Yes.

M: Then you are forced to admit that 2+(1+1)=(2+1)+1.

S: Yes I am.

M: But (2+1)+1 is the number after the number after 2, so it's 4.

S: OK I'm convinced.

The general idea of the above argument is that we have some precise definitions of concepts such as 2, 4, "the number after" and +, and one or two simple axioms concerning them, such as the associativity of addition, and from those it is easy to *prove* that 2+2=4. End of story.

That is, of course, roughly what I think, but before we move on I would nevertheless like to try to imagine a world in which it was natural to think that 2+2=5 and see what that tells us about our belief that 2+2=4.

In such a world, physical objects might have less clear boundaries than they do in ours, or vary more over time, and the following might be an observed empirical fact: that if you put two objects into a container, and then another two, and if you then look inside the container, you will find not four objects but five. A phenomenon like that, though strange, is certainly not a logical impossibility, though one does feel the need for more details: for example, if a being in this world holds up two fingers on one hand and two on the other, how many fingers is it holding up? If you put *no* apples into a bag and then put no further apples into the bag, do you have one apple? But then why not a tomato?

We are free to invent any answers we like to such questions if we can somehow remain logically consistent, so here is a simple suggestion. Perhaps in the strange world it is really the act of enclosing objects in a container that causes what seems to us to be peculiar consequences. It might be that this requires an expenditure of energy so that the container doesn't just explode the moment you put anything into it and there isn't the physical equivalent of what economists call arbitrage. And yet, the apparent duplication-machine properties of plastic bags and the like might be sufficiently common for 2+2=5 to seem a more natural statement than 2+2=4.

But what, one wants to ask, about our mental picture of numbers? If we just *think* of two apples and then think of another two, surely we are thinking of four apples, however you look at it.

But what should we say if we put that point to a being X from the other world, and X reacted as follows?

X: I don't know what you're talking about. Look, I'm thinking of two apples now. [Holds up three fingers from one hand.] Now I'm thinking of two more. [Holds up three fingers from the other hand making a row of six fingers.] The result - five apples.

Suppose that we were initially confused, but after a bit of discussion came to realize that X was associating the apples not with the fingers themselves but with the *gaps between them*. After all, between three fingers there are two gaps and between six fingers there are five. At any rate this might seem a good explanation to us. But perhaps X would be so used to a different way of thinking that it would resist our interpretation. To X, holding up three fingers and saying "two" could seem utterly natural: it might feel absolutely no need for a one-to-one correspondence between fingers and apples.

Faced with such a situation, it is a tempting to take the following line: what X is "really" doing is giving different names to the positive integers. When X says "2+2=5", what this actually means is "3+3=6", and more generally X's false sounding statement that "a+b=c" corresponds to our true statement "(a+1)+(b+1)=c+1".

But should we say this? Or is it better to say that what X means by addition is not our notion of addition but the more complicated (or so we judge) binary operation f(m,n)=m+n+1? Or is it enough simply to say that X uses a system of arithmetic that we can understand and explain in terms of ours in more than one way?

These questions are bothersome for a Platonist, particularly one who believes in direct reference, a philosophical doctrine I shan't discuss here. If the word "five", as used by us, directly refers to the number 5, then surely there ought to be a fact of the matter as to whether the same word used by X directly refers to 5 or 6 or something else. And yet there doesn't seem to be such a fact of the matter.

I think I will leave that question hanging since the world I have just attempted to describe is rather fanciful and there are many other ways of attacking Platonism.

2. The empty set

My next question is whether there is such a thing as the empty set. This question might seem more basic than the first, but if it does then I put it to you that your mind has been warped by a century of logicism, because there is, if you think about it, something rather odd about the concept of a set with no elements. What, after all, is a set? It is a collection of objects (whatever that means). And to say that you have a collection of objects, except that there are no objects, sounds like a contradiction in terms.

I have various questions like this, and normally I don't worry about them. But they do cause me small problems when I lecture on concepts such as sets, functions and the like. I will explain that a set is a collection of objects, usually mathematical, but will not go on to say what a collection is, or a mathematical object. The empty set, I recently told the first-year undergraduates to whom I am lecturing this term, is the set with no elements, but I made no attempt to justify that there was such a set, and I'm glad to report that I got away with it.

There are various ways that one could try to argue for the existence of the empty set. For example, if it doesn't exist then what is the intersection of the sets {1,2} and {3,4}? Or what is the set of all natural numbers n such that n=n+1? These arguments demonstrate that the empty set does indeed exist if one is prepared to accept natural statements such as that the sets {1,2} and {3,4} exist and that given any two sets A and B there is a set C consisting of exactly the elements that A and B have in common. So if you are going to doubt the existence of the empty set you will probably find yourself doubting the existence of any sets at all.

So let us consider the more general question: are there sets? What exactly are we doing to the numbers 1 and 2 if we separate them with a comma and enclose them in curly brackets? Similarly, what is the difference between the number 1 and the set whose sole element is 1?

Let us look at a simple problem that I recently set to the first-years. I asked them whether there could be a sequence of sets A_1, A_2, A_3,... such that for every n the intersection of the first n A_i is non-empty but yet the intersection of all the A_i is empty. The answer is yes and one example that shows it is to let $A_n=\{n,n+1,n+2,...\}$.

Why is this a suitable example? Well, the number n belongs to all the first n A_i but if m is any number then m does not belong to *all* A_i since it does not belong, for example, to A_{m+1}.

We could spell out this justification even more. Why, for example, does n belong to the first n A_i? Well, n belongs to A_i if and only if $n \geq i$, so n does indeed belong to all of $A_1,...,A_n$. Similarly, m does not belong to A_{m+1}.

So an equivalent way to describe the above example is to say that whatever (finite) number of conditions you impose on a number n of the form $n \geq i$, it will be possible for them to be all to be satisfied, but there is no n that is greater than or equal to *every* i. And the interesting thing about this formulation is that it makes no explicit mention of sets. What's more, it isn't just an artificial translation cooked up with the sole purpose of not talking about sets. Rather, it reflects quite accurately what actually goes on in our minds when we go about proving what we want to prove about the sets.

So could it be that all that matters about the empty set is something like this? Whenever you see the sentence "x is an element of the empty set" it is false. More generally, could it be that whenever you actually prove something about sets in a normal mathematical context, one of the first things you do is get rid of the sets. I had a good example of this recently when proving in lectures that the equivalence classes of an equivalence relation form a partition. If R is an equivalence relation on a set A and x belongs to A, then the equivalence class of x is the set E(x)={y in A: xRy}, but as the proof proceeded, every time I wrote down a statement such as "z is an element of E(x)" I immediately translated it into the equivalent and much simpler non-set-theoretic statement xRz.

3. Subsets of the natural numbers

I think it would be possible to defend a position that set theory could be dispensed with, at least when it involved sets defined by properties. We could regard the expression A={x:P(x)} not as actually denoting an object named A but as being a convenient piece of shorthand. The statement "z belongs to A", on this view, means nothing more than P(z). Similarly, if B={x:Q(x)} then the statement "A cap B = emptyset" means nothing more than that there is no x such that P(x) and Q(x).

But not all sets that crop up in mathematics, and I am still talking about "ordinary" mathematics rather than logic and set theory, are defined by properties. Often we talk about sets in a much more general way, using sentences like, "Let A be a set of natural numbers," and proving theorems such as that there are uncountably many such sets. In such contexts it is not as easy to dispense with the language of set theory. And yet the sets we are supposedly discussing, general sets of positive integers, are rather puzzling. My third question is this: *what is* an arbitrary set of positive integers? Here I have in mind the sort of utterly general set that cannot be defined, the infinite equivalent of a subset of the first thousand integers chosen randomly. We have a strong intuition that such sets exist, but why?

Let us look at the proof that there are uncountably many sets of positive integers, and see what it tells us about our attitude to sets in general. We start with an arbitrary sequence A_1,A_2,A_3,... of subsets of N, and from those construct a new set A according to the rule

n is an element of A if and only if n is not an element of A_n.

Then A is a subset of N not in the sequence. Since the sequence we looked at was arbitrary, *no* sequence of subsets of N exhausts all of them.

Why was A not in the sequence? Well, if it had been then there would have had to be some n such that A=A_n. But for each n we know that it belongs to A if and only if it does *not* belong to A_n, so A is not the same set as A_n.

This argument shows a very basic property of the two sets A and A_n - that they are not equal. And yet even here I did not really reason about the sets themselves and say some mathematical equivalent of, "Look, they're different." Instead, I used the standard criterion for when two sets are equal:

A=B if and only if every element of A is an element of B and every element of B is an element of A.

This tells me that in order to prove that A and A_n are distinct I must find a positive integer m and show that either

m belongs to A but not to A_n

or

m belongs to A_n but not to A.

So, once again, what I seem to be focusing on is not so much the sets themselves but statements such as "m is an element of A".

Does this mean that set-theoretic language is dispensable to an ordinary mathematician? I think it often is, but I wouldn't want to go too far - after all, I would certainly feel hampered if I couldn't use it myself. Here is an analogy that I have not had time to think about in any detail, so perhaps it won't stand up, but let me float it anyway. A central project in philosophy is to explain the notion of truth. What is it for a statement to be true? There are some who hold that the word "true" adds very little to our language: if we say that the sentence, "Snow is white" is true, then what we have said is that snow is white, and that is all there is to it. And yet the word "true" does seem to be hard to avoid in some contexts. For example, it isn't easy to think of a way to paraphrase the sentence, "Not all of what George Bush says in the next week will be true" without invoking some notion of a similar nature to that of truth. I think perhaps it is similar for the language of sets -that it makes it much easier to talk in generalities, but can be dispensed with when we make more particular statements.

I must press on, but before I ask my next question, let me tell you, or remind you, of three useful pieces of terminology.

4. Some terminology

The first is the phrase "ontological commitment", a phrase associated with and much used by Quine. One of the standard tricks that we do as mathematicians is "reduce" one concept to another - showing, for example that complex numbers can be "constructed" as ordered pairs of real numbers, or that positive integers can be "built out of sets". People sometimes use extravagant language to describe such constructions, sounding as though what they are claiming is that positive integers "really are" special kinds of sets. Such a claim is, of course, ridiculous, and probably almost nobody, when pressed, would say that they actually believed it.

But another position, taken by many philosophers, is more appealing. In describing the world, and in particular the rather problematic abstract world of mathematics, it makes sense to try to keep one's list of dubious beliefs to an absolute minimum. One example of a belief that might be thought dubious, or at least problematic, is that the number 5 actually exists. Questions about what exists belong to the branch of philosophy known as ontology, a word derived from the Greek for "to be"; and if what you say implies that something exists, then you are making an ontological commitment. One view, which I do not share, is that at least some ontological commitment is implicit in mathematical language. But those who subscribe to such a view will often seek to minimize their commitment by reducing concepts to others. Such people may, for example, be comforted by knowing that complex numbers can be thought of as ordered pairs of real numbers, so that we are not making any *further* worrying ontological commitments when we introduce them than we had already made when talking about the reals.

The second piece of terminology is the distinction between naive and abstract set theory. A professional set theorist does not spend time worrying about whether sets exist and what they are if they do. Instead, he or she studies *models* of set theory, which are mathematical structures containing things that we conventionally call sets - rather as a vector space contains things that we conventionally call vectors. And just as, when we do abstract linear algebra, we do not have to say what a vector is, or at least we do not have to say any more than that it is an element of a vector space, so, when we do set theory, we do not need to say what a set is, except to say that it belongs to a model. (In an appropriate metalanguage one could say that a model is a set, and sets, in the sense of the object language, are its elements. But this is confusing and it would be more usual to call the model a proper class.) And to pursue the analogy further, just as there are rules that tell you how to form new vectors out of old ones - addition and scalar multiplication - so there are rules that tell you how to form new sets out of old ones - unions, intersections, power sets, replacement and so on. To get yourself started you need to assume that there are at least some sets, so you want an axiom asserting the existence of the empty set, or perhaps of a set with infinitely many elements.

If you attempt to say what a set is, then you are probably doing naive set theory. What I have just described, where sets are not defined (philosophers would call the word "set" an undefined primitive), is abstract set theory. Notice that as soon as you do abstract set theory, you do not find yourself thinking about the actual existence or nature of sets, though you might, if you were that way inclined, transfer your worries into the meta-world and wonder about the existence of the model. Even so, when you were actually doing set theory, your activity would more naturally fit the formalist picture than the Platonist one.

The distinction between naive and abstract set theory gives one possible answer to the question "What is an arbitrary set of natural numbers?" The answer is, "Don't ask." Instead, learn a few rules that allow you to build new sets out of old ones (including unions, intersections and the diagonal process we have just seen) and make it all feel real by thinking from time to time about sets you can actually define such as the set of all primes - even if in the end the definition is more important than the set.

Another distinction, which I introduce because it may well have occurred to those here who have not previously come across it, is one made by Rudolf Carnap between what he called internal and external questions. Suppose I ask you whether you accept that there are infinitely many primes. I hope that you will say that you do. But if I then say, "Ah, but prime numbers are positive integers and positive integers are numbers and numbers are mathematical objets so you've admitted that there are infinitely many mathematical objects," you may well feel cheated. If you do, the chances are that you will want to say, with Carnap, that there are two senses of the phrase "there exists". One is the sense in which it is used in ordinary mathematical discourse - if I say that there are infinitely many primes I merely mean that the normal rules for proving mathematical statements license me to use appropriate quantifiers. The other is the more philosophical sense, the idea that those infinitely many primes "actually exist out there". These are the *internal* and *external* uses respectively. And it seems, though not all philosophers would agree, that this is a clear distinction, and that the answers you give in the internal sense do not commit you to any particular external and philosophical position. In fact, to many mathematicians, including me, it is not altogether clear what is even meant by the phrase "there exists" in the external sense.

5. Ordered pairs

I will not spend long over my next question, since I have discussed similar issues already, but it is one of the simplest examples of the slight difficulty I have when lecturing about basic mathematical concepts from the naive point of view. The question is, what is an ordered pair?

This is what I take to be the standard account that a mathematician would give. Let x and y be two mathematical objects. Then from a formal point of view the ordered pair (x,y) is defined to be the set $\{\{x\},\{x,y\}\}$, and it can be checked easily that

$$\{\{x\},\{x,y\}\}=\{\{z\},\{z,w\}\} \text{ if and only if } x=z \text{ and } y=w.$$

Less formally, the ordered pair (x,y) is a bit like the set $\{x,y\}$ except that "the order matters" and x is allowed to equal y.

Contrast this account with the way ordered pairs are sneaked in at a school level. There, the phrase "ordered pair" is not even used. Instead, schoolchildren are told that points in the plane can be represented by coordinates, and that the point (x,y) means the point x to the right and y up from the origin. It is then geometrically obvious that (x,y)=(z,w) if and only if x=z and y=w.

Pupils who are thoroughly used to this idea will usually have no difficulty accepting later on that they can form "coordinates" not just out of real numbers but also out of elements of more general sets. And because of their experience with plane geometry, they will take for granted that (x,y)=(z,w) if and only if x=z and y=w, whether or not you bother to spell this out as an axiom for ordered pairs. In other words, it is possible to convey the idea of an ordered pair in a way that is clearly inadequate from the formal point of view but that does not seem to lead to any problems. One can quite easily imagine an eminent physicist successfully using the language of ordered pairs without knowing how to formalize it.

It is clear that what matters in practice about ordered pairs is *just* the condition for when two of them are equal. So why does anybody bother to "define" the ordered pair (x,y) as $\{\{x\},\{x,y\}\}$? The standard answer is that if you want to adopt a statement such as

$$(x,y)=(z,w) \text{ if and only if } x=y \text{ and } z=w$$

as an axiom, then you are obliged to show that your axiom is consistent. And this you do by constructing a model that satisfies the axiom. For ordered pairs, the strange-looking definition $(x,y)=\{\{x\},\{x,y\}\}$ is exactly such a model. What this shows is that ordered pairs can be defined in terms of sets and the axiom for ordered pairs can then be deduced from the axioms of set theory. So we are not making new ontological commitments by introducing ordered pairs, or being asked to accept any new and unproved mathematical beliefs.

This account still leaves me wanting to ask the following question. Granted, the theory of ordered pairs can be reduced to set theory, but that is not quite the same as saying that an ordered pair is "really" a funny kind of set. (*That* view is obviously wrong, since there are many different set-theoretic constructions that do the job equally well.) And if an ordered pair isn't really a set, then what is it? Is there any way of doing justice to our pre-theoretic notion of an ordered pair other than producing this rather artificial translation of it into set theory?

I don't think there is, at least if you want to start your explanation with the words, "An ordered pair is". At least, I have never found a completely satisfactory way of defining them in lectures. To my mind this presents a pretty serious difficulty for Platonism. And yet, as I have said, it doesn't really seem to matter to mathematics. Why not?

I would contend that it doesn't matter because it *never* matters what a mathematical object is, or whether it exists. What *does* matter is the set of rules governing how you talk about it - or perhaps I should say, since that sounds as though "it" refers to something, what matters about a piece of *mathematical terminology* is the set of rules governing its use. In the case of ordered pairs, there is only one rule that matters - the one I have mentioned several times that tells us when two of them are equal (or, to rephrase again, the one that tells us when we are allowed to write down that they are equal, substitute one for another and so on).

I said earlier that I like to think about actual practice when I consider philosophical questions about mathematics. Another useful technique is to think what you would have to program into a computer if you wanted it to handle a mathematical concept correctly. If the concept was an ordered pair, then it would be ridiculous to tell your computer to convert the ordered pair (x,y) into the set $\{\{x\},\{x,y\}\}$ every time it came across it. Far more sensible, for almost all mathematical contexts, would be to tell it the axiom for equality of ordered pairs. And if it used that axiom without a fuss, we would be inclined to judge that it understood the concept of ordered pairs, at least if we had a reasonably non-metaphysical idea of understanding - something like Wittgenstein's, for example.

This point applies much more generally. I have sometimes read that computers cannot do the mathematics we can because they are finite machines, whereas we have a mysterious access to the infinite. Here, for example, is a quotation from the famous mathematician Alain Connes:

... this direct access to the infinite which characterizes Euclid's reasoning [in his proof that there are infinitely many primes], or in a more mature form Gödel's, is actually a trait of the living being that contradicts the reductionist's model.

But, just as it is not necessary to tell a computer what an ordered pair is, so we don't have to embed into it some "model of infinity". All we have to do is teach it some syntactic rules for handling, with care, the *word* infinity - which is also what we have to do when teaching undergraduates. And, just as we often try to get rid of set-theoretic language when talking about sets, so we avoid talking about infinity when justifying statements that are ostensibly about the infinite. For example, what Euclid's proof actually gives is a recipe for extending any finite list of primes. To take a simpler example, if I prove that there are infinitely many even numbers by saying, "2 is even and if n is even then so is n+2", have I somehow exhibited infinite mental powers? I think not: it would be easy to programme a computer to come up with such an argument.

6. Truth and provability

I can date my own conversion from an unthinking childhood Platonism from the moment when I learnt that the continuum hypothesis was independent of the other axioms of set theory. If as apparently concrete a statement as that can neither be proved nor disproved, then what grounds can there be for saying that it is true or that it is false? But if you think there is no fact of the matter either way with the continuum hypothesis, then why stop there? What about the axiom of infinity -that there is an infinite set? It doesn't follow from the other axioms of set theory, and nor, it seems, does its negation. So why should we believe it? Surely not because of some view that the universe is infinite in extent, or infinitely divisible. What would that show anyway? There would still be the problem of applying those funny curly brackets. As I have said, even the axiom that the empty set exists is hard to justify if one interprets it realistically.

So I am driven to the view that there isn't much to mathematical truth over and above our accepted procedures of justification - that is, formal proofs. But something in me still rebels against the intuitionists' idea that a statement could *become* true when a proof is found, and I'm sure most mathematicians agree with me. So my next question is why, and I would like to look at a few concrete examples.

Here is one that intuitionists like: consider the statement "somewhere in the decimal expansion of pi there is a string of a million sevens". Surely, one feels, there is a fact of the matter as to whether that is true or false, even if it may never be known which.

What is it that makes me want to say that the long string of sevens is definitely either there or not there - other than a general and question-begging belief in the law of the excluded middle? Well, actually I am tempted to go further and say that I believe that the long string of sevens *is* there, and I have a definite reason for that stronger belief, which is the following. All the evidence is that there is nothing very systematic about the sequence of digits of pi. Indeed, they seem to behave much as they would if you just chose a sequence of random numbers between 0 to 9. This hunch sounds vague, but it can be made precise as follows: there are various tests that statisticians perfom on sequences to see whether they are likely to have been generated randomly, and it looks very much as though the sequence of digits of pi would pass these tests. Certainly the first few million do. One obvious test is to see whether any given short sequence of digits, such as 137, occurs with about the right frequency in the long term. In the case of the string 137 one would expect it to crop up about one thousandth of the time in the decimal expansion of pi. If after examining several million digits we found that it had in fact occurred a hundredth of the time, or not at all, then we would be surprised and wonder whether there was an explanation.

But experience strongly suggests that short sequences in the decimal expansions of the irrational numbers that crop up in nature, such as pi, e or the square root of 2, *do* occur with the correct frequencies. And if that is so, then we would expect a million sevens to occur in the decimal expansion of pi about $10^{-1000000}$ of the time - and it is of course no surprise that we will not actually be able to check that directly. And yet, the argument that it does eventually occur, while not a proof, is pretty convincing.

This raises an interesting philosophical question. A number for which the short sequences of digits occur with the right frequencies is called *normal*. Artificial examples have been constructed of normal numbers, but there is no naturally occurring number that is known to be normal. Perhaps the normality of pi is not just an unsolved problem but actually an unprovable theorem. If so, then it is highly unlikely that we shall ever find an abstract argument that shows that the expansion of pi contains a million sevens in a row, and direct calculation of the number of digits that would be necessary to verify it "empirically" is out of the question. So what, then, is the status of the reasonable-sounding heuristic argument that pi contains a million sevens in a row, an argument that convinces me and many others?

This question raises difficulties for those who are too ready to identify truth and provability. If you look at actual mathematical practice, and in particular at how mathematical beliefs are formed, you find that mathematicians have opinions long before they have formal proofs. When I say that I think pi almost certainly has a million sevens somewhere in its decimal expansion, I am not saying that I think there is almost certainly a (feasibly short) *proof* of this assertion - perhaps there is and perhaps there isn't. So it begins to look as though I am committed to some sort of Platonism -that there is a fact of the matter one way or the other and that that is why it makes sense to speculate about which.

There is an obvious way to try to wriggle out of this difficulty, but I'm not sure how satisfactory it is. One could admit that a simple identification of truth with provability does not do justice to mathematical practice, but still argue that what really matters about a mathematical statement is not some metaphysical notion of truth, but rather the conditions that have to hold to make us inclined to assert it. By far the most important such condition is the existence of a formal proof, but it is not the only one. And if I say something like, "pi is probably normal", that is just a shorthand for, "there is a convincing heuristic argument, the conclusion of which is that pi is normal". Of course, a move like this leaves very much open the question of which informal arguments we find convincing and why. I think that is an important philosophical project, but not one I have carried out, or one that I would have time to tell you about now even if I had.

Actually, it is closely related to another interesting question, a mathematical version of the well-known philosophical problem of induction. A large part of mathematical research consists in spotting patterns, making conjectures, guessing general statements after examining a few specific instances, and so on. In other words, mathematicians practise induction in the scientific as well as mathematical sense. Suppose, for example, that f is a complicated function of the positive integers arising from some research problem and that the first ten values it takes are 2, 6, 14, 24, 28, 40, 42, 66, 70, 80. In the absence of any other knowledge about f, it is reasonable to guess that it always takes even values, or that it is an increasing function, but it would be silly to imagine that f(n) is always less than 1000. Why? I think the beginning of an answer is that any guess about f should be backed up by some sort of heuristic argument. In this case, if we have in the back of our minds some picture of a "typical function that occurs in nature" then we might be inclined to say that the likelihood of its first ten values being even or strictly increasing just by chance is small, whereas the likelihood of their all being less than 1000 is quite high.

Let me return to the question of why it seems so obvious that there is a fact of the matter as to whether the decimal expansion of pi contains a string of a million sevens. In the back of many minds is probably an argument like this. Since pi almost certainly *is* normal, if we look instead for shorter strings, such as 137, then we don't have to look very far before we find them. And in principle we could do the same for much longer strings - even if in practice we certainly can't. So the difference between the two situations is not mathematically interesting and should not have any philosophical significance.

Now let me ask a rather vague question: what is interesting about mathematical theorems that begin "for every natural number n"? There seem to me to be two attitudes one can take. One is that a typical number of the order of magnitude of, say, $10^{10^{100}}$, will be too large for us to specify and therefore isn't really anything more to us than a purely abstract n. So the instances of a theorem that starts "for all n" are, after a certain point, no more concrete than the general statement, the evidence for which consists of a certain manipulation of symbols, as a formalist would contend. So, in a sense the "real meaning" of the general theorem is that it tells us, in a succinct way, that the small "observable" instances of the theorem are true, the ones that we might wish to use in applications.

This attitude is not at all the one taken by most pure mathematicians, as is clear from a consideration of two unsolved problems. Goldbach's conjecture, that every even number over 4 is the sum of two primes, has been verified up to a very large number, but is still regarded as completely open. Conversely, Vinogradov's three-primes theorem, that every sufficiently large odd integer is the sum of three primes, is thought of as "basically solving" that problem, even though in the current state of knowledge "sufficiently large" means "at least 10^{13000}" which makes checking the remaining cases way beyond what a computer could do. This last example is particularly interesting since to date only 79 primes are known above 10^{13000} (or even a third of 10^{13000}) are known. So the theorem has almost no observable consequences.

In other words, there are two conclusions you can draw from the fact that very large integers are inaccessible to us. One is that what actually matters is small numbers, and the other is that what actually matters is abstract statements.

One small extra comment before I move to a completely new question. Another conjecture that seems almost certainly true is the twin-primes conjecture - that there are infinitely many primes p for which p+2 is also prime. This time the heuristic argument that backs up the statement is based on the idea that the primes appear to be "distributed randomly", and that a sensible-looking probabilistic model for the primes not only suggests that the twin-primes conjecture is true but also agrees with our observations about how often they occur. But I find that my feeling that there must be a fact of the matter one way or the other is less strong than it was for the sevens in the expansion of pi, because no amount of finite checking could, even in principle, settle the question. The difference is that the pi statement began with just an existential quantifier, whereas "there are infinitely many" gives us "for all" and then "there exists". On the other hand, there does seem to be a fact of the matter about whether there are at least n twin primes, for any n you might choose to specify. But now I am talking about very subjective feelings, so it is time to turn to my last question.

7. The axiom of choice

I mentioned earlier that the status of the continuum hypothesis convinced me that Platonist views of mathematical ontology and truth could not be correct. Instead of discussing that, let me ask a similar question. Is the axiom of choice true?

By now it probably goes without saying that I don't think that either it or its negation is true in any absolute, transcendent, metaphysical sense, but many philosophers disagree. I recently read an article by Hilary Putnam in which he ridiculed the idea that one could draw any philosophical conclusions from the independence of the continuum hypothesis. But, as nearly always happens in philosophy, I emerged with my beliefs intact, and will now try to do what he thinks I can't with the axiom of choice.

Consider the following statement, which is an infinitary analogue of a famous theorem of Ramsey.

Let A be a collection of infinite subsets of the natural numbers. Then there is an infinite set Z of natural numbers such that either all *its* infinite subsets belong to A or none of them do.

As it stands, that statement is false (or so one usually says) because it is quite easy to use the axiom of choice to build a counterexample. But there are many mathematical contexts in which the result could be applied if only it were true, and actually for those contexts - that is, for the specific instances of A that crop up "in nature", as mathematicians like to say - the result *is* true. In fact, there is a precise theorem along these lines, which comes close to saying that the only counterexamples are ridiculous ones cooked up using the axiom of choice. So in a way, the statement is "basically true", or at least true whenever you care about it. In this context the axiom of choice is a minor irritant that forces you to qualify your statement by putting some not very restrictive conditions on A.

There are many results like this. For example, not every function is measurable but all the ones that you might actually want to integrate are, and so on.

Now let's consider another statement, the infinite-dimensional analogue of a simple result of finite-dimensional linear algebra.

Let V be an infinite-dimensional vector space over R and let v be a non-zero vector in V. Then there is a linear map f from V to R such that f(v) is non-zero.

To prove this in a finite-dimensional context you take the vector v, call it v_1, and extend it to a basis $v_1,...,v_n$. Then you let $f(v_1)=1$ and $f(v_i)=0$ for all other i.

For an infinite-dimensional space, the proof is exactly analogous, but now when you extend v_1 to a basis, you have to continue transfinitely, and since each time you are choosing a v_i you are not saying how you did it, you have to appeal to the axiom of choice. And yet it seems unreasonable to say that you can't make the choices just because you can't specify them - after all, you can't specify the choices you make in the finite-dimensional context either, and for the same reason, that nothing has been told to you about the vector space V.

If you are given an explicit example of V then the picture changes, but there is still a close analogy between the two situations. Sometimes there is a fairly obvious choice of the function f, but sometimes there is no canonical way to extend v to a basis. Then, what rescues you if V is finite-dimensional may be merely that you have to make 10 trillion ugly non-canonical choices rather than infinitely many.

But by and large, for the vector spaces that matter, it is clear that a function f can be found, so this time it would be the *negation* of the axiom of choice that is a minor irritant, telling you that you have to apologize for your theorem by confessing that it depends on the axiom.

So for good reasons - and this is what I would like to stress - we sometimes dismiss the consequences of the axiom of choice and we sometimes insist on them. And in both cases our governing principle is nothing to do with anything like truth, but more a matter of convenience. It is as though when we talk about the world of the infinite, we think of it as a sort of idealization of the finite world we actually inhabit. If the axiom of choice helps to make the infinite world better reflect the finite one then we are happy to use it. If it doesn't then we describe its consequences as "bizarre" and not really part of "mainstream mathematics". And that is why I do not believe that it is "really" true or "really" false.

8. Concluding remarks

It may seem as though I have ignored the title of my talk, so here is what I mean when I say that mathematics both needs and does not need a philosophy.

Suppose a paper were published tomorrow that gave a new and very compelling argument for some position in the philosophy of mathematics, and that, most unusually, the argument caused many philosophers to abandon their old beliefs and embrace a whole new -ism. What would be the effect on mathematics? I contend that there would be almost none, that the development would go virtually unnoticed. And basically, the reason is that the questions considered fundamental by philosophers are the strange, external ones that seem to make no difference to the real, internal business of doing mathematics. I can't resist quoting Wittgenstein here:

> A wheel that can be turned though nothing else moves with it, is not part of the mechanism.

Now this is not a wholly fair comment about philosophers of mathematics, since much of what they do is of a technical nature -attempting to reduce one sort of discourse to another, investigating complicated logical systems and so on. This may not be of much relevance to mathematicians, but neither are some branches of mathematics relevant to other ones. That does not make them unrespectable.

But the point remains that if A is a mathematician who believes that mathematical objects exist in a Platonic sense, his outward behaviour will be no different from that of his colleague B who believes that they are fictitious entities, and hers in turn will be just like that of C who believes that the very question of whether they exist is meaningless.

So why should a mathematician bother to think about philosophy? Here I would like to advance a rather cheeky thesis: that modern mathematicians are formalists, even if they profess otherwise, and that it is good that they are.

This is the sort of evidence I have in mind. When mathematicians discuss unsolved problems, what they are doing is not so much trying to uncover the truth as trying to find proofs. Suppose somebody suggests an approach to an unsolved problem that involves proving an intermediate lemma. It is common to hear assessments such as, "Well, your lemma certainly looks true, but it is very similar to the following unsolved problem that is known to be hard," or, "What makes you think that the lemma isn't more or less equivalent to the whole problem?" The probable truth and apparent relevance of the lemma are basic minimal requirements, but what matters more is whether it forms part of a realistic-looking research strategy, and what that means is that one should be able to imagine, however dimly, an *argument* that involves it.

I think that most successful mathematicians are very much aware of this principle, even if they don't bother to articulate it. But I also think that it is a good idea to articulate it - if you are doing research, you might as well have as clear and explicit an idea as possible of what you are doing rather than groping about and waiting for that magic inspiration to strike. And it is a principle that sits more naturally with formalism than with Platonism.

I also believe that the formalist way of looking at mathematics has beneficial pedagogical consequences. If you are too much of a Platonist or logicist, you may well be tempted by the idea that an ordered pair is *really* a funny kind of set - the idea I criticized earlier. And if you teach that to undergraduates, you will confuse them unnecessarily. The same goes for many artificial definitions. What matters about them is the basic properties enjoyed by the objects being defined, and learning to use these fluently and easily means learning appropriate replacement rules rather than grasping the essence of the concept. If you take this attitude to the kind of basic undergraduate mathematics I am teaching this term, you find that many proofs write themselves - an assertion I could back up with several examples.

So philosophical, or at least quasi-philosophical, considerations do have an effect on the practice of mathematics, and therein lies their importance. I have mentioned some other questions I find interesting, such as the problem of non-mathematical induction in mathematics, and I would justify those the same way. And that is the sense in which mathematics needs philosophy.

Brief additions in response to the discussion after the talk

1. Thomas Forster informed me that Russell and Whitehead took roughly the view of ordered pairs that I advocate - treat them as an undefined primitive with a simple rule for equality. So perhaps I did logicists an injustice

(unless they felt that the construction of ordered pairs out of sets was significant progress).

2. It was pointed out by Peter Smith that I had blurred the distinction between a pure formalism - mere pushing around of symbols - and "if-then-ism", the view that what mathematicians do is explore the consequences of axioms to obtain conditional statements (if this set of axioms is true, then this follows, while this other set implies such and such else), but nevertheless statements with a definite content over and above the formalism. I don't know exactly where I stand, but probably a bit further towards the purely formal end than most mathematicians. See my page on *how to solve basic analysis exercises without thinking* for some idea of why.

11

How and Why Mathematics Is Unique as a Social Practice[1]

JODY AZZOUNI

I

I'm sympathetic to *many things* those who self-style themselves "mavericks" have to say about how mathematics is a *social* practice. I'll start with the uncontroversial point that mathematicians usually reassure themselves about their results by showing colleagues what they've done. But *many* activities are similarly (epistemically) social: politicians ratify commonly-held beliefs and behavior; so do religious cultists, bank tellers, empirical scientists, and prisoners.

Sociologists, typically, study methods of attaining *consensus* or *conformity*[2] since groups act in *concert*. And (after all) ironing out mathematical "mistakes" is suppressing a form of *deviant behavior*. One way to find genuine examples of socially-induced consensus is to limn the range of behaviors *possible* for such groups. One *empirically* studies, that is, how groups deviate

[1] This paper owes its existence to an invitation by Jean Paul Van Bendegem and Bart Van Kerkhove to give a talk at a conference on Perspectives on Mathematical Practices (October 24-25, 2002). I also subsequently gave this talk at Columbia University on November 21, 2002. I'm grateful to both audiences for their suggestions which enabled improvements in the paper. I want to single out in this regard: David Albert, Haim Gaifman, and Philip Kitcher. Thanks also to Isaac Levi, Michael D. Resnik and Robert Thomas. I've made a small—but significant—number of changes for the appearance of the article in this volume based on the usual run of second-thoughts, "oops"-reactions, and so on, that are so natural in philosophy (at least when *I* do it).

[2] Attaining "conformity" and "consensus" are mild-sounding phrases for what's often a pretty *brutal* process. Although what I say is intended to be understood generally, the reader does best *not* to think of the practice of torturing political deviants (in order to bring them and their kind into line), but of doting parents teaching children to count, to hold forks, to maneuver about in clothing, or to speak.

from one another in their (group) practices. Consider admissible eating behavior. The options that *exist* are virtually *unimaginable*: in *what's* eaten, *how* it's eaten (in what order, with what *tools*, over how much time), how it's cooked—*if* it's cooked—what's allowed to be said (or not) during a meal, and so on. To understand why a group (at a time) eats meals as it does, and why its members find variants inappropriate (even *revolting*), we see how consensus is determined by childhood training, how ideology crushes variations by making them *unimaginable* or viscerally *repulsive* (so that, say, when someone so trained imagines an otherwise innocent *cheeseburger*, what's felt is—nearly instinctive—disgust), and so on. Equally coercive social factors in conjunction with the ones just mentioned explain why we obey laws, respect property (in the *particular* ways we do), and so on: The threat of punishment, corporeal or financial.

Before turning specifically to mathematical practice, note two presuppositions of any empirical study of the social inducing of consensus (and which have been assumed in my sketchy delineation of the sociology of eating). First, such *social* inducing presupposes (empirical) evidence of the *possibility* of alternative behaviors. The best way (although not the only way) to verify that a kind of behavior *is* possible is to find a group engaged in it; but, in any case, if a behavior is biologically or psychologically impossible, or if the resources available to a group prevent it, we don't need social restraints to explain why individuals *uniformly* avoid that behavior.

The second presupposition is that the study of social mechanisms should uncover factors powerful enough to exclude (in a given population) the alternatives we otherwise know are possible; either the absence of such factors, or the presence of empirical reasons that show such factors can't *enforce* behavioral consensus, will motivate the hypothesis of *internal factors*—psychological, physiological, or both—in conforming individuals: consider, for example, the Chomskian argument that internal *dispositions* in humans strongly constrain the general form of the rules for natural languages.

II

Let's turn to mathematical practice. There are two striking ways it seems to differ from just about *any* other group practice humans engage in. One has been repeatedly noted by commentators on mathematics; the other, oddly, is (pretty much) overlooked.

It's widely observed that, unlike other cases of conformity, and where social factors *really are* the source of that conformity, one finds in mathematical practice *nothing like* the variability found in cuisine, clothing, or metaphysical doctrine. There *are* examples of deviant computational practices: Babylonian fractions or the one-two-many form of counting; but overall empirical evidence for the possibility of deviation from standard mathematical practice—at least until the twentieth century—isn't rich.

Two points. First, as Kripke and others have noted (in the wake of Wittgenstein),[3] it's easy to *design* thought experiments where people, impervious to correction, *systematically* follow rules differently from us. Despite the ease of *imagining* people like this, they're not *found* outside philosophical fiction. One *does* (unfortunately) meet people who can't grasp rules at all—but that's different. In rule-following thought-experiments someone is portrayed who seems to follow *a* rule but who also understands "similar," so that she "goes on" differently from us. (After being shown a finite number of examples of sums, she sums new examples as we would until she reaches a particular border (pairs of numbers both over a hundred, say) whereupon she sums differently—in some systematic way—while claiming she's still doing the same thing. This really is different from people who don't grasp generalization at all.)

Despite the absence of *empirically real* examples of alternative rule-following (in counting or summing), such *thought experiments* are often used to press the view that it's (purely) *social factors* that induce mathematical consensus. Given my remarks about *appropriate* empirical methods for recognizing *real* options in group practices, such a claim—to be empirically respectable, anyway—can't batten on thought experiment alone; it needs an analysis of social factors that arise in *every* society that counts or adds—and which force humans to agree to the same numerical claims. The social factors that are pointed to, however, for example childhood learning, are ones shared by almost every other group practice (diet, language, cosmetics, and so on) which—in contrast to mathematical practice—show great deviation across groups. That is, even when systematic algorithmic rules (such as the ones of languages or games) govern a practice, that practice still drifts over time—unlike, as it seems, the algorithmic rules of mathematics.[4]

One possible explanation for this[5] is that practical exigencies exclude deviant rule-following mathematics: someone who doesn't add as we do can be exploited—in business transactions, say. And so it's thought that deviant counting would die quickly. But this idea is sociologically naive because, even if the dangerousness of a practice did imply its quick demise, this wouldn't mean it couldn't emerge to begin with, and leave evidence in our historical record of its temporary stay among us; all sorts of idiotic and quite dangerous practices (medical ones, cosmetic ones, practices motivated by religious superstition) are *widespread*. Even *shallow* historical reading exposes a

---

[3] E.g.: Kripke 1982, Bloor 1983, and, of course, Wittgenstein 1953.

[4] I should make this clear: by "drift" I mean a change in the rules and practices which doesn't merely involve augmentation of such rules, but the elimination of at least some of them. Mathematics is always being augmented; the point of denying "drift" in its case is that such augmentation is overwhelmingly monotonically increasing.

[5] See e.g., Hersh 1997, p. 203.

plethora of, to speak frankly, pretty dumb activities that (i) allowed exploitation of all sorts (*and* helped shorten lives), and yet (ii) didn't require too much insight to *realize* were both pretty dumb and pretty dangerous. *There's* no shortage of such practices *today*—as the religious right and the raw-food movement, both in the United States, make clear. So it's hard to see why there can't have been really dumb counting practices that flourished (by, for example, exploiting the rich vein of number superstition we *know* existed), and then died out (along with the poor fools practicing them).

Another way around the apparent sociological uniqueness of mathematical practice is the blunt response that mathematical practice *isn't* unique; there *are* deviant mathematical practices; we just haven't looked in the right places for them. Consider, instead of *counting* variants, the development of alternative mathematics—intuitionism, for example, or mathematics based on alternative logics (e.g., paraconsistent logics). Aren't *these* examples of mathematical *deviancy* every bit as breathtakingly different as all the things people willingly put in their mouths (and claim tastes good and is good for them)?

Well, no. What should strike you about "alternative mathematics"—unless you're blinded by an a priori style of foundationalism, where a specific style of mathematical proof (and logic), and a specific subject matter, are definitional *of* mathematics—is that such mathematics is *mathematics as usual*. One mark of the ordinariness of the stuff is that contemporary mathematicians shift in what they prove results about: they practice one or another branch of "classical" mathematics, and then try something more exotic—if the mood strikes. Proof, informal or formal, looks like the same thing (despite principles of proof being *severely* augmented or diminished in such approaches).

Schisms among mathematicians, prior to the late nineteenth century, prove even *shallower* than this.[6] That differences in methodology historically prove divisive can't be denied: differences in the methodology of the calculus, in England and on the continent, for example, *retarded* mathematical developments in England for over a century. Nevertheless, one finds British mathematicians (eventually) adopting the continental approach to the calculus, and doing so because they (eventually) recognized that the results they wanted, and more generally, the development of the mathematics surrounding the calculus, were easier given continental approaches. British mathematicians didn't deny the cogency of such results on the grounds that the methods that yielded them occurred in a "different (incomprehensible) tradition."

Let's turn to the second (*unnoticed*) way that mathematics *shockingly* differs from other group practices. *Mistakes are ubiquitous in mathematics*. I'm not *just* speaking of the mistakes of professional—even brilliant—mathematicians although, notoriously, they make *many* mistakes;[7] I'm speaking

[6] One point of this paper is to provide an explanation for why this should be so; see what's forthcoming, especially section VI.

[7] This is especially stressed in the "maverick tradition" to repeatedly hammer home the point that proof doesn't confer "certainty."

of *ordinary* people: they find mathematics *hard*—harder, in fact, than just about any other intellectual activity they attempt. What makes mathematics difficult is (1) that it's *so easy* to blunder in; and (2) that it's *so easy* for others (or oneself) to see—when they're pointed out—that blunders *have* been made.[8]

So? This is where it gets cute. When the factors forcing behavioral consensus are genuinely social, *mistakes can lead to new practices*. This is for two reasons at least: first, because the social factors imposing consensus are often blind to details about the behaviors enforced—they're better at imposing uniformity of behavior than at pinning down *exactly which* uniform behavior the population is to conform to. If enough people make a certain mistake, and if enough of them pass the mistake on, the social factors enforcing consensus continue doing so despite the shift in content. Social mechanisms that impose conformity are good at synchronic enforcement; they're not as good at diachronic enforcement. (Thus what's sometimes described as a "generation gap.")[9]

The second reason is that the power of social factors to enforce conformity often turns on the successful *psychological internalization* of social standards; but if such standards are imperfectly internalized (and *any* standard—however mechanical, i.e., algorithmic—can be imperfectly internalized), then the *social standards themselves* can evolve, since, in certain cases, nothing else fixes them. Two examples are, first, the drift in natural languages over time: this is often because of systematic mishearings by speakers, or interference phenomena (among internalized linguistic rules), so that certain locutions or sounds drop out (or arise). The second example is when an external standard supplementing psychological internalization of social standards is operative, and is *taken to* prevent drift—for example the holy books of a religious tradition. Notoriously, such things are open to *hermeneutical drift*: the subject population reinterprets them (often inadvertantly) because of changes in language, "common sense," and therefore changes in their (collective) view of what a given law-maker (e.g., God) *obviously* had in mind.

In short, although every social practice is easy to blunder in, it's not at all easy to get people to recognize or accept *that* they've made mistakes (and therefore, if enough of them do so, it's impossible—nearly enough—to *restore* the practice as opposed to—often inadvertantly—starting a new one).

The foregoing remarks about mistakes *aren't* meant to imply that *conscious* attempts to change traditions aren't effective: of course they can be (and

---

[8] What makes mathematics *hard* is both how easy it is to make mistakes and how difficult it is *to hide them*. Contrast this with *poetry*. It's as easy to make mistakes in poetry—write stunningly bad poetry—as it is to blunder in mathematics. But it's much easier to cover up poetic blunders. *Why* that is is extremely interesting, but something I can't fully get into now.

[9] Consider school uniforms. All sorts of contingent accidents cause mutations in such uniforms; but that (at a time) the uniforms should be, um, *uniform*, is a requirement.

often are). But mathematical practice resists *willful* (deliberate) change too. A dramatic case of a conscious attempt to change mathematical practice which failed (in large part because of incompetence at the standard fare) is Hobbes.[10] Another informative failure is Brouwer, because Brouwer was *anything but* incompetent at the standard practice.

Notice the point: Brouwer wasn't interested in developing *more* mathematics, nor were (and are) the other kinds of constructivists that followed; he wanted to change the *practice*, including his own earlier practice. But he only succeeded in developing *more* mathematics, not in changing *that* practice (as a whole). This makes Hilbert's response to Brouwer's challenge, by the way, *misguided*, because Hilbert's response was also predicated on (the fear of) Brouwer inducing a change in the practice. This is common: fads in mathematics often arise because someone (or a group, e.g., Bourbaki) thinks that some approach can become *the* tradition of mathematics—the result, invariably, is just *more* (additional) mathematics. A related (sociological) phenomenon is the mathematical *kook*—there are enough of these to write *books* about.[11] Only a field in which the recognition of mistakes is extremely robust can (sociologically speaking) successfully marginalize so many otherwise competent people *without* standard social forms of coercion, e.g., prison.

So (to recap.) mistakes in mathematics are common, and yet mathematical culture doesn't splinter because of them, or *for any other reason* (for that matter)[12]; that is, permanent *competing* practices don't arise as they can with other socially-constrained practices. This makes mathematics (sociologically speaking) *very odd*. Mathematical standards—here's another way to put the point—are robust. Mistakes *do* persevere, of course; but mostly they're eliminated, even when *repeatedly* made. More importantly, mathematical practice

---

It's very common for a population to slowly evolve its culinary practices, dress, accent, religious beliefs, etc., *without realizing that it's doing so*.

[10] See Jesseph 1999.

[11] For example, Dudley 1987.

[12] Philip Kitcher, during the discussion period on November 21, 2002, urged otherwise—not with respect to mistakes, but with respect to conscious disagreement on method: he invoked historical cases where mathematicians found themselves arrayed oppositely with respect to methodology—and the suggestion is that this led to schisms which lasted as long (comparatively speaking) as those found among, say, various sorts of religious believers: one thinks (again) of the controversy over the calculus, or the disputes over Cantor's work in the late nineteenth century. What's striking—when the dust settles, and historians look over the episodes—is how nicely a distinction may be drawn between a dispute in terms of proof procedures and one in terms of admissible concepts. The latter sort of dispute allows a (subsequent) consistent pooling of the results from the so-called disparate traditions; the former does not. Thus there is a sharp distinction between the (eventual) outcome of disputes over the calculus, and (some of) those over Cantor's work. The latter eventually flowered into a dispute over proof procedures which proved irresolvable in one sense (the results cannot be pooled) but not in another. See VI.

is so robust that even if a mistake eludes detection for years, and even if many results are built on that mistake, this *won't* provide enough social inertia—once the error *is* unearthed—to resist changing the practice back to what it was originally: in mathematics, even after lots of time, the subsequent mathematics built on the "falsehood" is repudiated.[13]

This aspect of mathematical practice has been (pretty much) unnoticed, or rather, *misdescribed*; and it's easy to see why. If one focuses on *other* epistemic issues, scepticism say, one can confuse the rigidity of group standards in mathematics with the availability of *certainty*: one can claim that, if only one is *sufficiently careful*, *really* attends to each step in a proof, carefully analyzes proofs so that each step immediately follows from earlier ones, dutifully surveys the whole repeatedly until it can be intuited in a flash, then one can rig it so that—in mathematics, at least—one won't *ever* make any mistakes *to begin with*: one can be totally *certain*.[14]

But there's a number of, er, mistakes in this Cartesian line. First, it's a robust part of mathematical practice that mistakes are found and corrected. Even though the practice is therefore *fixed* enough to rule out deviant practices that would otherwise result from allowing such "mistakes" to change that practice, this *won't* imply that *psychologically-based* certainty is within reach. For it's compatible with the robustness of our (collective) capacity to correct mathematical mistakes that some mistakes are still undetected—even old ones.

Apart from this, the psychological picture the Cartesian recipe for certainty presupposes is inaccurate. It's very hard to correct your own mistakes, as you know, having proofread your work in the past. *And* yet, someone else often sees *your* mistakes at a glance. This shows that the Cartesian project of gaining certainty *all alone*, a strategy crucial for Descartes' demon-driven epistemic program, is quixotic.[15]

Notice, however, that the Cartesian view would explain, *if it were only true*, how individuals can disagree on an answer, look over each other's work, and then come to agree *on what the error is*. (They become CERTAIN of THE TRUTH, and THE TRUTH is, after all, THE SAME.) Without this story, we

[13] Contrast this with our referential practices: Evans (1973, p. 11) mentions that (a corrupt form of) the term "Madagascar," applied to the African mainland, was mistakenly taken to apply to an island (indeed, *the* island we currently use the term to refer to). Our discovery of this error doesn't affect our current use of the term "Madagascar"—the social inertial of our current referential practice trumps any social mechanisms for correcting dated mistakes in that practice.

[14] And *then* one can make this an epistemic requirement on *all* knowledge (and offer recipes on how to carry it off). *Entire* philosophical traditions start this way.

[15] *One* of the ways Newton is *so* remarkable is that he did so much totally on his own, by *obsessively* going over his own work. (See Westfall 1980.) Newton's work is an impressive example of what heroic individualistic epistemic practice can *sometimes* look like. Despite this, Newton made *mistakes*.

need to know what practitioners have internalized (psychologically) to allow such an unnaturally agreeable social practice to arise.[16]

To summarize: What seems odd about mathematics as a social practice is the presence of substantial conformity on the one hand, and yet, on the other, the absence of (sometimes brutal) social tools to induce conformity that routinely appear among us *whenever* behavior really is socially constrained. Let's call this "the benign fixation of mathematical practice."

III

The benign fixation of mathematical practice *requires* an explanation. And (it should be said) Platonism is an appealing one: mathematical objects have their properties necessarily, and we perceive these properties (somehow). Keeping our (inner) eye firmly on mathematical objects keeps mathematical practice robust (enables us to find mistakes). The problem with this view—as the literature makes clear—is that we can't explain our epistemic access to the objects so posited.[17]

One might try to finesse things: demote Platonic objects to socially-constructed items (draw analogies between numbers and laws, language, banks, or Sherlock Holmes). Address the worry that socially-constructed nonmathematical objects like languages, Mickey Mouse, or laws, *evolve over time* (and that their properties look conventional or arbitrary) by invoking the content of mathematics (mathematical rules have content; linguistic rules are only a "semitransparent transmission medium" without content). And, claim that such content makes mathematical rules "necessary."[18]

---

[16] Unlike politics, for example, or any of the other numerous group activities we might consider, mathematical agreement isn't coerced. Individuals can see who's wrong; at least, if someone is stubborn, others (pretty much *all* the competent others) see it. Again, see Jesseph 1999 for the Hobbesian example. Also recall Leibniz's fond hope that this genial aspect of mathematical practice could be grafted onto other discourses, if we learned to "calculate together." By contrast, Protestantism, with all its numerous sects—in the United States especially—is what results when coercion isn't possible (because deviants can, say, move to Rhode Island). And much of the history of the Byzantine empire with its unpleasant treatment of "heretics" is the normal course of events when there's no Rhode Island to escape *to*. It's sociologically very surprising that conformity in mathematics isn't achieved as in *these* group practices. Imagine—here's a dark Wittgensteinian fable—we *tortured* numerical deviants to force them to add as we do. (Recall, for that matter, George Orwell's *1984*.)

[17] Current metaphysics robs Platonism of respectability. Judiciously sprinkle mysticism among your beliefs, and the perceptual analogy looks better; surreptitiously introduce deities to imprint *true* mathematical principles in our minds, and the approach also looks appealing. Explicitly deny all this, and Platonism looks *bizarre.*

[18] See Hersh 1997, p. 206. I deny that (certain) socially-constructed objects, mathematical objects and fictional objects, in particular, exist *in any sense at all.* See my 2004a. Nominalism, though, won't absolve me of the need to explain the benign fixation of mathematical practice. On the contrary.

As this stands, it won't work: we can't bless necessity upon whatever we'd like by chanting "content." Terms that refer to fictional objects have content too—that doesn't stop the properties attributed to such "things" from *evolving* over time; socially-constructed objects are *our* objects—if we take their properties to be fixed, that's something we've (collectively) imposed on them. It's a good question *why* we did this with mathematical terms, and not with other sorts of terms.

If socially-constructed objects are stiffened into "logical constructions" of some *fixed* logic plus set theory (say), this doesn't solve the problem: one still must explain how logic (of whatever sort) and set theory accrue social rigidity (why won't we let our set theory and logic *change*?).

There is no simple explanation for the benign fixation of mathematical practice because, as with any group practice, even if that practice retains its properties over time, that doesn't show that it has those properties (at different times) for the same reasons. Mathematical practice, despite its venerable association with *unchanging objects*, is an historical entity with a long pedigree, and so the reasons for why the correction of mistakes, for example, is robust in early mathematics, are *not* the same reasons for that robustness *now*.

IV

So now I'll discuss a number of factors, social and otherwise, and speculate how (and when) they contributed to the benign fixation of mathematical practice. The result, interestingly, is that if I'm right, benign fixation is historically contingent (and complex). That's a surprise, I suppose, for apriorists, but not for those of us who long ago thought of mathematics as something *humans* do over time.

Let's start (as it were) at the beginning: the historical emergence of mathematical practice (primarily counting and sums), and as that practice appears today among people with little or no other mathematics. Here it's appropriate to consider the role of "hard-wired" psychological dispositions. There seem two such relevant kinds of disposition. The first is a capacity to carry out algorithms, and—it's important to stress—this is a species-wide capacity: we can carry out algorithms, and teach each other to carry out (specific) algorithms *in the same way*. That's why we can play games *with* each other (as opposed to *past* each other), and why we can teach each other games that we play alone in the same way (e.g., one or another version of solitaire).

I can't say what it is about us—neurophysiologically I mean—that enables us to carry out algorithms the same way—no one can (yet); it's clear that some of us are better at some algorithms than others (think of games, and how our abilities to play them varies)—but what's striking is that those of us who are better aren't, by virtue of *that*, in *any* danger of being regarded as *doing something else*.

In describing us as able to "carry out" the same algorithms, I don't mean to say that we're *executing* the same algorithms (otherwise our abilities to carry out algorithms wouldn't *differ* in the so many *ways* that they do). I understand "executing an algorithm," as doing (roughly) what a Turing machine does when it operates. Perhaps humans do something like that with *some* algorithm(s) or other (but, surely, different humans execute different algorithms). In any case, when we learn arithmetic, for example, we're *not* learning to *execute* any of the numerous (but equivalent) algorithms that officially characterize arithmetic operations—instead what we're learning is what a particular algorithm *is*, and how to *imitate* its result—or at least *some* of its results—by actions of our own. So when I describe us as "carrying out" an algorithm **A**, I mean that we're imitating it by doing something else **B**, not by executing *it*.

I mean this. Add two numbers fifteen times, and you do *something different* each time—you do fifteen *different* things that (if you don't blunder) are the same *in the respect needed*: the sum you write down at the end of each process is the same (right) one. *We* can't do *anything* twice; it's only, as it were, *parts* of our behavior (at a time) that occur repeatedly. Proof? We *remember* much of what we do, and we're never but never *just* imitating a numerical algorithm when we do so; we're squirming in our chairs, taking in some of the passing scene (through our *ears*, if not through our eyes), etc. *Machines* execute algorithms and can do so by doing some things *twice*. We're (I hope this isn't news) *animals*.

Another way to make the same point about how we imitate rather than execute the algorithms that we're officially working on is that our learning such algorithms enjoys an interesting flexibility: we not only (apparently) acquire and learn new algorithms, but we can get better at the algorithms we've already learned by practicing them.

Finally, it should be noted that, usually, mathematicians don't execute the algorithms they're officially deriving results from; they *short-cut* them.[19]

These considerations—phenomenological ones, it's true—suggest that we don't—probably can't—execute the official algorithms we're carrying out; we're executing other algorithms instead that imitate the target algorithm (and over time, no doubt, different ones are used to do this); and this neatly explains why we can improve our abilities, by practice, to add sums, carry out other mathematical algorithms, and win games (for that matter).

Having said this, I must stress that I'm speculating about something that must ultimately be established empirically. So (of course) it could turn out that I'm just *wrong*, that we really do execute (some of) these algorithms (or, at least, some subpersonal part of us does), and that we don't imitate them

[19] Our ability to imitate algorithms flourishes into mathematical genius (in some individuals, anyway); for the mathematician, as I said, never (or almost never) figures out what an algorithm (proof procedure, say) will yield by executing that algorithm directly. Ordinary mathematical proof—its form, I mean—already shows this. See my forthcoming for more details.

via other algorithms that we execute. The neurophysiologists, in the end, will tell us what's what (if it's possible for *anyone* to tell us this, I mean): I'm betting, however, on my story—it explains our algorithmic flexibility, and our capacity to make and correct mistakes, in a way that a story that requires us to actually execute such algorithms doesn't.[20]

I'm also unable to say—because this too is ultimately a matter of neurophysiology—how *general* our capacity to mimic algorithms is; that we can now (since Turing and others) formulate in full generality the notion of mechanized practice—algorithm—doesn't mean that we have the *innate* capacity to "carry out"—imitate—the results of *any* such algorithm whatsoever. Our capacity to imitate algorithms may be, contrary to (introspective) appearance, more restricted than we realize.

A (species-wide) capacity to imitate the execution of algorithms *in the same way* doesn't explain the benign fixation of mathematical practice. This is because that robustness turns on *conserving* the official rules governing mathematical objects, and a group ability to imitate algorithms the same way won't explain why a practice *doesn't* evolve by *changing* the applicable algorithms altogether—in just the way that languages, which involve algorithms too, evolves.

A second innate capacity I'm willing to attribute to us is a disposition to *execute* certain *specific* algorithms.[21] I'm still thinking here primarily of our (primitive) ability to count and handle small sums. My suggestion is that why, wherever primitive numerical practices emerge, they're (pretty much) the same isn't because of sociological factors that constrain psychologically possible variants—rather, it's because of fixed *innate dispositions*.

Don't read too much into this second set of dispositions since they're also too weak to explain the benign fixation of mathematical practice: they don't extend far enough. They're *not* rules that apply to, say, any counting number whatsoever. These dispositions—I suspect—are very specific: they may facilitate handling certain small sums by visualizing them, or manipulating tokens in certain ways. I'm *not* claiming that such dispositions enable the execution of (certain) algorithms so that we can count as high as we like, add arbitrarily large sums, and so on.[22]

---

[20] The last four paragraphs respond to a line of questioning raised by David Albert; my thanks for this.

[21] Caveat: Given my earlier remarks about the empirical nature of my speculations about how we imitate the execution of algorithms, I'm not sure I've succeeded in describing two *distinct* capacities: Our ability to imitate algorithms—in general—needn't be a general ability to execute algorithms because, as I've said, we don't "carry out" algorithms by directly executing *them*. What we do, perhaps, is apply a quite specific algorithm or set of algorithms *to* official algorithms that we want to carry out, a process which enables us to extract (some) information about *any* algorithm (once we've psychologically couched it a certain way).

[22] Thus I *haven't* (entirely) deflected Kripkean attacks on the dispositionalist approach to rule-following. *And*: On my view, dispositions have only a *partial* role in the benign fixation of mathematical practice.

V

Such innate dispositions as I've described, although they explain why the independent emergence of counting and summing among various populations always turns out the same, won't explain why, when mathematics becomes professionalized—in particular, when *informal deduction* is hit upon by the ancient Greeks—benign fixation continues, rather than mathematical practice splintering.

I introduce something of a sociological idealization, which I'll call "mature mathematics," and which I'll describe as emerging somewhat before Euclid and continuing until the beginning of the twentieth century.[23] I claim that several factors conspire to benignly fix mature mathematical practice.

The first is that, *pretty much until the twentieth century*, mathematics came with intended *empirical domains of application* (from which mathematical concepts so applied largely arose). Arithmetic and geometry, in particular, come with *obviously* intended domains of application. These fixed domains of application prevent, to some extent, drift in the rules governing terms of mathematics—in these subjects so applied, anyway. This is because *successful application* makes us loathe to change successfully applied theorems—if that costs us applicability.[24]

But something more must be going on with *mathematics*, as a comparison with empirical science indicates. For the history of empirical science (physics, in particular) proves that drift *can* occur and yet the intended application of the concepts and theories not vanish as a result. Newtonian motion, *strictly speaking*, occurs only when objects *don't move*. But its *approximate* correctness suffices for successful application. Furthermore, the application of mathematics—geometry especially—always involves (some) approximation because of the nature of what geometric concepts are applied to (in particular, fuzzily-drawn figures).

---

[23] Twentieth-century mathematics isn't *mature*? Well, *of course it is*, but I'm arguing that it's different in important respects that require distinguishing it (sociologically, anyway) from what I've called "mature mathematics." Maybe—taking a nomenclatural tip from literature studies—we can call it "post-mature mathematics." On the other hand, maybe we better not. I'll contrast "mature" mathematics with "contemporary" mathematics.
This idealization is artificial because aspects of mathematical practice present in "mature" mathematics (they're in Euclid), and which continue to play an important role in contemporary mathematics, don't fit my official characterization of mature mathematics. I'll touch on this in due course.

[24] The ancient Greeks, it's pointed out more than once, were disdainful of "applied mathematics." Yes, but that disdain is compatible with what I've just written. The view, for example, that the empirical realm is a copy of the mathematical realm both determines the intended empirical domain in the sense I mean, *while simultaneously* demeaning the intellectual significance of that domain.

At work fixing mathematical practice beyond the drift allowed by successful (but approximate) application is a crucial factor, the essential role of *informal proof*, or deduction. It's no doubt debatable exactly what's involved in informal proof, but in mature mathematics it can be safely described as this:[25] a canonization of logical principles, and an (open-ended) set of additional (mathematical) principles and concepts which, (1) (partially) characterize subject matters with intended domains of application, (2) are more or less tractable insofar as we can, by means of them, informally prove new unanticipated results, and (3) which grow monotonically over time.

The need for tractable informal proofs drives the existential commitments of mathematics, and in particular, drives such commitments *away from* the objects characterized (empirically) in the intended domains of application. I've described this process in two case studies elsewhere, and won't dwell on it now.[26] But, the particular form mathematical posits take, itself now contributes several ways to benign fixation.

First, ordinary folk practices with empirical concepts allow those concepts to drift in what we can claim about them, and what they refer to, without our taking ourselves, as a result, to be referring to something new. If we discover that gold is actually blue, we describe that discovery in exactly those words (and not as a discovery that there is *no* gold).[27] By taking mathematical posits as empirically uninstantiated items, we detach mathematical language from this significant source of drift in what we take to be true of them.

Second, once mathematical posits are taken to be *real* but sensorily unavailable items which provide truths successfully applicable to empirical domains, mathematical practice opens itself to philosophical concerns both about the nature of such truths and how such truths *are established*. For a number of reasons—mostly involving various philosophical prejudices about *truth*[28]—the conventionalist view that mathematical truths are stipulated, and that mathematical objects exist in no sense at all, isn't seen as tenable (or even

---

[25] Recall, however, the second paragraph of footnote 23.

[26] See my 2004a and my 2004b. A discussion of the special qualities that a set of concepts, and principles governing them, needs for amenability to mathematical development—qualities that empirically derived notions, and truths about them, *usually don't have*—may be found in my 2000a.

[27] I'm alluding here to the sorts of thought experiments Putnam gave in his 1975. See my 2000b, especially Parts III and IV. There are subtleties and complications with this view of empirical terms, of course; but they don't affect points made in this paper.

[28] I have in mind claims like: (i) we can't have truths about things that don't exist, and (ii) even if we could, such truths wouldn't prove as empirically useful as mathematics is. Such prejudices are hardly restricted to ancient philosophers—e.g., Plato and Aristotle. They are standard fare among contemporaries *too*. See my 2004a for what things look like once we purge ourselves of them.

considered), and a view of eternal and unchanging mathematical objects carries the day instead.[29]

Of course, such an eternalist view of mathematical objects doesn't, all on its own, eliminate the possibility of a mathematical practice which allows drift in what we take to be true of mathematical objects: we could (in principle) still allow ourselves to be wrong about mathematical objects, and to be willing to change basic axioms governing them as a result. Imagine this thought experiment: a possible world much like ours, except that we discover nonEuclidean geometry centuries earlier, and due to the curvature of space in that world, its applicability is much more evident than in the actual world. In that world, we decide that Euclidean geometry is wrong; that is, we take ourselves to have been wrong about geometric *abstracta*—there are no abstracta that obey Euclidean axioms. This attitude is compatible with a view of mathematical abstracta as eternal, unchanging, etc. What prevented such a view from emerging among *us*, I claim, is the relative *late* discovery of non-Euclidean geometry (in the actual world). I touch on this later, but my view is that had (one or another) nonEuclidean geometry emerged in, say, ancient times, and had it been the case that Euclidean geometry proved useless in its intended domain of application (in comparison to nonEuclidean geometry), it would have been supplanted by nonEuclidean geometry—we would have taken ourselves to have been wrong about geometry and would have changed the basic axioms of what we called *Geometry* to suit.[30]

I've stressed how intended domains of application helped to benignly fix mathematical practice; the (implicit) canonization of logical principles is just as essential. Had there been shifts in the (implicit) logic, then we would have found ourselves—when considering early mathematics—in exactly the same

[29] Philosophical views about what positions are sensible or not *can't be ignored* in any sociological analysis of why a group practice develops as it does. There are *some*, no doubt, who take philosophical views as mere ideology, as advertising for other more substantial social motives (e.g., professional or class interests). I can't see how to take such a position seriously, especially if it's the sociology of knowledge-practices (of one sort or another) that's under study. What looked philosophically respectable, or not, I claim, had (and has) a profound impact on mathematical practice. It may be a mistake to search for that effect in the theorem-proving practices of the ordinary mathematician, but, in any case, as this paper illustrates, I locate it in, as it might be described, the general framework of how mathematics operates as a subject-matter (in particular, in how it's allowed to change over time).

[30] This would have happened, in part, because of an implicit metaphysical role for mathematical objects in the explanation for why that mathematics applied to its intended empirical domain—recall the resemblence doctrine mentioned in footnote 24. But in part I attribute the late emergence of nonEuclidean geometry *not* having a supplanting effect on Euclidean geometry as due to the already in place change of "mature" mathematics into "contemporary" mathematics, as I characterize the latter shortly. I guess I'm hypothesizing a "paradigm shift" although I don't much like this kind of talk. It seems that Kline (1980) is sensitive to some of the changes from mature to contemporary mathematics, although he takes a rather darker view of the shift than I do.

position that modern Greeks find themselves if they try to read ancient Greek on the basis of their knowledge of the contemporary stuff: incomprehension. In addition, shifts in the implicit logical principles utilized would have led to incompatible branchings in mathematical practice because of (irresolvable) disagreements about the implications of axioms and the validity of proofs.

VI

In order to motivate my discussion of how twentieth-century mathematics differs from what came before, I need to amplify my claim about the (tacit) canonization of logical principles in mature mathematics. Contemporary discussions of Frege's logicist program, and the *Principia* program that followed it, often dwell—quite melodramatically—on paradox; and the maverick animus towards such projects focuses on the set-theoretic foundationalism that's taken to have undergirded both the ontological concerns and the obsession with rigor proponents of such programs expressed.[31] But this focus obscures what those projects *really showed*: Nothing about the (real) subject-matter of mathematics (I rush to say), for *that's* proven to be elusive in any case—ways of embedding systems of mathematical posits in other systems is so unconstrained, ontologically speaking, that it's inspired structuralist views of that ontology.[32]

However, a very good case can be made that the *logic* of mature mathematics *was* something (more or less) equivalent to the first-order predicate logic, and that this was a nontrivial thing to have shown.[33] What *proves* this

---

[31] In describing the complex history this way, I'm not necessarily agreeing either with the depiction or with the attribution of these motives to later proponents of set-theoretical foundationalism.

[32] See, e.g., Resnik 1997.

[33] Why fix on *first-order* logic, and not a higher-order (classical) logic, especially since it was a higher-order logic that historically arose first? Well, there are a *number* of reasons; but the one most pertinent to the topic of this paper is that first-order logic and higher-order logics are nicely distinguishable because first-order logic is a *canonization* of reasoning principles without a subject matter, but the higher-order stuff (with standard semantics—where e.g., second order quantifiers range over *all* the subsets of the domain) is implicitly saddled with a set-theoretic subject-matter. See my 1994, Part I, § 3, where the point is made that second-order logic is equivalent to what are there called truncated first-order logics which require an additional logical constant which introduces set-theoretic resources. Notice the point: the intrusion of set-theoretic facts into the implication relation of higher-order logics constrains them in ways that families of axiom systems of first-order systems are not so constrained.
There are also grammatical considerations that suggest that the implicit logic of ordinary mathematics is first-order: the direct way that higher-order quantifiers quantify into predicate positions must be imitated in natural languages (at least in English) via nominalization. This suggests (again, at least in English) that predicates must be objectified, and containment relations stipulated between such objectified predicates and the objects the original predicates hold of. But such is the way of first-order idioms.

is that the project of characterizing (classical) mathematics axiomatically in first-order classical systems succeeded.[34] What shows it isn't a trivial point is that, in fact, much of twentieth-century mathematics *can't* be so axiomatically characterized.

What's *especially* striking about this success is that the classical logic which is the algorithmic skeleton behind informal proof remained tacit until its (late) nineteenth-century *uncovery* (I coin this word deliberately). But, as the study of ever-changing linguistic rules shows, implicit rules have a slippery way of mutating; in particular, what seems to be a general rule (at a time) can subsequently divide into a set of *domain-specific* rules, only *some* of which are retained.[35] The logical principles implicit in mathematical practice—until the twentieth century, however—remained the *same* topic-neutral ones (at least relative to mathematical subject-matters). Such uniformity of logical practice suggests, as does the uniformity of counting and summation practices I discussed earlier, a "hard-wired" disposition to reason in a particular way.[36]

This brings us to the points I want to make about twentieth-century mathematics. *Contemporary mathematics*, I claim, breaks away from the earlier practice in two extremely dramatic respects. First, it substitutes for classical logic (the tacit canon of logical principles operative in "mature" mathematics), proof procedures of *any* sort (of logic) *whatsoever* provided only that they admit of the (in principle) mechanical *recognition* of completely explicit proofs. That is, not only are alternative logics, and the mathematics based on such things, *now* part of contemporary mathematics; but various sorts of

---

[34] Hersh (1997) and other mavericks deny this but offer only the (weak) argument that the project hasn't been carried out *in detail* for *all* the mathematics it was supposed to apply to. But why is *that* needed? (The same grounds show, I suppose, that Gödel's second incompleteness theorem hasn't been shown *either*; and there are other examples *in mathematics* as well.) By the way, notice that it's irrelevant that the ordinary mathematician neither now, nor historically, couched any of his or her reasoning *in* such a formalism. This is because—as mentioned earlier—*nobody* carries out an algorithm by executing that algorithm—especially not gifted mathematicians who strategize proofs (and their descriptions) *routinely*, if they give proofs *at all*. Consider a similar argument that because ordinary speakers don't introspectively have access to the rules of natural language, they (of course) aren't implementing such rules (or recognizing the grammaticality of sentences by means of such rules).

[35] See, e.g., Anderson 1988, especially pp. 334-335.

[36] There is a complication that (potentially) mars this otherwise appealing view of the implicit role of first-order logic in mathematics: the "logic" of ordinary language looks much richer than what the first-order predicate calculus can handle—notoriously, projects of canonizing the logic of *anything other than* mathematics using (even enrichments of) the first-order predicate calculus have proved stunningly unsuccessful. This leaves us without a similar argument that the tacit logic of natural language is (something similar to) the first-order predicate calculus. But it would be very surprising if the tacit logic of mathematics were different from that of ordinary language—especially given the apparent topic-neutrality of that logic. I can't get further into this very puzzling issue now.

diagrammatic proof procedures are part of it as well; such (analogue) proof procedures, which involve conventionalized moves in the construction of diagrams, need not be proofs easily replicated in language-based axiomatic systems of any sort.[37]

One factor that accelerated the generalization of mathematical practice beyond the tacit classical logic employed up until the twentieth century was the explicit formalization of that very logic. For once (a version of) the logic in use was made explicit, mathematicians could *change it*. Why? because what's *conceptually central* to the notion of *formal proof*, and had been all along (as it had been operating in mature mathematics), isn't the presence of any particular *logic*, or logical axioms of some sort, but only the unarticulated idea that something "follows from" something else. This is neat: since (until the late nineteenth century) the logic *was* tacit, its particular principles couldn't have been seen as essential to mathematical proof since they weren't seen *at all*. What *was* seen clearly by mathematicians and fellow-travelers (recall footnote 16) was the benign fixation of mathematical practice; but *that's* preserved by generalizing proofs to anything algorithmically recognizable, *regardless* of the logic used.[38]

---

[37] There's lots more to say about diagrammatic proofs, but not now. I've now discharged the promissory note of the second paragraph of footnote 23, however: Diagrammatic proofs are in Euclid's elements (see my 2004b), and they continued to appear in mathematics during its entire mature phase—even though practices using them are only awkwardly canonized in a language-based theorem-proving picture of mathematical practice. The discussion of such items in contemporary mathematics is showing up in the literature on mathematical method. See, e.g., Brown 1999. I should stress again, however, that such practices require mechanical recognition of proofs; so they nicely fit within *my* (1994) characterization of mathematics as an interlocking system of algorithmic systems. I should also add that diagrammatic practices within classical mathematics are clearly compatible with the tacit standard logic used there—they provide consistent extensions of the axiomatic systems they accompany (or so I conjecture)—something not true of the more exotic items (e.g., logics) invented in the twentieth century.

[38] Haim Gaifman (November 21, 2002) has raised a challenge to the idea that contemporary mathematics can (genuinely) substitute algorithmic recognizability for the implicit logic of mature mathematics. For given the fact—aired previously—that mathematicians don't execute the actual algorithmic systems they prove results from, it must be that they rely on methods (modeling, adopting a metalanguage vantage point, etc.) which incorporate, or are likely to, the classical logic mathematicians naturally (implicitly) rely on. This suggests that if a mathematician were to attempt to *really* desert the classical context (and not merely avoid a principle or two—as intuitionists do), he or she would have to execute such algorithms mechanically—any other option would endanger the validity of the results (because the short-cuts used could presuppose inadmissible logical principles).

It may be that this is correct: an adoption of a seriously deviant nonclassical logic for (some) mathematics *requires* formalization. I'm simply not sure: none of the factors that distinguish mathematical proof from formal derivation that I discuss in my forthcoming seem to require any particular logic but my discussion there hardly exhausts such differences; and so I could easily be wrong about this.

The second way that contemporary mathematics bursts out from the previous practice is that it allows pure mathematics *such* a substantial life of its own that areas of mathematics can be explored and practiced without even a *hope* (as far as we can tell) of empirical application.[39] This, coupled with the generalization of mathematical proof to mechanical recognition procedures (of one sort or another) allows a *different* way to benignly fix mathematical practice. For now branches of mathematics can be individuated by families of algorithmic systems: by (tacit) stipulation, one doesn't *change* mathematical practice; *new* mathematics is created by the introduction of new algorithmic systems (i) with rules different from all the others, and (ii) which aren't augmentations of systems already in use. Should such an invention prove empirically applicable, and should it supplant some other (family of) system(s) previously applied to that domain, this doesn't cause a change in *mathematics*: the old family of systems is still mathematics, and is still something that can be profitably practiced (from the pure mathematical point of view). All that changes is the mathematics applied (and perhaps, the mathematics *funded*).

Notice that these reasons for the benign fixation of mathematical practice differ from those at work during mature mathematics. In particular, recall my thought experiment about the much earlier discovery of nonEuclidean geometry in a nearby possible world; its discovery *in our world*, given when it happened, *spurred on* the detachment of mathematics (as a practice) from intended domains of application; but that was hardly something that it *started*. Mathematical development had already started to explode (in complex analysis, especially)—but although intended domains of application were still exerting a strong impact on the direction of mathematical research, the introduction of mathematical concepts was no longer solely a matter of abstracting and idealizing empirical notions, as the notion of the square root of −1 makes clear all on its own. I claim (but this is something only historians of mathematics can evaluate the truth of) that this, coupled with a more sophisticated view of how mathematical posits could prove empirically valuable (not just by a "resemblance" to what they're applied to), and both of these coupled with the emergence of a confident mathematical profession not directly concerned with the application of said mathematics, allowed the birth of *mathematical liberalism*: the side-by-side noncompetitive existence of (logically incompatible) mathematical systems. And what a nice outcome *that* was!

[39] What about classical number theory? Well, I'm *not* claiming that "mature" mathematics didn't have subject matters, the exploration of (some) of which wasn't expected to yield empirical application; but *numbers* aren't the best counterexample to my claim since *they* were clearly perceived to have (intended) empirical applications. The contemporary *invention* and *exploration* of whole domains of abstracta without (any) empirical application *whatsoever* is a different matter. Consider, e.g., *most* of the explorations of set-theoretic exotica or (all of) degree theory. (None of this is to say, of course, that empirical applications can't arise *later*.)

References

Anderson, Stephen R. 1988. Morphological change. In *Linguistics: the Cambridge survey, Vol.I: Linguistic theory: foundations*. Cambridge: Cambridge University Press, 324–362.

Azzouni, Jody 1994. *Metaphysical myths, mathematical practice: the ontology and epistemology of the exact sciences*. Cambridge: Cambridge University Press.

—— 2000a. Applying mathematics: an attempt to design a philosophical problem, *The Monist* 83(2):209–227.

—— 2000b. *Knowledge and reference in empirical science*. London: Routledge.

—— 2004a. *Deflating existential commitment: a case for nominalism*. Oxford: Oxford University Press.

—— 2004b. Proof and ontology in mathematics. In (eds. Tinne Hoff Kjeldsen, Stig Andur Pedersen, Lise Mariane Sonne-Hansen) *New Trends in the History and Philosophy of Mathematics*. Denmark: University Press of Southern Denmark, pp. 117–133.

—— forthcoming. The derivation-indicator view of mathematical practice. *Philosophia Mathematica*.

Bloor, David 1983. *Wittgenstein: a social theory of knowledge*. New York: Columbia University Press.

Brown, James Robert 1999. *Philosophy of mathematics: an introduction to the world of proofs and pictures*. London: Routledge.

Descartes, René 1931a. Rules for the direction of the mind. In *The philosophical works of Descartes*, trans. by Elizabeth S. Haldane and G.R.T.Ross (pp. 1–77). Cambridge: Cambridge University Press.

—— 1931b. Meditations on first philosophy. In *The philosophical works of Descartes*, trans. by Elizabeth S. Haldane and G.R.T.Ross (pp. 131–99). Cambridge: Cambridge University Press.

Dudley, Underwood 1987. *A Budget of trisections*. New York: Springer-Verlag.

Evans, Gareth 1973. The causal theory of names. In *Collected papers*, 1-24. Oxford: Oxford University Press (1985).

Hersh, Reuben 1997. *What is mathematics, really?* Oxford: Oxford University Press.

Jesseph, Douglas M. 1999. *Squaring the circle: the war between Hobbes and Wallis* Chicago: University of Chicago Press.

Kline, Morris 1980. *Mathematics: the loss of certainty*. Oxford: Oxford University Press.

Kripke, Saul A. 1982. *Wittgenstein on rules and private language*. Cambridge, Massachusetts: Harvard University Press.

Orwell, George 1949. *1984*. New York: Harcourt, Brace and Company, Inc.

Putnam, Hilary 1975. The meaning of 'meaning'. In *Mind, language and reality: philosophical papers* (Vol. 2, pp. 139–52). Cambridge: Cambridge University Press.

Resnik, Michael D. 1997. *Mathematics as a science of patterns*. Oxford: Oxford University Press.

Westfall, Richard S. 1980. *Never at rest: a biography of Isaac Newton*. Cambridge: Cambridge University Press.

Wittgenstein, Ludwig 1953. *Philosophical investigations*. Trans. G.E.M. Anscombe. New York: The Macmillan Company.

12

The Pernicious Influence of Mathematics upon Philosophy

GIAN-CARLO ROTA

The Double Life of Mathematics

ARE MATHEMATICAL IDEAS INVENTED or discovered? This question has been repeatedly posed by philosophers through the ages and will probably be with us forever. We will not be concerned with the answer. What matters is that by asking the question, we acknowledge that in the first of its lives mathematics deals with facts, like any other science. It is a fact that the altitudes of a triangle meet at a point; it is a fact that there are only seventeen kinds of symmetry in the plane; it is a fact that there are only five non-linear differential equations with fixed singularities; it is a fact that every finite group of odd order is solvable. The work of a mathematician consists of dealing with such facts in various ways. When mathematicians talk to each other, they tell the facts of mathematics. In their research, mathematicians study the facts of mathematics with a taxonomic zeal similar to a botanist studying the properties of some rare plant.

The facts of mathematics are as useful as the facts of any other science. No matter how abstruse they may first seem, sooner or later they find their way back to practical applications. The facts of group theory, for example, may appear abstract and remote, but the practical applications of group theory have been numerous, and have occurred in ways that no one could have anticipated. The facts of today's mathematics are the springboard for the science of tomorrow.

In its second life, mathematics deals with proofs. A mathematical theory begins with definitions and derives its results from dearly agreed-upon rules of inference. Every fact of mathematics must be ensconced in an axiomatic theory and formally proved if it is to be accepted as true. Axiomatic exposition is indispensable in mathematics because the facts of mathematics, unlike the facts of physics, are not amenable to experimental verification.

The axiomatic method of mathematics is one of the great achievements of our culture. However, it is only a method. Whereas the facts of mathematics once discovered will never change, the method by which these facts are verified has changed many times in the past, and it would be foolhardy to expect that changes will not occur again at some future date.

The Double Life of Philosophy

The success of mathematics in leading a double life has long been the envy of philosophy, another field which also is blessed- or maybe we should say cursed - to live in two worlds but which has not been quite as comfortable with its double life.

In the first of its lives, philosophy sets itself the task of telling us how to look at the world. Philosophy is effective at correcting and redirecting our thinking, helping us do away with glaring prejudices and unwarranted assumptions. Philosophy lays bare contradictions that we would rather avoid facing. Philosophical descriptions make us aware of phenomena that lie at the other end of the spectrum of rationality that science will not and cannot deal with.

The assertions of philosophy are less reliable than the assertions of mathematics but they run deeper into the roots of our existence. Philosophical assertions of today will be the common sense of tomorrow.

In its second life, philosophy, like mathematics, relies on a method of argumentation that seems to follow the rules of some logic. But the method of philosophical reasoning, unlike the method of mathematical reasoning, has never been clearly agreed upon by philosophers, and much philosophical discussion since its Greek beginnings has been spent on method. Philosophy's relationship with Goddess Reason is closer to a forced cohabitation than to the romantic liaison which has always existed between Goddess Reason and mathematics.

The assertions of philosophy are tentative and partial. It is not even clear what it is that philosophy deals with. It used to be said that philosophy was "purely speculative," and this used to be an expression of praise. But lately the word "speculative" has become a *bad word.*

Philosophical arguments are emotion-laden to a greater degree than mathematical arguments and written in a style more reminiscent of a shameful admission than of a dispassionate description. Behind every question of philosophy there lurks a gnarl of unacknowledged emotional cravings which act as a powerful motivation for conclusions in which reason plays at best a supporting role. To bring such hidden emotional cravings out into the open, as philosophers have felt it their duty to do, is to ask for trouble. Philosophical disclosures are frequently met with the anger that we reserve for the betrayal of our family secrets.

This confused state of affairs makes philosophical reasoning more difficult but far more rewarding. Although philosophical arguments are blended with emotion, although philosophy seldom reaches a firm conclusion, although the method of philosophy has never been clearly agreed upon, nonetheless the assertions of philosophy, tentative and partial as they are, come far closer to the truth of our existence than the proofs of mathematics.

The Loss of Autonomy

Philosophers of all times, beginning with Thales and Socrates, have suffered from recurring suspicions about the soundness of their work and have responded to them as well as they could.

The latest reaction against the criticism of philosophy began around the turn of the twentieth century and is still very much with us.

Today's philosophers (not all of them) have become great believers in mathematization. They have recast Galileo's famous sentence to read, "The great book of philosophy is written in the language of mathematics."

"Mathematics calls attention to itself," wrote Jack Schwartz in a famous paper on another kind of misunderstanding. Philosophers in this century have suffered more than ever from the dictatorship of definitiveness. The illusion of the final answer, what two thousand years of Western philosophy failed to accomplish, was thought in this century to have come at last within reach by the slavish imitation of mathematics.

Mathematizing philosophers have claimed that philosophy should be made factual and precise. They have given guidelines based upon mathematical logic to philosophical argument. Their contention is that the eternal riddles of philosophy can be definitively solved by pure reasoning, unencumbered by the weight of history. Confident in their faith in the power of pure thought, they have cut all ties to the past, claiming that the messages of past philosophers are now "obsolete."

Mathematizing philosophers will agree that traditional philosophical reasoning is radically different from mathematical reasoning. But this difference, rather than being viewed as strong evidence for the heterogeneity of philosophy and mathematics, is taken as a reason for doing away completely with non-mathematical philosophy.

In one area of philosophy the program of mathematization has succeeded. Logic is nowadays no longer part of philosophy. Under the name of mathematical logic it is now a successful and respected branch of mathematics, one that has found substantial practical applications in computer science, more than any other branch of mathematics.

But logic has become mathematical at a price. Mathematical logic has given up all claims of providing a foundation to mathematics. Very few logicians of our day believe that mathematical logic has anything to do with the way we think. Mathematicians are therefore mystified by the spectacle of philosophers pretending to re-inject philosophical sense into the language of mathematical logic. A hygienic cleansing of every trace of philosophical reference had been the price of admission of logic into the mathematical fold. Mathematical logic is now just another branch of mathematics, like topology and probability. The philosophical aspects of mathematical logic are qualitatively no different from the philosophical aspects of topology or the theory of functions, aside from a curious terminology which, by chance, goes back to the Middle Ages.

The fake philosophical terminology of mathematical logic has mis led philosophers into believing that mathematical logic deals with the truth in the philosophical sense. But this is a mistake. Mathematical logic deals not with the truth but only with the game of truth. The snobbish symbol-dropping found nowadays in philosophical papers raises eyebrows among mathematicians, like someone paying his grocery bill with Monopoly money.

Mathematics and Philosophy: Success and Failure

By all accounts mathematics is mankind's most successful intellectual undertaking. Every problem of mathematics gets solved, sooner or later. Once solved, a mathematical problem is forever finished: no later event will disprove a correct solution. As mathematics progresses, problems that were difficult become easy and can be assigned to schoolchildren. Thus Euclidean geometry is taught in the second year of high school. Similarly, the mathematics learned by my generation in graduate school is now taught at the undergraduate level, and perhaps in the not too distant future, in the high schools.

Not only is every mathematical problem solved, but eventually every mathematical problem is proved trivial. The quest for ultimate triviality is characteristic of the mathematical enterprise.

Another picture emerges when we look at the problems of philosophy. Philosophy can be described as the study of a few problems whose statements have changed little since the Greeks: the mind-body problem and the problem of reality, to mention only two. A dispassionate look at the history of philosophy discloses two contradictory features: first, these problems have in no way been solved, nor are they likely to be solved as long as philosophy survives; and second, every philosopher who has ever worked on any of these problems has proposed his own "definitive solution," which has invariably been rejected by his successors.

Such crushing historical evidence forces us to conclude that these two paradoxical features must be an inescapable concomitant of the philosophical enterprise. Failure to conclude has been an outstanding characteristic of philosophy throughout its history.

Philosophers of the past have repeatedly stressed the essential role of failure in philosophy. José Ortega y Gasset used to describe philosophy as "a constant shipwreck." However, fear of failure did not stop him or any other philosopher from doing philosophy.

The failure of philosophers to reach any kind of agreement does not make their writings any less relevant to the problems of our day. We reread with interest the mutually contradictory theories of mind that Plato, Aristotle, Kant and Comte have bequeathed to us, and find their opinions timely and enlightening, even in problems of artificial intelligence.

But the latter day mathematizers of philosophy are unable to face up to the inevitability of failure. Borrowing from the world of business, they have embraced the ideal of success. Philosophy had better be successful, or else it should be given up.

The Myth of Precision

Since mathematical concepts are precise and since mathematics has been successful, our darling philosophers infer - mistakenly that philosophy would be better off, that is, would have a better chance of being successful, if it utilized precise concepts and unequivocal statements.

The prejudice that a concept must be precisely defined in order to be meaningful, or that an argument must be precisely stated in order to make sense, is one of the most insidious of the twentieth century. The best known expression of this prejudice appears at the end of Ludwig Wittgenstein's *Tractatus.* The author's later writings, in particular *Philosophical Investigations,* are a loud and repeated retraction of his earlier gaffe.

Looked at from the vantage point of ordinary experience, the ideal of precision seems preposterous. Our everyday reasoning is not precise, yet it is effective. Nature itself, from the cosmos to the gene, is approximate and inaccurate.

The concepts of philosophy are among the least precise. The mind, perception, memory, cognition are words that do not have any fixed or clear meaning. Yet they do have meaning. We misunderstand these concepts when we force them to be precise. To use an image due to Wittgenstein, philosophical concepts are like the winding streets of an old city, which we must accept as they are, and which we must familiarize ourselves with by strolling through them while admiring their historical heritage. Like a Carpathian dictator, the advocates of precision would raze the city and replace it with the straight and wide Avenue of Precision.

The ideal of precision in philosophy has its roots in a misunderstanding of the notion of rigor. It has not occurred to our mathematizing philosophers that philosophy might be endowed with its own kind of rigor, a rigor that philosophers should dispassionately describe and codify, as mathematicians did with their own kind of rigor a long time ago. Bewitched as they are by the success of mathematics, they remain enslaved by the prejudice that the only possible rigor is that of mathematics and that philosophy has no choice but to imitate it.

Misunderstanding the Axiomatic Method

The facts of mathematics are verified and presented by the axiomatic method. One must guard, however, against confusing the presentation of mathematics with the *content* of mathematics. An axiomatic presentation of a mathematical fact differs from the fact that is being presented as medicine differs from food. It is true that this particular medicine is necessary to keep the mathematician at a safe distance from the self-delusions of the mind. Nonetheless, understanding mathematics means being able to forget the medicine and enjoy the food. Confusing mathematics with the axiomatic method for its presentation is as preposterous as confusing the music of Johann Sebastian Bach with the techniques for counterpoint in the Baroque age.

This is not, however, the opinion held by our mathematizing philosophers. They are convinced that the axiomatic method is a basic instrument of discovery. They mistakenly believe that mathematicians use the axiomatic method in solving problems and proving theorems. To the misunderstanding of the role of the method they add the absurd pretense that this presumed method should be adopted in philosophy. Systematically confusing food with medicine, they pretend to replace the food of philosophical thought with the medicine of axiomatics.

This mistake betrays the philosophers' pessimistic view of their own field. Unable or afraid as they are of singling out, describing and analyzing the structure of philosophical reasoning, they seek help from the proven technique of another field, a field that is the object of their envy and veneration. Secretly disbelieving in the power of autonomous philosophical reasoning to discover truth, they surrender to a slavish and superficial imitation of the truth of mathematics.

The negative opinion that many philosophers hold of their own field has caused damage to philosophy. The mathematician's contempt for the philosopher's exaggerated estimation of a method of mathematical exposition feeds back onto the philosopher's inferiority complex and further decreases the philosopher's confidence.

"Define Your Terms!"

This old injunction has become a platitude in everyday discussions. What could be healthier than a clear statement right at the start of what it is that we are talking about? Doesn't mathematics begin with definitions and then develop the properties of the objects that have been defined by an admirable and infallible logic?

Salutary as this injunction may be in mathematics, it has had disastrous consequences when carried over to philosophy. Whereas mathematics *starts* with a definition, philosophy *ends* with a definition. A clear statement of what it is we are talking about is not only missing in philosophy, such a statement would be the instant end of all philosophy. If we could define our terms, then we would gladly dispense with philosophical argument.

The "define your terms" imperative is flawed in more than one way. When reading a formal mathematical argument we are given to believe that the "undefined terms," or the "basic definitions," have been whimsically chosen out of a variety of possibilities. Mathematicians take mischievous pleasure in faking the arbitrariness of definition. In fact no mathematical definition is arbitrary. The theorems of mathematics motivate the definitions as much as the definitions motivate the theorems. A good definition is "justified" by the theorems that can be proved with it, just as the proof of the theorem is "justified" by appealing to a previously given definition.

There is, thus, a hidden circularity in formal mathematical exposition. The theorems are proved starting with definitions; but the definitions themselves are motivated by the theorems that we have previously decided ought to be correct.

Instead of focusing on this strange circularity, philosophers have pretended it does not exist, as if the axiomatic method, proceeding linearly from definition to theorem, were endowed with definitiveness. This is, as every mathematician knows, a subtle fakery to be debunked.

Perform the following thought experiment. Suppose you are given two formal presentations of the same mathematical theory. The definitions of the first presentation are the theorems of the second, and vice versa. This situation frequently occurs in mathematics. Which of the two presentations makes the theory "true?" Neither, evidently: what we have are two presentations of the *same* theory.

This thought experiment shows that mathematical truth is not brought into being by a formal presentation; instead, formal presentation is only a technique for displaying mathematical truth. The truth of a mathematical theory is distinct from the correctness of any axiomatic method that may be chosen for the presentation of the theory.

Mathematizing philosophers have missed this distinction.

The Appeal to Psychology

What will happen to the philosopher who insists on precise statements and clear definitions? Realizing after futile trials that philosophy resists such a treatment, the philosopher will proclaim that most problems previously thought to belong to philosophy are henceforth to be excluded from consideration. He will claim that they are "meaningless," or at best, can be settled by an analysis of their statements that will eventually show them to be vacuous.

This is not an exaggeration. The classical problems of philosophy have become forbidden topics in many philosophy departments. The mere mention of one such problem by a graduate student or by a junior colleague will result in raised eyebrows followed by severe penalties. In this dictatorial regime we have witnessed the shrinking of philosophical activity to an impoverished *problématique,* mainly dealing with language.

In order to justify their neglect of most of the old and substantial questions of philosophy, our mathematizing philosophers have resorted to the ruse of claiming that many questions formerly thought to be philosophical are instead "purely psychological" and that they should be dealt with in the psychology department.

If the psychology department of any university were to consider only one tenth of the problems that philosophers are palming off on them, then psychology would without question be the most fascinating of all subjects. Maybe it is. But the fact is that psychologists have no intention of dealing with problems abandoned by philosophers who have been derelict in their duties.

One cannot do away with problems by decree. The classical problems of philosophy are now coming back with a vengeance in the forefront of science.

Experimental psychology, neurophysiology and computer science may turn out to be the best friends of traditional philosophy. The awesome complexities of the phenomena that are being studied in these sciences have convinced scientists (well in advance of the philosophical establishment) that progress in science will depend on philosophical research in the most classical vein.

The Reductionist Concept of Mind

What does a mathematician do when working on a mathematical problem? An adequate description of the project of solving a mathematical problem might require a thick volume. We will be content with recalling an old saying, probably going back to the mathematician George Polya: "Few mathematical problems are ever solved directly."

Every mathematician will agree that an important step in solving a mathematical problem, perhaps *the* most important step, consists of analyzing other attempts, either those attempts that have been previously carried out or attempts that he imagines might have been carried out, with a view to discovering how such "previous" approaches failed. In short, no mathematician will ever dream of attacking a substantial mathematical problem without first becoming acquainted with the *history* of the problem, be it the real history or an ideal history reconstructed by the gifted mathematician. The solution of a mathematical problem goes hand in hand with the discovery of the inadequacy ofprevious attempts, with the enthusiasm that sees through and gradually does away with layers of irrelevancies which formerly clouded the real nature of the problem. In philosophical terms, a mathematician who solves a problem cannot avoid facing up to the historicity of the problem.

Mathematics is nothing if not a historical subject *par excellence,* Every philosopher since Heraclitus with striking uniformity has stressed the lesson that all thought is constitutively historical. Until, that is, our mathematizing philosophers came along, claiming that the mind is nothing but a complex thinking machine, not to be polluted by the inconclusive ramblings of bygone ages. Historical thought was dealt a *coup de grace* by those who today occupy some of the chairs of our philosophy departments. Graduate school requirements in the history of philosophy were dropped, together with language requirements, and in their place we find required courses in mathematical logic. It is important to uncover the myth that underlies such drastic revision of the concept of mind, that is, the myth that the mind is some sort of mechanical device. This myth has been repeatedly and successfully attacked by the best philosophers of our century (Husserl, John Dewey, Wittgenstein, Austin, Ryle, Croce, to name a few).

According to this myth, the process of reasoning functions like a vending machine which, by setting into motion a complex mechanism reminiscent of Charlie Chaplin's *Modern Times,* grinds out solutions to problems. Believers in the theory of the mind as a vending machine will rate human beings by "degrees" of intelligence, the more intelligent ones being those endowed with bigger and better gears in their brains, as may of course be verified by administering cleverly devised 1. Q. tests. Philosophers believing in the mechanistic myth assert that the solution of a problem is arrived at in just one way: by thinking hard about it. They will go so far as to assert that acquaintance with previous contributions to a problem may bias the well-geared mind. A blank mind, they insist, is better geared to complete the solution process than an informed mind.

This outrageous proposition originates from a misconception of the working habits of mathematicians. Our mathematizing philosophers are failed mathematicians. They gape at the spectacle of mathematicians at work in wide-eyed admiration. To them, mathematicians are superminds who spew out solutions of one problem after another by dint of pure brain power, simply by staring long enough at a blank piece of paper.

The myth of the vending machine that grinds out solutions may appropriately describe the way to solve the linguistic puzzles of today's impoverished philosophy, but this myth is wide of the mark in describing the work of mathematicians, or any kind of serious work.

The fundamental error is an instance of reductionism. The process by which the mind works, which may be of interest to physicians but is of no help to working mathematicians, is confused with the progress of thought that is required in the solution of any problem. This catastrophic misunderstanding of the concept of mind is the heritage of one hundred-odd years of pseudo-mathematization of philosophy.

The Illusion of Definitiveness

The results of mathematics are definitive. No one will ever improve on a sorting algorithm which has been proved best possible. No one will ever discover a new finite simple group, now that the list has been drawn after a century of research. Mathematics is forever.

We could order the sciences by how close their results come to being definitive. At the top of the list we would find sciences of lesser philosophical interest, such as mechanics, organic chemistry, botany. At the bottom of the list we would find more philosophically inclined sciences such as cosmology and evolutionary biology.

The problems of philosophy, such as mind and matter, reality, perception, are the least likely to have "solutions." We would be hard put to spell out what kind of argument might be acceptable as a "solution to a problem of philosophy." The idea of a "solution" is borrowed from mathematics and tacitly presupposes an analogy between problems of philosophy and problems of science that is fatally misleading.

Philosophers of our day go one step further in their mis-analogies between philosophy and mathematics. Driven by a misplaced belief in definitiveness measured in terms of problems solved, and realizing the futility of any program that promises definitive solutions, they have been compelled to get rid of all classical problems. And where do they think they have found problems worthy of them? Why, in the world of facts!

Science deals with facts. Whatever traditional philosophy deals with, it is not facts in any known sense of the word. Therefore, traditional philosophy is meaningless.

This syllogism, wrong on several counts, is predicated on the assumption that no statement is of any value unless it is a statement of fact. Instead of realizing the absurdity of this vulgar assumption, philosophers have swallowed it, hook, line and sinker, and have busied themselves in making their living on facts.

But philosophy has never been equipped to deal directly with facts, and no classical philosopher has ever considered facts to be any of his business. Nobody will ever turn to philosophy to learn facts. Facts are the business of science, not of philosophy.

And so, a new slogan had to be coined: philosophy *should* be dealing with facts.

This "should" comes at the end ofa long normative line of "shoulds." Philosophy should be precise; it should follow the rules of mathematical logic; it should define its terms carefully; it should ignore the lessons of the past; it should be successful at solving its problems; it should produce definitive solutions.

"Pigs should fly," the old saying goes.

But what is the standing of such "shoulds," flatly negated as they are by two thousand years of philosophy? Are we to believe the not so subtle insinuation that the royal road to right reasoning will at last be ours if we follow these imperatives?

There is a more plausible explanation of this barrage of "shoulds." The reality we live in is constituted by a myriad contradictions, which traditional philosophy has taken pains to describe with courageous realism. But contradiction cannot be confronted by minds who have put all their eggs in the basket of precision and definitiveness. The real world is filled with absences, absurdities, abnormalities, aberrances, abominations, abuses, with *Abgrund.* But our latter-day philosophers are not concerned with facing up to these discomforting features of the world, nor to any relevant features whatsoever. They would rather tell us what the world *should* be like. They find it safer to escape from distasteful description of what is into pointless prescription of what isn't. Like ostriches with their heads buried in the sand, they will meet the fate of those who refuse to remember the past and fail to face the challenges of our difficult present: increasing irrelevance followed by eventual extinction.

NOTES

1. J. T. Schwartz, "The Pernicious Influence of Mathematics Upon Science" in Mark Kac, Gian-Carlo Rota, J. T. Schwartz, "Discrete Thoughts. Essays on Mathematics, Science, and Philosophy," Birkhauser Boston, 1992, Chapter 3
2. Ludwig Wittgenstein "Tractatus logico-philosophicus," Kegan, Trench, Trubner & Co. Ltd., London, 1922; Proposition 7.00
3. Ludwig Wittgenstein, "Philosophische Untersuchungen," Basil Blackwell, Oxford, 1953

13

The Pernicious Influence of Mathematics on Science

JACK SCHWARTZ

I wish to confine myself to the negative aspects, leaving it to others to dwell on the amazing triumphs of the mathematical method; and also to comment not only on physical science but also on social science, in which the characteristic inadequacies which I wish to discuss are more readily apparent.

Computer programmers often make a certain remark about computing machines, which may perhaps be taken as a complaint: that computing machines, with a perfect lack of discrimination, will do any foolish thing they are told to do. The reason for this lies of course in the narrow fixation of the computing machine "intelligence" upon the basely typographical details of its own perceptions–its inability to be guided by any large context. In a psychological description of the computer intelligence, three related adjectives push themselves forward: single-mindedness, literal-mindedness, simple-mindedness. Recognizing this, we should at the same time recognize that this single-mindedness, literal-mindedness, simple-mindedness also characterizes theoretical mathematics, though to a lesser extent.

It is a continual result of the fact that science tries to deal with reality that even the most precise sciences normally work with more or less ill-understood approximations toward which the scientist must maintain an appropriate skepticism. Thus, for instance, it may come as a shock to the mathematician to learn that the Schrodinger equation for the hydrogen atom, which he is able to solve only after a considerable effort of functional analysis and special function theory, is not a literally correct description of this atom, but only an approximation to a somewhat more correct equation taking account of spin, magnetic dipole, and relativistic effects; that this corrected equation is itself only an ill-understood approximation to an infinite set of quantum field-theoretical equations; and finally that the quantum field theory, besides diverging, neglects a myriad of strange-particle interactions whose strength and form are largely unknown. The physicist, looking at the original Schrodinger equation, learns to sense in it the presence of many invisible terms, integral, integrodifferential, perhaps even more complicated types of operators, in addition to the differential terms visible, and this sense inspires an entirely appropriate disregard for the purely technical features of the equation which he sees. This very healthy self-skepticism is foreign to the mathematical approach.

Mathematics must deal with well-defined situations. Thus, in its relations with science mathematics depends on an intellectual effort outside of mathematics for the crucial specification of the approximation which mathematics is to take literally. Give a mathematician a situation which is the least bit ill-defined–he will first of all make it well defined. Perhaps appropriately, but perhaps also inappropriately. The hydrogen atom illustrates this process nicely. The physicist asks: "What are the eigenfunctions of such-and-such a differential operator?" The mathematician replies: "The question as put is not well defined. First you must specify the linear space in which you wish to operate, then the precise domain of the operator as a subspace. Carrying all this out in the simplest way, we find the following result..." Whereupon the physicist may answer, much to the mathematician's chagrin: "Incidentally, I am not so much interested in the operator you have just analyzed as in the following operator, which has four or five additional small terms–how different is the analysis of this modified problem?" In the case just cited, one may perhaps consider that nothing much is lost, nothing at any rate but the vigor and wide sweep of the physicist's less formal attack. But, in other cases, the mathematician's habit of making definite his literal-mindedness may have more unfortunate consequences. The mathematician turns the scientist's theoretical assumptions, i.e., convenient points of analytical emphasis, into axioms, and then takes these axioms literally. This brings with it the danger that he may also persuade the scientist to take these axioms literally. The question, central to the scientific investigation but intensely disturbing in the mathematical context–what happens to all this if the axioms are relaxed–is thereby put into shadow.

In this way, mathematics has often succeeded in proving, for instance, that the fundamental objects of the scientist's calculations do not exist. The sorry history of the Dirac Delta function should teach us the pitfalls of rigor. Used repeatedly by Heaviside in the last century, used constantly and systematically by physicists since the 1920's, this function remained for mathematicians a monstrosity and an amusing example of the physicists' naiveté until it was realized that the Dirac Delta function was not literally a function but a generalized function. It is not hard to surmise that this history will be repeated for many of the notions of mathematical physics which are currently regarded as mathematically questionable. The physicist rightly dreads precise argument, since an argument which is only convincing if precise loses all its force if the assumptions upon which it is based are slightly changed, while an argument which is convincing though imprecise may well be stable under small perturbations of its underlying axioms.

The literal-mindedness of mathematics thus makes it essential, if mathematics is to be appropriately used in science, that the assumptions upon which mathematics is to elaborate be correctly chosen from a larger point of view, invisible to mathematics itself. The single-mindedness of mathematics reinforces this conclusion. Mathematics is able to deal successfully only with the simplest of situations, more precisely, with a complex situation only to the extent that rare good fortune makes this complex situation hinge upon a few dominant simple factors. Beyond the well-traversed path, mathematics loses its bearings in a jungle of unnamed special functions and impenetrable combinatorial particularities. Thus, the mathematical technique can only reach far if it starts from a point close to the simple essentials of a problem which has simple essentials. That form of wisdom which is the opposite of single-mindedness, the ability to keep many threads in hand, to draw for an argument from many disparate sources, is quite foreign to mathematics. This inability accounts for much of the difficulty which mathematics experiences in attempting to penetrate the social sciences. We may perhaps attempt a mathematical economics–but how difficult would be a mathematical history! Mathematics adjusts only with reluctance to the external, and vitally necessary, approximating of the scientists, and shudders each time a batch of small terms is cavalierly erased. Only with difficulty does it find its way to the scientist's ready grasp of the relative importance of many factors. Quite typically, science leaps ahead and mathematics plods behind.

Related to this deficiency of mathematics, and perhaps more productive of rueful consequence, is the simple-mindedness of mathematics–its willingness, like that of a computing machine, to elaborate upon any idea, however absurd; to dress scientific brilliancies and scientific absurdities alike in the impressive uniform of formulae and theorems. Unfortunately however, an absurdity in uniform is far more persuasive than an absurdity unclad. The very fact that a theory appears in mathematical form, that, for instance, a theory has provided the occasion for the application of a fixed-point theorem, or of a result about difference equations, somehow makes us more ready to take it seriously. And the mathematical-intellectual effort of applying the theorem fixes in us the particular point of view of the theory with which we deal, making us blind to whatever appears neither as a dependent nor as an independent parameter in its mathematical formulation.

The result, perhaps most common in the social sciences, is bad theory with a mathematical passport. The present point is best established by reference to a few horrible examples. In so large and public a gathering, however, prudence dictates the avoidance of any possible faux pas. I confine myself therefore, to the citation of a delightful passage form Keynes' *General Theory*, in which the issues before us are discussed with a characteristic wisdom and wit: "It is the great fault of symbolic pseudomathematical methods of formalizing a system of economic analysis...that they expressly assume strict independence between the factors involved and lose all their cogency and authority if this is disallowed; whereas, in ordinary discourse, where we are not blindly manipulating but know all the time what we are doing and what the words mean, we can keep 'at the back of our heads' the necessary reserves and qualifications and adjustments which we shall have to make later on, in a way in which we cannot keep complicated partial differentials "at the back" of several pages of algebra which assume they all vanish. Too large a proportion of recent 'mathematical' economics are mere concoctions, as imprecise as the initial assumptions they rest on, which allow the author to lose sight of the complexities and interdependencies of the real world in a maze of pretentious and unhelpful symbols."

The intellectual attractiveness of a mathematical argument, as well as the considerable mental labor involved in following it, makes mathematics a powerful tool of intellectual prestidigitation–a glittering deception in which some are entrapped, and some, alas, entrappers. Thus, for instance, the delicious ingenuity of the Birkhoff ergodic theorem has created the general impression that it must play a central role in the foundations of statistical mechanics. (This dictum is promulgated, with a characteristically straight face, in Dunford-Schwartz, *Linear Operators*, Vol. 1, Chapter 7.) Let us examine this case carefully, and see. Mechanics tells us that the configuration of an isolated system is specified by choice of a point p in its phase surface, and that after t seconds a system initially in the configuration represented by p moves into the configuration represented by M_t p. The Birkhoff theorem tells us that if f is any numerical function of the configuration p (and if the mechanical system is metrically transitive), the time average tends (as $t \to \infty$) to a certain constant; at any rate for all initial configurations p not lying in a set e in the phase surface whose measure μ (e) is zero; μ here is the (natural) Lebesgue measure in the phase surface. Thus, the familiar argument continues, we should not expect to observe a configuration in which the long-time average of such a function f is not close to its equilibrium value. Here I may conveniently use a bit of mathematical prestidigitation of the very sort to which I object, thus paradoxically making an argument serve the purpose of its own denunciation. Let v(e) denote the probability of observing a configuration in the set e; the application of the Birkhoff theorem just made is then justified only if μ (e) = 0 implies that v(e) = 0. If this is the case, a known result of measure theory tells us that v(e) is extremely small wherever μ(e) is extremely small. Now the functions f of principal interest in statistical mechanics are those which, like the local pressure and density of a gas, come into equilibrium, i.e., those functions for which f(M_t p) is constant for long periods of time and for almost all initial configurations p. As is evident by direct computation in simple cases, and as the Birkhoff theorem itself tells us in these cases in which it is applicable, this means that f(p) is close to its equilibrium value except for a set e of configurations of very small measure μ. Thus, not the Birkhoff theorem but the simple and generally unstated hypothesis "μ (e)= 0 implies v(e) = 0" necessary to make the Birkhoff theorem relevant in any sense at all tells us why we are apt to find f(p) having its equilibrium value. The Birkhoff theorem in fact does us the service of establishing its own inability to be more than a questionably relevant superstructure upon this hypothesis.

The phenomenon to be observed here is that of an involved mathematical argument hiding the fact that we understand only poorly what it is based on.

This shows, in sophisticated form, the manner in which mathematics, concentrating our attention, makes us blind to its own omission–what I have already called the single-mindedness of mathematics. Typically, mathematics knows better what to do than why to do it. Probability theory is a famous example. An example which is perhaps of far greater significance is the quantum theory. The mathematical structure of operators in Hilbert space and unitary transformations is clear enough, as are certain features of the interpretation of this mathematics to give physical assertions, particularly assertions about general scattering experiments. But the larger question here, a systematic elaboration of the world-picture which quantum theory provides, is still unanswered. Philosophical questions of the deepest significance may well be involved. Here also, the mathematical formalism may be hiding as much as it reveals.

14

What Is Philosophy of Mathematics Looking for?*

ALFONSO C. ÁVILA DEL PALACIO

1. Dialogue of the deaf

There have been many different and opposing answers to the question, what is mathematics? I think that one problem is that it is not clear what would be the most adequate point of view to search for an answer, because of the vagueness of the question itself.

In addition to diversity of opinions about the nature of mathematics, which we might resolve by presenting facts or reasons to support one of these opinions or to oppose the others, I affirm that there is a "dialogue of the deaf".[1]

There are mathematicians and philosophers who think that they are talking about the same subject.

> ...it is characteristic of mathematical theories that they can themselves become the subject matter of mathematical theories. It is thus in principle possible for mathematical theories and philosophical theories about mathematics to be incompatible. (Körner [1967], p. 118)

It is generally accepted that mathematics is recursive, as Körner says; but not all thinkers accept that philosophy is formed by theories. Wittgenstein [1918], for instance, says that '*Philosophy aims at the logical clarification of thoughts. Philosophy is not one doctrine but an activity.*' (4.112)

It seems that Gödel's mathematical work has generated many philosophical reflections. Rodriguez-Consuegra [1992], for example, said:

---

* *This paper is an updated version of one that I published in Syntesis No. 3 (1997), Revista de la Universidad Autónoma de Chihuahua, México. I am very grateful to Profs. R. Hersh and R. Morado for their help in the translation into English.*

[1] Jody Azzouni [1994] affirms that the philosophical point of view frequently is prone to linguistic pitfalls and misperceptions; and it seems that he suggests a dialogue of the deaf too.

> *Gödel proved the existence of propositions true but undemonstrated in a formal system sufficiently rich for containing arithmetic... It seems to me that the more relevant philosophical consequences are the following: Once they proved that truth and demonstrability are different things, then the truth of certain propositions is directly intuitive.* (p. 446)

Körner [1967], going far beyond this, concludes that,

> ...the metamathematical discoveries of the present century imply the falsehood of the common doctrines shared by the classical philosophies of non-competitive mathematical theories. (p. 132)

But, against these opinions, Wittgenstein [1967] affirms:

> *A philosophical problem has the form: 'I don't know my way about' (123). 'It leaves everything as it is. It also leaves mathematics as it is, and no mathematical discovery can advance it.' (124)*

We agree with Wittgenstein that the philosophy of mathematics is different from mathematics or metamathematics. But we must make precise how philosophical work is performed, and what is its difference with mathematical work. Wittgenstein [1918] says that '*A philosophical work consists essentially of elucidations*' (4.112). But that which is not clear for one person, can be clear for somebody else. Putnam [1967] affirms that 'The fact that philosophers all agree that a notion is 'unclear' doesn't mean that it *is* unclear' (p. 296). In this difference of opinions between philosophers of mathematics, we can see that philosophers do not agree unanimously, maybe because they conceive of mathematics and philosophy in accordance with their own philosophical perspective. For this reason, most mathematicians reject the philosophical attempts to characterize mathematical activity.

In addition, mathematicians are not interested in these philosophical attempts because they are almost always external to mathematics itself. Santiago Ramirez [1990] says:

> *They have conceived traditionally the relation between philosophy and mathematics as that in which philosophy, whatever its metaphysical foundation, tries to subject mathematics to philosophical discourse, or philosophical norm. From Pythagoras to Analytical Philosophy the question is to exhibit mathematics as a discipline, discourse, or special kind of knowledge where philosophical or epistemological hypothesis about existence, about truth, and about method are confirmed.* (p. 419)

But against these pretensions, Ramirez [1989] had said following Cavailles: '*The essence of mathematics is a problem, among others, which philosophy can not resolve*' (p. 318). In the same direction, Courant and Robbins [1941] conclude that 'it is not philosophy but active experience in mathematics itself that alone can answer the question: What is mathematics?' (p. 7)

Perhaps the difficulty of resolving this question is in philosophy itself, as Ramirez, Cavailles, Courant and Robbins seem to affirm; that is, it is not philosophy that can answer it. Or maybe, as Hersh [1979] says, the reason is the fact that 'There do not seem to be many professional philosophers who know functional analysis or algebraic topology or stochastic processes' (p. 34). This is reaffirmed by Amor [1981] when he says, commenting on Hersh's work: '*this is a reflection of an active mathematician, and not of a philosopher non mathematician, and for this reason it is an authentic reflection about the real mathematics.*' (p. II)

But, how much mathematics we must know in order to do a philosophical reflection about mathematics? Maurice Frechet [1955] said: '*mathematicians do not know, for example, the whole mathematical analysis*' (p. 21). On the other hand, some mathematicians say that the number of mathematicians who really could understand completely the arguments of Wiles about Fermat's theorem could meet in a meeting hall. Therefore, what is the meaning of the expression 'to know mathematics'? Perhaps it is not enough to be a mathematician. Maybe it would be necessary to be a creator of part of mathematics. Maybe even this would not be enough, because of the growing and unfinished complexity of mathematics.

Then, who or which discipline should answer the question, what is mathematics? What kind of elements are required for this enterprise? Some philosophers, such as Plato, Aristotle, and Kant, were not mathematicians, but they wrote important ideas about philosophy of mathematics. Other philosophers, such as Pythagoras, Descartes, and Leibniz, were philosophers and great mathematicians too. Others, such as Frege and Wittgenstein had mathematical training. But, it seems that these circumstances are not conclusive for the importance of their philosophical ideas about mathematics.

On the other hand, some mathematicians, such as Cantor, Poincaré, and Frechet, among others, have reflected in different ways about their own work, giving us their rich historical and psychological experiences. But, would that be doing philosophy? Hersh [1979] says in this connection:

> But the art of philosophical discourse is not well developed today among mathematicians, even among the most brilliant. Philosophical issues just as much as mathematical ones deserve careful arguments, fully developed analysis, and due consideration of objections. A bald statement of one's own opinion is not an argument, even in philosophy. (pp. 34-35)

In consequence, in order to know how we can do philosophy of mathematics, and how we can understand different affirmations about mathematics, I propose that we must first make precise, what is a philosopher looking for? what is a mathematician looking for? when either of them pose questions about mathematics. That is, how one and the other understand the question, what is mathematics?

2. Mathematical reasoning about mathematics

Mathematicians, in general, do not pay much attention to the question, what is mathematics? Only a few of them pay attention to it. Nevertheless, it seems that almost all of them think that it is an internal affair to mathematics itself. In word of Cavailles [1938]: '*There is no definition, nor justification of the mathematical objects, except mathematics itself.*' (p. 172)

But if this is the case, how is it possible to meditate about mathematics from mathematics itself? It seems that there have been, at least, three ways: 1) doing metamathematics; 2) doing history of mathematics; and 3) practicing mathematics.

2.1. *Metamathematics*

In general it is accepted that mathematics is recursive; that is, that mathematics can rework mathematically its own results. Notwithstanding, if we do not know what mathematics is, it would be difficult to make precise how we can do mathematics of mathematics. In any case, I believe that we can say, at least, that it is possible to do mathematics in an axiomatic way, and in a non-axiomatic way. Polya [1957] says,

> Mathematics has two faces... Mathematics presented in the Euclidean way appears as a systematic, deductive science; but mathematics in the making appears as an experimental, inductive science. (p. vii)

In the words of Maurice Frechet [1955]:

> *Mathematics is not a completely logic theory... In spite of the fact that most of the mathematical works consist in doing logical transformations from propositions admitted as truthful,... it is not hard to admit that intuition guides the work in a specific direction. (p. 21-22)*

Let us remember that the use of axioms is a technique, based on ideas of Plato and Aristotle, which consists in ordering a certain body of knowledge (mathematical or non mathematical), by finding some affirmations (axioms) from which we can deduce all the other affirmation in that body of knowledge. The knowledge ordered in that way can, in theory, substitute for the former knowledge, gaining in clarity and precision.

In that sense, there could be axiomatic and non axiomatic metamathematics. It is characteristic of those two forms that both start from a certain mathematical theory and build another mathematical theory. It seems that the difference consists in the circumstance that the axiomatic one is born with the claim of substituting for the former and being clearer and more precise; while, the non axiomatic one does not substitute for the primitive theory, but only subsumes the former in a larger body.

With respect to the axiomatic theories (called foundations), I believe that all axiomatic mathematics is by its very nature metamathematics, and it tries to make clear the nature and relations of mathematical entities. In the words of Gödel:

> The so-called logical or set-theoretical 'foundation' for number-theory, or of any other well established mathematical theory, is explanatory, rather than really foundational, exactly as in physics where the actual function of axioms is to *explain* the phenomena described by the theorems of this system rather than to provide a genuine 'foundation' for such theorems (in Lakatos [1978], p. 27)

The limitations of the formal systems in the axiomatic work are well known. For this reason, the axiomatic theories, in general, have not substituted for the primitive mathematical theories, and both subsist. Hersh [1979] says,

> The common presupposition was that mathematics must be provided with an absolutely reliable foundation. The disagreement was on strategy, on what had to be sacrificed for the sake of the agreed-on goal. But the goal was never attained, and there are few who still hope for its attainment. (p. 38)

With respect to non axiomatic metamathematics, we can mention the theory of groups in its early stages; and, in general, that which Cavailles [1938] called 'theme'; that is, '*a transformation of one operation in element of one superior operational field: example, the topology of the topological transformations*' (p. 173). If we see the primitive mathematical theories as structures, this metamathematical work would build other structures formed by simplest structures; that is, it would build complex structures by means of which they can study and clarify the properties and relations of some primitive structures and their elements.

Summarizing, metamathematics is a development of mathematics itself in the direction of becoming more unitary and precise. In any case, we can say that both metatheories, intending to substitute or subsume primitive structures, in fact, explain and clarify the primitive mathematical theories, and tend to make mathematics more homogeneous. In consequence, it seems that mathematicians who do metamathematics understand the question, what is mathematics? as asking for the nature and relations of the elements which constitute mathematics.

2.2. Historical analysis

In opposition to metamathematics, above all to the axiomatic one, some mathematicians, such as R. Thom [1980], hold that these works are not enough because,

> *Formalism denies the status of mathematical to most of that which they commonly have understood as mathematics, and it does not say something new about its development.* (p. 27)

For this reason, some people think that an answer that takes into account more completely the fruitfulness of mathematical work, must be done from a historical analysis that clarifies the facts and methods which have permitted its growth.

Among the historical works there are at least two important varieties about "what is mathematics?": a) those which study the origin and development of the mathematical elements; and b) those which are the result of an analysis or a heuristic-psychological self reflection.

With respect to the former, Garciadiego [1996] says,

> *As historians of mathematics and sciences we are interested in knowing the origins of problems which men have tried to solve in the past time, the ideas they used as a start-*

> *ing point, and what they expected as an answer...; that is, how an idea is born, grows and changes in order to conform to the science's field.* (p. 14)

All history is an intelligible reconstruction of data which we consider relevant. This implies an interpretation of those data. 'History emerges when chronology is selected, organized, related and explained' (May [1974], p. 28). So history of mathematics is an explanation, which give us a picture that takes in count not only the final result of one mathematical theory, but its origin and development too. And that, I believe, lets us understand better, how mathematics was born and grew.

On the other hand, with respect to the heuristic-psychological works, we can mention Pappus, at the end of the third century A. D., Polya [1957] and Velleman [1994] in our days, passing by Descartes, Leibniz, Bolzano and Poincaré, among others. They deal, in general, with the conscious and even unconscious ways which mathematicians have followed trying to solve problems: regressive reasoning masterfully exposed by Pappus, the method of analysis-synthesis used in book XIII of Euclid's *Elements*, reduction ad absurdum used by Eleatics, mathematical induction, analogous reasoning, and the recourse of drawing a figure,[2] among others. It seems that in all these works, mathematicians are guided always by an '*esthetic feeling which all true mathematicians know..., because the useful combinations are, just, the more beautiful*' (Poincaré [1908], p. 52). Occasionally they are helped by the unconscious work too, as Poincaré and Polya say. With all that, they try to describe how mathematicians work, and so, how has mathematics became what it is now.

Briefly, history of mathematics gives us an approach to the subjective elements which have been present in the conformation of the objective elements of mathematics. In consequence, it seems that historians of mathematics understand the question, what is mathematics? as asking for the causes or motors of mathematical development.

2.3. Mathematical practice

Finally, for many mathematicians, such as Courant and Robbins [1941],

> What points, lines, numbers "actually" *are* cannot and need not be discussed in mathematical science. What matters and what corresponds to "verifiable" facts is structure and relationship... For scholars and laymen alike it is not philosophy [and maybe history not either], but active experience in mathematics itself that alone can answer the question: What is mathematics? (p. iv, v)

For example, that which has preoccupied mathematicians about numbers is their generalization, the *mathematical* existence of some numbers, such as the irrational, the transcendent, the ideal, and so on; and some problems

[2] That which De Lorenzo [1993] calls 'figural work'.

derived from there, such as the continuum problem or the infinity problem (see Frechet [1955], pp. 417-449).

Perhaps, we could say, as Newton Da-Costa said one day: '*mathematics is all that which is in mathematical books and reviews*'. This is, of course, an insufficient characterization; but, in its defense, we could say that mathematics is an interminable field, and '*with respect to the motor of progress, it seems to escape to all investigation.*' (Cavailles [1938], p. 175)

There are mathematicians, such as Cavailles himself in some of his works, who do not stop here but try to characterize mathematics based on their own mathematical practice. This is the case of J. De Lorenzo [1992] when he says:

> *It is a myth that all mathematical work is a logic syntactic work... In mathematical practice, axioms are not the starting point, they are not the key of the knowledge process, but nuclear concepts are, and some times, hypothesis or conjectures.* (pp. 447-448)

Thus it seems that some mathematicians understand the question, what is mathematics? as questioning about the mathematical doing; that is, questioning about that which is done by mathematicians.

3. Philosophical reasoning about mathematics

Philosophers, even when they appreciate and use the reflections mentioned above, in general are not satisfied with it. They can agree with certain metamathematical, or historical versions; but it seems that philosophers[3] ask other questions too.[4] But, what is it that philosophers, or mathematicians acting as philosophers, are looking for? That is, what could be for them an acceptable, or even controvertible answer to the question, what is mathematics?

3.1. What is philosophy?

When I meditate about philosophy, I ought to confess that it is a genuine question. I do not know a definitive answer, and I believe that we could not find one easily. Nevertheless, because the present investigation needs to make precise what we mean by 'philosophy', I will expound my point of view about that.

---

[3] This distinction, of course, is not exact. In fact, there have been mathematicians and historians who worked on central philosophical subjects. We can mention, for instance, the strong philosophical concerns of Gödel in [1949]; which, it seems, guided him to his technical works. On the other hand, I believe that it would be unjust to classify Wilbur Knorr's work as only historic, because I believe it is philosophical too.

[4] As Shapiro [1994] says, 'Philosophers have their own interests, beyond those of their colleagues in other departments' (p. 157).

I think we can see philosophy as a Socratic dialogue of a certain community[5], which starts with doubts about the inherited knowledge, and grows through history,[6] trying to make clear the concepts, or building new concepts, with which that community thinks its world. As Hegel said, 'it is to think about thinking,' and I add, by means of which, man has questioned about thought itself, about its limits and capacities, and about its fruits. Maybe we can sum up these questions this way: how is thought possible? Or, what is its place in the world?

This characterization is, of course, tentative, and only intends to pick out some of the more general characteristics of philosophy in order to make clear the philosophy of mathematics. There are, of course, other characterizations. It seems that there are not so many, because only a few philosophers have dedicated their attention to that question. I will comment on two of these characterizations, which are very common.

The first says that philosophy is the mother of the sciences. For instance, Cornman, Lehrer and Pappas [1992] say,

> Philosophy was once construed so broadly as to cover any field of theoretical inquiry. Any subject matter for which some general explanatory theory might be offered would have been a branch of philosophy. However, once a field of study came to be dominated by some main theory and developed standard methods of criticism and confirmation, then the field was cut off from the mother country of philosophy and become independent (p. 5).

Supporting this thesis, we can mention the fact that in the past centuries, many philosophers were scientists too, such as Aristotle and Descartes; and, on the other hand, some scientists, such as Newton for example, called their scientific works 'natural philosophy.' Nevertheless, even though since Thales philosophical work has been very close to scientific work, that does not give us license to say that philosophy and science have the same subjects and try to discover the same things.

I believe that we can explain from my perspective why some people see philosophy as the mother of science. In fact, I believe that it would not be hard to accept that the critical and dialogical work of philosophy about inherited knowledge can provoke new researches in some scientific fields, or can even provoke the beginning of new fields. For example, the critical work of Berkeley on Analysis provoked Weierstrass' mathematical work. Most of the time, mathematicians do not care about the philosophical work, because it leaves mathematics as it is; as Analysis continued the same in spite of Berkeley's

---

[5] I understand philosophy here as a dialogue of the western culture. Other cultures, such as that of China or that of India have their own internal dialogues. That circumstance makes difficult, but not impossible, the dialogue between different cultures.

[6] Aristotle, for instance, discusses with Plato and Pre-Socratics about innate ideas. In the same way, Descartes, Locke, Leibniz and Kant discuss with each other and with Aristotle and Plato about the same subject.

criticism. But sometimes the philosophical work has effects on the scientific work, even when this is not its goal. In the same manner, mathematical work can provoke new philosophical works. I believe that there is a mutual relation between both works; but each one of those is looking for a different thing, and has its own road.

Another common idea about philosophy has its roots in phenomenology and existentialism. We can see it, for example, in the teachings of Ortega y Gasset [1973] where he says: '*The radical problem of philosophy is defining that mode of being, that primary reality which we call our life*' (P. 177). This idea is close to what many people think philosophy must be: something like a view of the world, or a personal cosmological view.

Nowadays, people accept that Physics, Chemistry, Astronomy, Psychology, and so on, are the disciplines that can say how is the world, and philosophy does not. But these disciplines could not answer, what is the meaning of our life? Or, as Heidegger would say, why existence is, and nonexistence is not? They think that philosophy could answer that; they think that philosophy, helped by the particular sciences, must give us that general view about our world.

Nevertheless, there is a great problem in this conception, because philosophers must know all the sciences and that is practically impossible in our days. Philosophy, from my point of view, is not a superior judge for sciences, and it is not a super-science. I believe that any one philosopher can only know deeply one scientific discipline, or perhaps, only one theory in that science. Then, the singular thing which one philosopher can do about that is to extract from that theory its suppositions or the view of the world which it implies. This task, surely, is not very different to my characterization of philosophical activity. I said that philosophy tries to make clear the concepts, and when we extract the suppositions or the general view of one theory, I think we can see most clearly the true meaning of the concepts which that theory uses. In that way, philosophy does not propose a cosmological view, but it only lets us know the cosmological view which is implicit in the discourses which it analyzes. Philosophy does that when it analyzes the ontology and epistemology of one theory, for instance.

3.2. Philosophy of mathematics

Consistent with the characterization of philosophy which I proposed above, I think that we can see philosophy of mathematics as the dialogue about mathematics by means of which we try to make clear the ontology and epistemology of mathematics; or, in other words, how mathematics is possible.

Maybe the first philosopher of mathematics was Pythagoras when he said: 'number is the principle both as matter for things and as constituting their attributes and permanent states' (Aristotle [c-IV BC], A.5, 986-a-16). In the words of an important member of the Neo-Pythagorean group, Nicomachus [c-I, A. D.],

Which of these four methods must we first learn?... this is arithmetic, not solely because we said that it existed before all the others in the mind of the creating God like some universal and exemplary plan, relying upon which as a design and archetypal example the creator of the universe sets in order his material creations and makes them attain to their proper ends; but also because it is naturally prior in birth. (Book 1, Chap. IV)

For these reasons, it seems that Pythagoras himself started the systematic study of numbers, and Nicomachus wrote the important textbook *Introduction to Arithmetic*, which was used throughout the Middle Ages. These Pythagorean ideas about the place of the mathematical entities in the entire world are similar to those of Galileo [1623]: '*The universe is written in mathematical language, being its characters triangles, circles and figures*' (6). These ideas are similar to those of Descartes too when he says that the entire world is composed only of two substances: extension and thought. He adds that the first must be studied by means of Geometry. For this reason he studied Geometry itself. In fact, we can say that almost all empirical sciences which use mathematics rest on the Pythagorean belief that the world is known only by means of mathematics, or at least it is better known using mathematics. I believe that this conception about mathematics implies in actual terms that mathematical entities and its relations can be seen as the general structure of the world. In words of Bigelow [1988], '*Mathematics is the theory of universals*' (p. 13). That means that mathematics is possible as the study of that which persists under changes. We can agree with this conception or not; but certainly, it presents an ontology and epistemology which give an answer to the questions, how is mathematics possible? Or, what is its place in the world?

Throughout the history of mathematics, there have been mutual influences between mathematics and philosophy. The first, and maybe the most important, influence of philosophy on mathematics was, according to Szabó [1967], the transformation of mathematics into a deductive science:

Deductive mathematics is born when knowledge acquired by practice *alone* is no longer accepted as true...this change was due to the impact of philosophy, and more precisely of Eleatic dialectic, upon mathematics. (pp. 1-2)

It is not certain that axiomatics came from Eleatic dialectic, because, according to Proclus [c-V A. D.], Thales and Pythagoras were the first who started to prove theorems. But, certainly, the pre-Greek mathematics was different from Greek mathematics which was deductive and was born at the same time as Greek philosophy. Maybe for these reasons, Plato affirmed in *The Republic* that mathematics was a hypothetical-deductive science. Additionally, we ought not forget the logic of Aristotle and later of Frege as a contribution from philosophy to mathematics. The whole of mathematics is not axiomatic or deductive, but one part of it certainly is.

With respect to the influence of mathematics on philosophy, we can mention the works of such important philosophers as Kant, Frege, Witggenstein, among others. The mathematical and axiomatic Physics of Newton puzzled

Kant, who thought that mathematics expresses universal and necessary knowledge because it refers not to the changeable world, but to fixed forms of our sensibility: space for geometry, and time for arithmetic. To Kant, Mathematics proves that a-priori synthetic judgments are possible, and shows the way which other sciences, such as metaphysics, must follow. I think that it is not hard to accept that mathematics is in the basis of Kantian philosophy; and, perhaps, in the basis of almost all contemporary Analytical philosophy which started with Frege.

There have been not only mutual influences between mathematics and philosophy, but parallel works too. This is the case of some studies on natural numbers. In mathematics, Peano [1889], Dedekind [1893], and Hilbert [1900] defined number in terms of classes, systems (series), or the mutual relations among numbers themselves. For these authors, a number is defined by means of its relations with other numbers; it is only a place in an infinite series. For philosophers, nevertheless, that is not enough, as Russell [1919] said: 'We want our numbers to be such as can be used for counting common objects, and this requires that our numbers should have a definite meaning, not merely that they should have certain formal properties' (p. 10). Frege, for instance, defined number in relation with concepts. According to this, an assertion about numbers is an assertion about concepts and their relations. In this form, he gave an ontology for numbers, which refers to something extra-mathematical. Other work in this line is Ávila [1993]. Apparently, all these authors were looking for an answer to the same question: What is a number? But, really, they were looking for different answers. Mathematicians, such as Peano, Dedekind and Hilbert were looking for the mathematical structure of numbers; meanwhile, philosophers, such as Frege, Russell and Ávila were looking for, how are numbers possible? Or, what is their place in the world? It is symptomatic that for mathematicians, in general, Peano is more important than Frege; while, for philosophers it is the contrary.

It seems that philosophy of mathematics went into an internal crisis with the works of Gödel. But his theorems about incompleteness and consistency of arithmetic are about formal systems in the meaning of Hilbert. They are important to axiomatic metamathematics (or foundations). They show that axiomatic pictures of arithmetic do not pick out the whole of arithmetic. This is interesting for mathematicians and metamathematicians because it let them examine the place and utility of formal systems in the whole of mathematics. It is an internal business about the interrelations of mathematical entities. The crisis was really internal to Hilbert's program and similar programs, such as Logicism and Intuitionism. Mathematics itself and philosophy of mathematics continued on their own path.

Nevertheless, due to confusion, metamathematics forgot its philosophical aspirations, and became a part of mathematics itself (see Lakatos [1978]); and philosophy kept silent for a while. After Gödel, we can mention almost only the important works of Putnam and Quine until the two famous papers of Benacerraf [1965] and [1973]. These papers gave new life to the philosophical discussion because, I believe, they focus on the proper field of philosophy. In fact, with the purpose of solving Benacerraf's dilemma,

> They have urged that the central issue in the philosophy of mathematics is to find a way to identify an ontology for mathematics that is compatible with an epistemology that does not invoke mysterious faculties. (Kitcher [1988], p. 397)

In other words, how is mathematics possible?

Once it was clear what is the goal for the philosophy of mathematics, there has been a revival in the philosophical discussion. There have even been meetings on this subject (see, for instance, Hersh [1991]). On one hand, a renaissance of empiricism with P. Kitcher [1984] and others; the realism of Maddy [1990] and Bigelow [1988]; or the structuralism of Shapiro and Resnik. On the other hand, we can find the modal mathematics of Hellman and Putnam or the nominalism of H. Field, among other interesting conceptions.

4. Conclusion

I believe that the distinction established here lets us understand, at least partially, the limits and capacity of either a philosophical, or mathematical, or historical view when it faces the question, what is mathematics? The explanation of why, on certain occasions, the answers of some are not important or satisfactory for others, can be given by saying that these different views ask different things, with the same question: The mathematical view inquires about the character and connections of the mathematical entities; the historical view inquires about the origin and growth of these entities; and the philosophical view inquires about how these entities are possible. Of course, these points of view are complementary; but misunderstanding arises when they believe that they are talking about the same thing when they use the same terms. At that moment, a dialogue of the deaf emerges.

References

Amor; J. A. [1981]: 'Prólogo' in Spanish edition of Hersh [1979] published by Facultad de Ciencias, Universidad Nacional Autónoma de México.

Aristotle [c-IV, BC]: *Metaphysics*. Translated by W. D. Ross. London: Encyclopaedia Britannica, Inc.

Ávila, A. [1993]: 'Existen números fuera de la matemática?' *Theoria* VIII, 19: 89-112.

Azzouni, Jody [1994]: *Metaphysical myths, mathematical practice*. New York: Cambridge University Press.

Benacerraf, P. [1965]: 'What numbers could not be' in Paul Benacerraf and Hilary Putnam, eds., *Philosophy of Mathematics*. New York: Cambridge University Press, pp. 272-294 (1983).

—— [1973]: 'Mathematical Truth' in P. Benacerraf and H. Putnam, eds., *Philosophy of Mathematics*. New York: Cambridge University Press, pp. 403-420 (1983).

Bigelow, J. [1988]: *The Reality of Numbers. A Physicalist's Philosophy of Mathematics*. New York: Oxford University Press.

Cavailles, J. [1938]: *Méthode Axiomatique et Formalisme*. Paris: Hermann Editeurs des Scienceset des Arts.

Cornman, J.W.; K. Lehrer; G.S. Pappas: [1992]: *Philosophical Problems and Arguments: An introduction*. Fourth edition, Hackett Publishing Co.

Courant, R. and H. Robbins [1941]: *What is mathematics?* Second edition, New York: Oxford University Press, 1996.

Dedekind, R. [1893]: 'The nature and meaning of numbers' in *Essays on the Theory of Numbers*. Translated by W. W. Beman. New York: Dover Publications, Inc. (1963).

Euclid [c-III, BC]: *The Thirteen Books of the Elements*. Translated by Sir Thomas L. Heath. Dover Publications (1956).

Frechet, M. [1955]: *Les mathématiques et le concret*, Paris: Presses Universitaires de France.

Frege, G. [1884]: *Die Grundlagen der Arithmetik*. Translated by Austin in *The Foundations of Arithmetic*. Oxford: Blackwell and Mott (1950).

Galileo, Galilei [1623]: *Il Sassiatore* nel quale con bilancia esquisita e guista si ponderano le core contenute nella libra astronomica e filosofica di Lotario Sorsi Sigensario.

Garciadiego, A. [1996]: 'Historia de las ideas matemáticas: un manual introductorio de investigación'. *Mathesis* XII, 1 (3-113).

Gödel, K. [1949]: 'A remark about the relationship between relativity theory and idealistic philosophy'. In *Albert Einstein, Philosopher-Scientist* edited by Paul Schipp, Evanston, Illinois: Library of Living Philosophers (555-562).

Hersh, Reuben [1979]: 'Some proposals for reviving the philosophy of mathematics', *Advances in Mathematics,* (31) 1, January 1979, pp. 31-50.

—— (ed.) [1991]: 'New directions in philosophy of mathematics', *Synthese* 88, 2.

Hilbert, D. [1900]: 'Über den Zahlbegriff', *Jahresbericht der Deutschen Mathematiker-Vereinigung* 8, pp. 180-194.

Knorr, W. R. [1986]: *The Ancient Tradition of Geometric Problems*. Boston: Birkhauser.

Kitcher, P. [1984]: *The Nature of Mathematical Knowledge*. New York: Oxford University Press.

—— [1988]: 'Introduction' *Revue Internationale de Philosophie* 42, 167: 397-399.

Körner, Stephan [1967]: 'On the relevance of Post-Gödelian Mathematics to Philosophy', in Lakatos [1967], pp. 118-137.

Lakatos, I. (ed.) [1967]: *Problems in the philosophy of mathematics*. Proceeding of the International Colloquium in the Philosophy of Science, London 1965. Amsterdam: North-Holland Publishing Company.

Lakatos, I. [1978]: *Mathematics, science and epistemology: Philosophical Papers, vol. 2 Edited by J. Worrall and G. Currie*. Cambridge University Press.

Lorenzo, J. de [1992]: 'La matemática, (incompleta, aleatoria, experimental?', *Theoria* VII, 16-17-18 (423-450).

—— [1993]: 'La razón constructiva matemática y sus haceres', *Mathesis* IX, 2 (129-153).

Maddy, P. [1990]: *Realism in Mathematics*. New York: Oxford University Press.

May, K. [1974]: *Bibliography and Research Manual of the History of Mathematics*. Toronto: University of Toronto Press.

Nicomachus [c-I, A.D.]: *Introduction to Arithmetic*. Translated by M. L. D'Ooge. London: Encyclopaedia Britannica, (1952).
Ortega y Gasset, J. [1973]: *&$$$;Qué es la filosofía?* Madrid: Espasa Calpe.
Peano, G. [1889]: 'The principles of arithmetic, presented by a new method' in J. v. Heijenoort, ed., *From Frege to Godel*. Cambridge Massachusetts: Harvard University Press, pp. 83-97 (1967).
Plato [c-IV, BC]: 'Republic' in *The Dialogues of Plato*. 5 vols. Edited by B. Jowett. Oxford: Oxford University Press.
Poincaré, H. [1908]: *Science et Méthode*. Paris: Flammarion
Pólya, G. [1957]: *How to solve it*. Second Edition, Princeton: Princeton University press.
Proclus de Lycie [c-V, A.D.]: *Les commentaires sur le premier livre des Eléments d'Euclide*, Brujas (1948).
Putnam, Hilary [1967]: 'Mathematics without foundations', *The Journal of Philosophy* LXIV, I (19 January).
Ramirez, Santiago [1989]: *El desconocido número 5*. Dissertation for Ph.D. in Universidad Nacional Autónoma de México.
—— [1990]: 'El reencuentro del cosmos: filosofía del caos', *Mathesis* VI, 4: 419-431.
Resnik, M. D. [1997]: *Mathematics as a Science of Patterns*. Oxford: Oxford University Press.
Rodriguez-Consuegra, F. [1992]: 'El convencionalismo y la analogía entre matemática y física', *Mathesis* VIII, 4: 441-458.
Russell, B. [1919]: *Introduction to Mathematical Philosophy*. London: Allen & Unwin.
Shapiro, S. [1994]: 'Mathematics and Philosophy of Mathematics' *Philosophia Mathematica* 2, 3: 148-160.
Shapiro, S. [1997]: *Philosophy of Mathematics: Structure and Ontology*. New York: Oxford University Press.
Szabo, A. [1967]: 'Greek dialectic and Euclid's axiomatic' in Lakatos [1967]: 1-26.
Thom, R. [1980]: *Parabole e Catastrofi. Intrevista su matemática, science e filosofia*. Milano: Il Sagiatore.
Velleman, D. J. [1994]: How to prove it. New York: Cambridge University Press
Wittgenstein, Ludwig [1918]: *Logisch-philosophische Abhandlung*.
Wittgenstein, Ludwig [1967]: *Philosophische Untersuchunggen*. Third edition.

15

Concepts and the Mangle of Practice Constructing Quaternions

ANDREW PICKERING

> Similarly, by surrounding √-1 by talk about vectors, it sounds quite natural to talk of a thing whose square is −1. That which at first seemed out of the question, if you surround it by the right kind of intermediate cases, becomes the most natural thing possible.
>
> Ludwig Wittgenstein,
> *Lectures on the Foundations of Mathematics* (p. 226).

> How can the workings of the mind lead the mind itself into problems? . . . How can the mind, by methodical research, furnish itself with difficult problems to solve?
>
> This happens whenever a definite method meets its own limit (and this happens, of course, to a certain extent, by chance).
>
> Simone Weil,
> *Lectures on Philosophy* (p. 116).

An asymmetry exists in our accounts of scientific practice: machines are located in a field of agency but concepts are not.[1] Thus while it easy to appreciate

[1] Questions of agency in science have been thematised most clearly and insistently in the actor-network approach developed by Michel Callon, Bruno Latour and John Law. See, for example, Callon and Latour, 'Don't Throw the Baby Out with the Bath School! A Reply to Collins and Yearley,' in *Science as Practice and Culture*, ed. Andrew Pickering (Chicago, 1992). But Latour is right to complain about the dearth of studies and analyses of conceptual practice in science: 'almost no one,' as he puts it, 'has had the courage to do a careful anthropological study' (*Science in Action* [Cambridge, MA, 1987], 246). Whether failure of nerve is quite the problem, I am less sure. Much of the emphasis on the material dimension of science in recent science studies must be, in part, a reaction against the theory-obsessed character of earlier history and philosophy of science. In any event, Eric Livingston's *The Ethnomethodological Foundations of Mathematics* (Boston, 1986) is a counter-example to Latour's claim, and the analysis of conceptual practice that follows is a direct extension of my own earlier analysis of the centrality of modelling to theory development in elementary-particle physics: Pickering, 'The Role of Interests in High-Energy Physics: The Choice Between Charm and Colour', in *The Social Process of Scientific Investigation. Sociology of the Sciences, Vol. 4, 1980*, eds

that dialectics of resistance and accommodation can arise in our dealings with machines – I have argued elsewhere that the contours of material agency only emerge in practice[2] – it is hard to see how the same could be said of our dealings with concepts. And this being the case the question arises of why concepts are not mere putty in our hands. Why is conceptual practice difficult? 'How can the workings of the mind lead the mind itself into problems?' It seems to me that one cannot claim to have a full analysis of scientific practice until one can suggest answers to questions like these, and my aim in this chapter is to argue, first in the abstract then via an example, that a symmetrising move is needed. We should think of conceptual structures as themselves located in fields of agency, and of the transformation and extension of such structures as emerging in dialectics of resistance and accommodation within those fields, dialectics which, for short, I call *the mangle of practice*.

In section 1 I develop a general understanding of agency in science appropriate to the analysis of conceptual practice, and I explain its relation to the mangle. I then turn to my example, which, for reasons discussed below, is taken from the history of mathematics rather than from the history of science proper. Section 2 provides some technical background, and section 3 is the heart of this essay, offering a reconstruction of Sir William Rowan Hamilton's construction of a system of 'quaternions' in 1843. Section 4 generalises from the example in a discussion of temporal emergence and posthumanist decentring in conceptual practice, and of the mangling and interactive stabilisation of conceptual structures, disciplines and intentions. Section 5 summarises the overall image of scientific practice and culture that emerges when the findings of the present essay are taken in conjunction with similar analyses of material practice in science. Finally, in an attempt to delineate more clearly what is at stake in my analysis, section 6 reviews David Bloor's non-(or quasi-)emergent and humanist account of Hamilton's metaphysics as an instance of the sociology of scientific knowledge in its empirical application. Bloor argues that Hamilton's metaphysics was fixed by the social; I argue instead that metaphysics, like everything else, is subject to mangling in practice.

1. Disciplinary Agency

When we think, we are conscious that a connection between feelings is determined by a general rule, we are aware of being governed by a habit. Intellectual power is nothing but facility in taking habits and in following them in cases essentially

---

Karin Knorr, Roger Krohn and Richard Whitley (Dordrecht, 1981) and *Constructing Quarks: A Sociological History of Particle Physics* (Chicago, 1984). Nevertheless, resistance and accommodation are not thematised in my earlier analyses, and it might be that this exemplifies the lack of which Latour complains.

[2] Pickering, *The Mangle of Practice: Time, Agency and Science* (Chicago, forthcoming), chs 2, 3 and 5. The present essay is a slightly revised version of ch. 4 of that book. I thank Barbara Herrnstein Smith for her editorial suggestions.

> analogous to, but in non-essentials widely remote from, the normal cases of connections of feelings under which those habits were formed.
>
> Charles Sanders Peirce, *Chance, Love and Logic* (p. 167).

> The student of mathematics often finds it hard to throw off the uncomfortable feeling that his science, in the person of his pencil, surpasses him in intelligence.
>
> Ernst Mach, quoted by Ernest Nagel, *Teleology Revisited* (p. 171)

My analysis of conceptual practice depends upon and elaborates three central ideas: first, that cultural practices (in the plural) are disciplined and machine-like; second, that practice, as cultural extension, is centrally a process of open-ended modelling; and third, that modelling takes place in a field of cultural multiplicity and is oriented to the production of associations between diverse cultural elements. I can take these ideas in turn.

Think of an established conceptual practice – elementary algebra, say. To know algebra is to recognise a set of characteristic symbols *and how to use them.* As Wittgenstein put it: 'Every sign *by itself* seems dead. *What* gives it life? – In use it is *alive.*'[3] And such uses are disciplined; they are machine-like actions, in Harry Collins' terminology.[4] Just as in arithmetic one completes '3 + 4 =' by writing '7' without hesitation, so in algebra one automatically multiplies out 'a(b + c)' as 'ab + ac.' Conceptual systems, then, hang together with specific disciplined patterns of human agency, particular routinised ways of connecting marks and symbols with one another. Such disciplines – acquired in training and refined in use – carry human conceptual practices along, as it were, independently of individual wishes and intents. The scientist is, in this sense, passive in disciplined conceptual practice. This is a key point in what follows and, in order to mark it and to symmetrise the formulation I want to redescribe such human passivity in terms of a notion of *disciplinary agency*. It is, I shall say, the agency of a discipline – elementary algebra, for example – that leads disciplined practitioners through a series of manipulations within an established conceptual system.[5]

---

[3] The Wittgenstein quotation is taken from Michael Lynch, 'Extending Wittgenstein: The Pivotal Move from Epistemology to the Sociology of Science', in *Science as Practice and Culture*, ed. Pickering (Chicago, 1992), 289. Lynch's commentary continues: 'If the "use" is the "life" of an expression, it is not as though a meaning is "attached" to an otherwise lifeless sign. We first encounter the sign in use or against the backdrop of a practice in which it has a use. It is already a meaningful part of the practice, even if the individual needs to learn the rule together with the other aspects of the practice. It is misleading to ask "how we attach meaning" to the sign, since the question implies that each of us separately accomplishes what is already established by the sign's use in the language game. This way of setting up the problem is like violently wresting a cell from a living body and then inspecting the cell to see how life would have been attached to it.'

[4] Harry Collins, *Artificial Experts: Social Knowledge and Intelligent Machines* (Cambridge, MA, 1990).

[5] The notion of discipline as a performative agent might seem odd to those accustomed to thinking of discipline as a constraint upon human agency, but I want (like Foucault) to recognise that discipline is productive. There could be no conceptual practice without the kind of discipline at issue; there could only be marks on paper.

I will return to disciplinary agency in a moment, but now we can turn from disciplined practices to the practice of cultural extension. A point that I take to be established about conceptual practice is that it proceeds through a process of modelling. Just as new machines are modelled on old ones, so are new conceptual structures modelled upon their forebears.[6] And much of what follows takes the form of a decomposition of the notion of modelling into more primitive elements. As it appears in my example, at least, it is useful to distinguish three stages within any given modelling sequence, which I will describe briefly in order to sketch out the overall form of my analysis. Modelling, I think, has to be understood as an open-ended process, having in advance no determinate destination, and this is certainly true of conceptual practice. Part of modelling is thus what I call *bridging*, or the construction of a *bridgehead*, that tentatively fixes a vector of cultural extension to be explored. Bridging, however, is not sufficient to efface the openness of modelling: it is not enough in itself to define a new conceptual system on the basis of an old one. Instead it marks out a space for *transcription* – the copying of established moves from the old system into the new space fixed by the bridgehead (hence my use of the word 'bridgehead'). And, if my example is a reliable guide, even transcription can be insufficient to complete the modelling process. What remains is *filling*, completing the new system in the absence of any clear guidance from the base model.

Now, this decomposition of modelling into bridging, transcription and filling is at the heart of my analysis of conceptual practice, and I will be able to clarify what these terms mean when we come to the example. For the moment, though, I want to make a general remark about how they connect to issues of agency. As I conceive them, bridging and filling are activities in which scientists display choice and discretion, the classic attributes of human agency. Scientists are active in these phases of the modelling process, in Fleck's sense.[7] Bridging and filling are *free moves*, as I shall say. In contrast, transcription is where discipline asserts itself, where the disciplinary agency just discussed carries scientists along, where scientists become passive in the face of their training and established procedures. Transcriptions, in this sense, are disciplined *forced moves*. Conceptual practice therefore has, in fact, the

[6] Pickering, 'The Role of Interests' and *Constructing Quarks*; for the literature on metaphor and analogy in science more generally, see Barry Barnes, *T. S. Kuhn and Social Science* (London, 1982), David Bloor, *Knowledge and Social Imagery* (Chicago, 1991, 2nd ed.), Mary Hesse, *Models and Analogies in Science* (Notre Dame, 1966), Karin Knorr-Cetina, *The Manufacture of Knowledge: An Essay on the Constructivist and Contextual Nature of Science* (Oxford, 1981) and Thomas Kuhn, *The Structure of Scientific Revolutions* (Chicago, 1970, 2nd ed.).

[7] Ludwik Fleck, *Genesis and Development of a Scientific Fact* (Chicago, 1979).

form of a *dance of agency*, in which the partners are alternately the classic human agent and disciplinary agency. And two points are worth emphasising here. First, this dance of agency, which manifests itself at the human end in the intertwining of free and forced moves in practice, is not optional. Practice has to take this form. The *point* of bridging as a free move is to invoke the forced moves that follow from it. Without such invocation, conceptual practice would be empty. Second, the intertwining of free and forced moves implies what Gingras and Schweber refer to (rather misleadingly) as a certain 'rigidity' of conceptual 'networks.'[8] I take this reference as a gesture towards the fact that scientists are not fully in control of where passages of conceptual practice will lead. Conceptual structures, one can say, relate to disciplinary agency much as do machines to material agency. Once one begins to tinker with the former, just as with the latter, one has to find out in practice how the resulting conceptual machinery will perform. It is precisely in this respect that dialectics of resistance and accommodation can arise in conceptual practice. To see how, though, requires some further discussion.

The constitutive role of disciplinary agency in conceptual practice is enough to guarantee that its end-points are temporally emergent. One simply has to play through the moves that follow from the construction of specific bridgeheads and see where they lead. But this is not enough to explain the emergence of resistance, to get at how the workings of the mind lead the mind itself into problems. To get at this, one needs to understand what conceptual practice is for. I do not suppose that any short general answer to this question exists, but all of the examples that I can think of lead to themes of cultural multiplicity and the making and breaking of associations between diverse cultural elements. Let me give just two examples to illustrate what I have in mind.

In science, one prominent object of conceptual practice is bringing theoretical ideas to bear upon empirical data, to understand or explain the latter, to extract supposedly more fundamental information from them, or whatever. In *Constructing Quarks*, I argued that this process was indeed one of modelling, and now I would add four remarks. First, this process points to the multiplicity (and heterogeneity) of scientific culture. Data and theory have no necessary connection to one another; such connections as exist between them have to be made. Hence my second point: conceptual practice aims at making associations (translations, alignments) between such diverse elements – here data and theory. Third, just because of the presence of the disciplinary partner in the dance of agency in conceptual practice, resistances can arise in the making of such associations. Because the destinations of conceptual practice cannot be known in advance, the pieces do not necessarily fit together as intended. And fourth, these resistances precipitate dialectics of resistance and accommodation, tentative revisions of modelling vectors,

[8] Yves Gingras and S. S. Schweber, 'Constraints on Construction', *Social Studies of Science* 16 (1986): 380.

manglings that can bear upon conceptual structures as well the form and performance of material apparatus.[9]

To exemplify these ideas, an obvious strategy would be to document how disciplines structure practice in theoretical science, but I will not take that route here because the disciplines and conceptual structures at stake in all of the interesting cases that I know about – largely in recent theoretical elementary-particle physics – are sufficiently esoteric to make analysis and exposition quite daunting. As mentioned already, I propose instead to concentrate on mathematics, and in particular on an example from the history of mathematics which is at once intellectually and historically interesting and simple, in that it draws only upon relatively low-level and already familiar disciplines and structures in basic algebra and geometry. I hope thus to find an example of the mangle in action in conceptual practice that is accessible while being rich enough to point to further extensions of the analysis, in science proper as well as in mathematics. I will come to the example in a moment, but first some remarks are needed on mathematics in general.

Physics might be said to seek, amongst other things, somehow to describe the world; but what is mathematics for? Once more, I suppose that there is no general answer to this question, but I think that Latour makes some important and insightful moves. In his discussion of mathematical formalisms, Latour continually invokes metaphors of joining, linking, association and alignment, comparing mathematical structures to railway turntables, crossroads, clover-leaf junctions and telephone exchanges.[10] His idea is, then, that such structures themselves serve as multipurpose translation devices, making connections between diverse cultural elements. And, as we shall see, this turns out to be the case in our example. The details follow, but the general point can be made in advance. If cultural extension in conceptual practice is not fully under the control of active human agents, due to the constitutive role of disciplinary agency, then the making of new associations – the construction of new telephone exchanges linking new kinds of subscribers – is nontrivial. Novel conceptual structures need to be tuned if they are to stand a chance of performing cooperatively in fields of disciplinary agency; one has to expect

[9] A point of clarification in relation to my earlier writings might be useful. In 'Living in the Material World: On Realism and Experimental Practice' (in *The Uses of Experiment: Studies in the Natural Sciences*, eds David Gooding, Trevor Pinch and Simon Schaffer [Cambridge, 1989]) I discussed the open-ended extension of scientific culture in terms of a metaphor of 'plasticity.' I said that cultural elements were plastic resources for practice. The problem with this metaphor is that, if taken too seriously, it makes scientific practice sound too easy – one just keeps moulding the bits of putty until they fit together. The upshot of the present discussion of disciplinary agency is that, unlike putty, pieces of conceptual culture keep transforming themselves in unpredictable ways after one has squeezed them (and evidently the same can be said of machinic culture: one can tinker with the material configuration of apparatus, but that does not determine how it will turn out to perform). This is why achieving associations in practice is really difficult (and chancy).

[10] Latour, *Science in Action*, 239, 241 and 242.

that resistances will arise in the construction of new conceptual associations, precipitating continuing dialectics of resistance and accommodation, manglings of modelling vectors – of bridgeheads and fillings, and even of disciplines themselves.[11]

This is the process that we can now explore in an example taken from the history of mathematics. In the next two sections we will be concerned with the work of the great Irish mathematician, Sir William Rowan Hamilton, and in particular with a brief passage of his mathematical practice that culminated on 16 October 1843 in the construction of his new mathematical system of quaternions. Before we turn to the study, however, I want to remark on the selection of this example. As indicated above, it recommends itself on several accounts. The disciplinary agency manifest in Hamilton's work has a simple and familiar structure, which makes his work much easier to follow than that of present-day mathematicians or scientists. At the same time, Hamilton's achievement in constructing quaternions is of considerable historical interest. It marked an important turning point in the development of mathematics, involving as it did the first introduction of non-commuting quantities into the subject matter of the field, as well as the introduction of an exemplary set of new entities and operations, the quaternion system, that mutated over time into the vector analysis central to modern physics. And further, detailed documentation of Hamilton's practice is available.[12]

---

[11] I should mention one important aspect of mathematics that distinguishes it from science and to which I cannot pay detailed attention here, namely mathematical *proof*. Here Imre Lakatos' (1976) account of the development of Euler's theorem points once more to the mangle in conceptual practice: *Proofs and Refutations: The Logic of Mathematical Discovery* (Cambridge, 1976). The exhibition of novel counterexamples to specific proofs of the theorem counts, in my terminology, as the emergence of resistances, and Lakatos describes very nicely the revision of proof procedures as open-ended accommodation to such resistances, with interactive stabilisation amounting to the reconciliation of such procedures to given counterexamples. Other work in the history and philosophy of mathematics that points towards an understanding of practice as the mutual adjustment of cultural elements includes that of Philip Kitcher (*The Nature of Mathematical Knowledge* [Oxford, 1983] and 'Mathematical Naturalism,' in *History and Philosophy of Modern Mathematics* [Minneapolis: 1988], eds William Aspray and Kitcher), who argues that every mathematical practice has five components ('Mathematical Naturalism,' 299), Michael Crowe ('Ten Misconceptions about Mathematics and Its History,' in Aspray and Kitcher , ibid., 'Duhem and History and Philosophy of Mathematics', *Synthese* 83 (1990): 431-447, see note 29 below), and Gaston Bachelard, who understands conceptual practice in terms of 'resistances' (his word) and 'interferences' between disjoint domains of mathematics (see Mary Tiles, *Bachelard: Science and Objectivity* [Cambridge, 1984]).

[12] I would have been entirely unaware of this, were it not for the work of Adam Stephanides, then a graduate student nominally under my supervision. Stephanides brought Hamilton's work to my attention by writing a very insightful essay emphasising the open-endedness of Hamilton's mathematical practice, an essay which eventually turned into Pickering and Stephanides 'Constructing Quaternions: On the Analysis of Conceptual Practice,' in Pickering ed., *Science as Practice*.

Hamilton himself left several accounts of the passage of practice that led him to quaternions, especially a notebook entry written on the day of the discovery and a letter to John T. Graves dated the following day.[13] As Hamilton's biographer puts it: 'These documents make the moment of truth on Dublin bridge [where Hamilton first conceived of the quaternion system] one of the best-documented discoveries in the history of mathematics.'[14] On this last point, some discussion is needed.

Hamilton's discovery of quaternions is not just well-documented, it is also much written about. Most accounts of Hamilton's algebraic researches contain some treatment of quaternions, and at least five accounts in the secondary literature rehearse to various ends Hamilton's own accounts more or less in their entirety.[15] I should therefore make it clear that what differentiates my account from others is that, as already indicated, I want to show that Hamilton's work can indeed be grasped within the more general understanding of agency and practice that I call the mangle. Together with the discussion of free and forced moves and disciplinary agency, the open-endedness of modelling is especially important here, and in the narrative that follows I seek to locate free moves in Hamilton's eventual route to quaternions by setting that trajectory in relation to his earlier attempts to construct systems of 'triplets.'

2. From Complex Numbers to Triplets

The early 19th century was a time of crisis in the foundations of algebra, centring on the question of how the 'absurd' quantities – negative numbers and their square-roots – should be understood.[16] Various moves were made in the

---

[13] Hamilton, 'Quaternions', Note-book 24.5, entry for 16 Oct. 1843, and 'Letter to Graves on Quaternions; or on a New System of Imaginaries in Algebra', dated 17 Oct 1843, published in *Phil. Mag.*, 25 (1843): 489-95, both reprinted in Hamilton, *The Mathematical Papers of Sir William Rowan Hamilton. Vol. III, Algebra* (Cambridge, 1967), 103-5, 106-10. I cite these below as NBE and LTG; all page number citations to these and other writings of Hamilton are to the 1967 reprint of his papers. I should note that my primary source of documentation is a first-person narrative written after the event. This has to be understood as an edited rather than a complete account (whatever the latter might mean) but it is sufficient to exemplify the operation of the mangle in conceptual practice, which is my central concern.

[14] Thomas L. Hankins, *Sir William Rowan Hamilton* (Baltimore, 1980), 295.

[15] Hankins, ibid., 295-300, J. O'Neill, 'Formalism, Hamilton and Complex Numbers,' *Studies in History and Philosophy of Science* 17 (1986): 351-72, Helena Pycior, *The Role of Sir William Rowan Hamilton in the Development of Modern British Algebra* (Cornell University, unpublished PhD dissertation, 1976), ch. 7, B. L. van der Waerden, 'Hamilton's Discovery of Quaternions,' *Mathematics Magazine* 49 (1976): 227-234 and E. T. Whittaker, 'The Sequence of Ideas in the Discovery of Quaternions,' *Royal Irish Academy, Proceedings* 50, Sect. A, No. 6 (1945): 93-98.

[16] Hankins, *Sir William Rowan Hamilton*, 248, Pycior, *The Role of Sir William Rowan Hamilton*, ch. 4.

debate over the absurd quantities, only one of which bears upon our story, and which serves to introduce the themes of cultural multiplicity and association as they will figure there. This was the move to construct an association between algebra and an otherwise disparate branch of mathematics, geometry, where the association in question consisted in establishing a *one-to-one correspondence* between the elements and operations of complex algebra and a particular geometrical system.[17] I need to go into some detail about the substance of this association, since it figured importantly in Hamilton's construction of quaternions.[18]

The standard algebraic notation for a complex number is $x + iy$, where x and y are real numbers and $i^2 = -1$. Positive real numbers can be thought of as representing measurable quantities or magnitudes – a number of apples, the length of a rod – and the foundational problem in algebra was to think what -1 and i (and multiples thereof) might stand for. What sense can one make of $\sqrt{-1}$ apples? How many apples is that? The geometrical response to such questions was to think of x and y not as quantities or magnitudes, but as coordinates of the end-point of a line-segment terminating at the origin in some 'complex' two-dimensional plane. Thus the x-axis of the plane measured the real component of a given complex number represented as such a line-segment, and the y-axis the imaginary part, the part multiplied by i in the algebraic expression (fig. 1a). In this way the entities of complex algebra were set in a one-to-one correspondence with geometrical line-segments. Further, it was possible to put the operations of complex algebra in a similar relation with suitably defined operations upon line-segments. Addition of line-segments was readily defined on this criterion. In algebraic notation, addition of two complex numbers was defined as

$$(a + ib) + (c + id) = (a + c) + i(b + d),$$

and the corresponding rule for line-segments was that the x-coordinate of the sum should be the sum of the x-coordinates of the segments to be summed, and likewise for the y-coordinate (fig. 1b). The rule for subtraction could be obtained directly from the rule for addition – coordinates of line-segments were to be subtracted instead of summed.

The rules for multiplication and division in the geometrical representation were more complicated, and we need only discuss that for multiplication, since

[17] Michael J. Crowe, *A History of Vector Analysis: The Evolution of the Idea of a Vectorial System* (New York, 1985), 5-11.

[18] At this point my analysis starts to get technical. This is inevitable if one's aim is to understand technical practice, but readers with limited familiarity with mathematics might try skimming this and the following section before moving on to section 4, and returning to them if aspects of the subsequent discussion seem obscure. The important thing is to grasp the overall form of the analysis of sections 2 and 3 rather than to follow all of Hamilton's mathematical manoeuvres in detail.

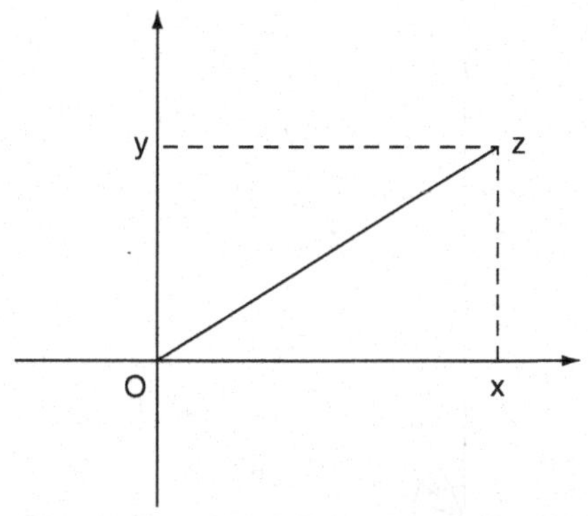

FIGURE 1A. Geometrical representation of the complex number $z = x + iy$ in the complex plane. The projections onto the x- and y-axes of the endpoint of the line Oz measure the real and imaginary parts of z, respectively.

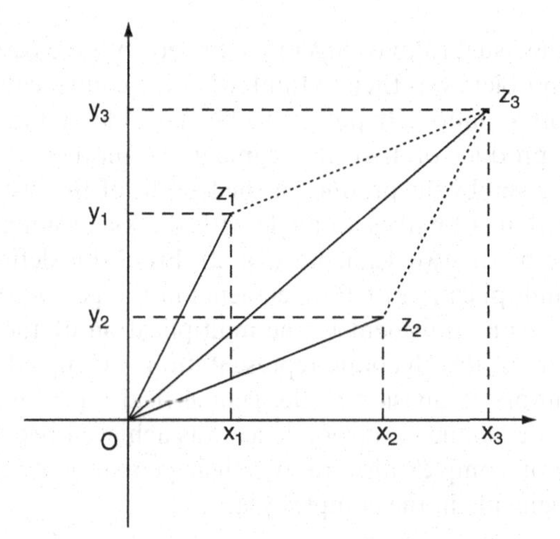

FIGURE 1B. Addition of complex numbers in the geometrical representation: $z_3 = z_1 + z_2$. By construction, the real part of z_3 is the sum of the real parts of z_1 and z_2 ($x_3 = x_1 + x_2$), and likewise the imaginary part ($y_3 = y_1 + y_2$).

this was the operation that became central in Hamilton's development of quaternions. The rule for algebraic multiplication of two complex numbers,

$$(a + ib)(c + id) = (ac - bd) + i(ad + bc),$$

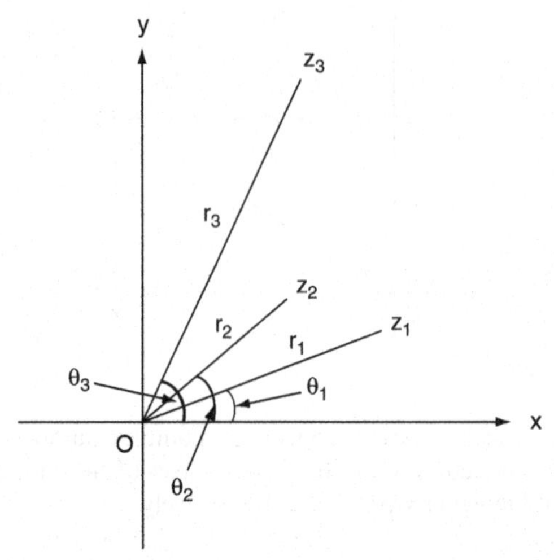

FIGURE 2. Multiplication of complex numbers in the geometrical representation: $z_3 = z_1 \times z_2$. Here the lengths of line-segments are multiplied ($r_3 = r_1 \times r_2$), while the angles subtended with the x-axis by line-segments are added ($\theta_3 = \theta_1 + \theta_2$).

followed from the usual rules of algebra, coupled with the peculiar definition of $i^2 = -1$. The problem was then to think what the equivalent might be in the geometrical representation. It proved to be stateable as the conjunction of two rules. The product of two line-segments is another line-segment that (a) has a length given by the product of the lengths of the two segments to be multiplied, and that (b) makes an angle with the x-axis equal to the sum of the angles made by the two segments (fig. 2). From this definition, it is easy to check that multiplication of line-segments in the geometrical representation leads to a result equivalent to the multiplication of the corresponding complex numbers in the algebraic representation.[19] Coupled with a suitably contrived definition of division in the geometrical representation, then, an association of one-to-one correspondence was achieved between the entities and operations of complex algebra and their geometrical representation in terms of line-segments in the complex plane.

[19] The easiest way to grasp these rules is as follows. In algebraic notation, any complex number $z = x + iy$ can be reexpressed as $r(\cos\theta + i\sin\theta)$, which can in turn be reinterpreted geometrically as a line-segment of length r, subtending an angle θ with the x-axis at the origin. The product of two complex numbers z_1 and z_2 is therefore $r_1r_2(\cos\theta_1 + i\sin\theta_1)(\cos\theta_2 + i\sin\theta_2)$. When the terms in brackets are multiplied out and rearranged using standard trigonometric relationships, one arrives at $z_1z_2 = r_1r_2[\cos(\theta_1 + \theta_2) + i\sin(\theta_1 + \theta_2)]$, which can itself be reinterpreted geometrically as a line-segment having a length which is the product of the lengths of the lines to be multiplied (part a of the rule) and making an angle with the x-axis equal to the sum of angles made by the lines to be multiplied (part b).

At least three important consequences for 19th-century mathematics flowed from this association. First, it could be said (though it could also be disputed) that the association solved the foundational problems centred on the absurd numbers. Instead of trying to understand negative and imaginary numbers as somehow measures of quantities or magnitudes of real objects, one should think of them geometrically, in terms of the orientation of line-segments. A negative number, for example, should be understood as referring to a line-segment lying along the negative (rather than positive) x-axis, a pure imaginary number as lying along the y-axis, and so on (fig. 3). Thus for an understanding of the absurd numbers one could appeal to an intuition of the possible differences in length and orientation of rigid bodies – sticks, say – in any given plane, and hence the foundational problem could be shown to be imaginary rather than real (so to speak).

Second, more practically, the geometrical representation of complex algebra functioned as a switchyard. Algebraic problems could be reformulated as geometrical ones, and thus perhaps solved using geometric techniques, and

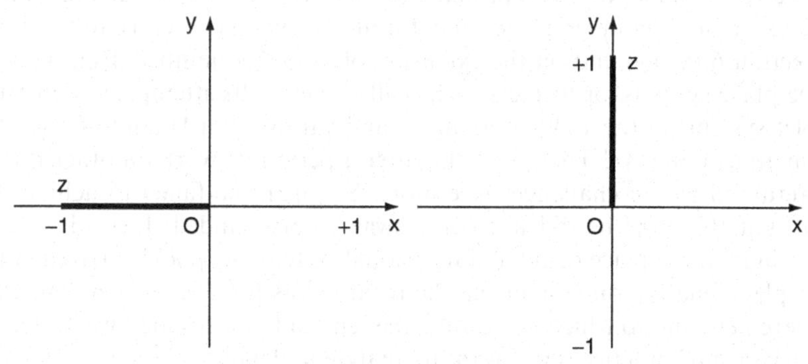

FIGURE 3A. Geometrical representations of absurd numbers: (a) $z = -1$.

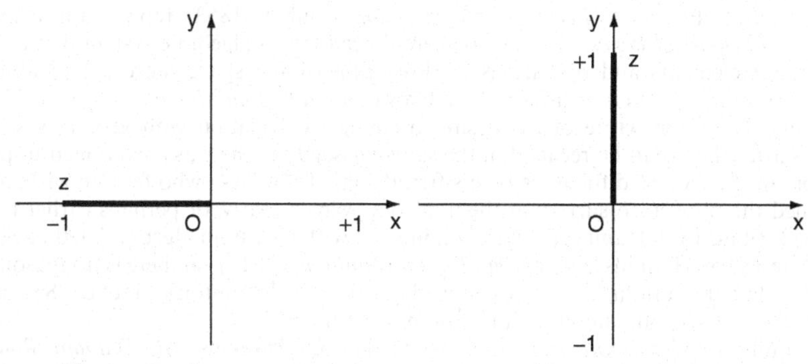

FIGURE 3B. $z = i = \sqrt{-1}$.

vice versa. The third consequence of this association of algebra with geometry was that the latter, more clearly than the former, invited extension. Complex algebra was a self-contained field of mathematical practice; geometry, in contrast, was by no means confined to the plane. The invitation, then, was to extend the geometrical representation of complex-number theory from a two- to a three-dimensional space, and to somehow carry along a three-place algebraic equivalent with it, maintaining the association already constructed in two dimensions. On the one hand, this extension could be attempted in a spirit of play, just to see what could be achieved. On the other, there was a promise of utility. The hope was to construct an algebraic replica of transformations of line-segments in three-dimensional space, and thus to develop a new and possibly useful algebraic system appropriate to calculations in three-dimensional geometry, 'to connect, in some new and useful (or at least interesting way) *calculation* with *geometry*, through some *extension* [of the association achieved in two dimensions], to *space of three dimensions*,' as Hamilton put it.[20]

Hamilton was involved in the development of complex algebra from the late 1820s onwards. He worked both on the foundational problems just discussed (developing his own approach to them via his 'Science of Pure Time' and a system of 'couples' rather than through geometry; I return to this topic in section 6 below) and on the extension of complex numbers from two- to three-place systems, or 'triplets' as he called them. His attempts to construct triplet systems in the 1830s were many and various, but Hamilton regarded them all as failures.[21] Then, in 1843, after a period of work on other topics, he returned to the challenge once more. Yet again he failed to achieve his goal, but this time he did not come away empty handed. Instead of constructing a three-place or three-dimensional system, he quickly arrived at the four-place quaternion system that he regarded as his greatest mathematical achievement and to which he devoted the remainder of his life's work. This is the passage of practice that I want to analyse in detail.

---

[20] Hamilton, Preface to *Lectures on Quaternions* (Dublin, 1853), reprinted in Hamilton, *Mathematical Papers*, 135.The perceived need for an algebraic system that could represent elements and operations in three-dimensional space more perspicuously than existing systems is discussed in Crowe, *Vector Analysisector Analysis*, 3-12. Though Hamilton wrote of his desire to connect calculation with geometry some years after the event, he recalled in the same passage that he was encouraged to persevere in the face of difficulties by his friend John T. Graves, 'who felt the wish, and formed the project, to surmount them in some way, as early, or perhaps earlier than myself' (ibid, 137). Hamilton's common interest with Graves in algebra dated back to the late 1820s (Hankins, *Sir William Rowan Hamilton*, ch. 17), so there is no reason to doubt that this utilitarian interest did play a role in Hamilton's practice. See also O'Neill, 'Formalism, Hamilton and Complex Numbers.'

[21] Hamilton, *Mathematical Papers*, 3-100, 117-42, Hankins, *Sir William Rowan Hamilton*, 245-301, Pycior, *The Role of Sir William Rowan Hamilton*, chs 3-6.

3. Constructing Quaternions

On 16 October 1843, Hamilton set down in a notebook his recollection of his path to quaternions. The entry begins:[22]

> I, this morning, was led to what seems to me a theory of *quaternions*, which may have interesting developments. *Couples* being supposed known, and known to be representable by points in a plane, so that √-1 is perpendicular to 1, it is natural to conceive that there may be another sort of √-1, perpendicular to the plane itself. Let this new imaginary be j; so that $j^2 = -1$, as well as $i^2 = -1$. A point x, y, z in space may suggest the triplet x + iy + jz.

I can begin my commentary on this passage by noting that a process of modelling was constitutive of Hamilton's practice. As is evident from these opening sentences, he did not attempt to construct a three-place mathematical system out of nothing. Instead he sought to move into the unknown from the known, to find a creative extension of the two-place systems already in existence. Further, as will become evident as we go along, the process of cultural extension through modelling was, in this instance as in general, an open-ended one: in his work on triplet systems that culminated in the construction of quaternions Hamilton tried out a large number of different extensions of complex algebra and geometry. Now I need to talk about how Hamilton moved around in this open-ended space, a discussion that will lead us into the tripartite decomposition of modelling mentioned in section 1.

In his reference above to 'points in a plane,' Hamilton first invokes the geometrical representation of complex algebra, and the extension that he considers is to move from thinking about line-segments in a plane to thinking about line-segments in a three-dimensional space. In so doing, I say that he established a *bridgehead* to a possible three-dimensional extension of complex algebra. As already stated and as discussed further below, the significance of such a bridging operation is that it marks a particular destination for modelling; at the moment I want to emphasise two points that I suspect are general about bridging. First, however natural Hamilton's move here from the plane to three-dimensional space might seem, it is important to recognise that it was by no means forced upon him. In fact, in his earlier attempts at triplet systems, he had proceeded differently, often working first in terms of an algebraic model and only towards the end of his calculations attempting to find geometrical representations of his findings, representations which were quite dissimilar from that with which he begins here.[23] In this sense, the act of fixing a bridgehead is an active or free move that serves to cut down the indefinite openness of modelling. My second point follows on from this. Such free moves need to be seen as tentative and

---

[22] NBE, 103.

[23] Hamilton, *Lectures on Quaternions*, 126-32. In such attempts, the intention to preserve any useful association of algebra and geometry does not seem to be central: Hamilton's principal intent was simply to model the development of a three-place

revisable trials that carry with them no guarantee of success. Just as Hamilton's earlier choices of bridgeheads had, in his own estimation, led to failure, so might this one. His only way of assessing this particular choice was to work with it and on it – to see what he could make of it. Similar comments apply to the second model that structured Hamilton's practice. This was the standard algebraic formulation of complex numbers, which he extends in the above quotation to a three-place system by moving from the usual $x + iy$ notation to $x + iy + jz$. This seems like another natural move to make. But again, when set against Hamilton's earlier work on triplets, it is better seen as the establishment of a bridgehead in a tentative free move.[24]

One more remark before returning to Hamilton's recollections. I noted above that complex algebra and its geometrical representation were associated with one another in a relation of one-to-one correspondence, and an intent to preserve that association characterised the passage of Hamilton's practice presently under discussion. In the quotation, he sets up a one-to-one correspondence between the *elements* defined in his two bridging moves – between the algebraic notation $x + iy + jz$ and suitably defined three-dimensional line-segments. In the passage that follows, he considers the possibility of preserving the same association of mathematical *operations* in the two systems. This is where the analysis of modelling becomes interesting, where disciplinary agency comes into play and the possibility of resistance in conceptual practice thus becomes manifest. Hamilton's notebook entry continues:[25]

> The square of this triplet [$x + iy + jz$] is on the one hand $x^2 - y^2 - z^2 + 2ixy + 2jxz + 2ijyz$; such at least it seemed to me at first, because I assumed $ij = ji$. On the other hand, if this is to represent the third proportional to 1, 0, 0 and x, y, z, considered as *indicators of lines*, (namely the lines which end in the points having these coordinates, while they begin at the origin) and if this third proportional be supposed to have its length a third proportional to 1 and $\surd(x^2 + y^2 + z^2)$, and its distance twice as far removed from 1, 0, 0 as x, y, z; then its real part ought to be $x^2 - y^2 - z^2$ and its two imaginary parts ought to have for coefficients $2xy$ and $2xz$; thus the term $2ijyz$ appeared de trop, and I was led to assume at first $ij = 0$. However I saw that this difficulty would be removed by supposing that $ji = -ij$.

This passage requires some exegesis. Here Hamilton begins to think about mathematical operations on the three-place elements that his bridgeheads

---

algebraic system on his existing two-place system of couples. Because the construction of associations in a multiple field plays a key role in my analysis, I should note that attention to this concept illuminates even these principally algebraic attempts. Hamilton found it necessary to transcribe parts of his development of couples *piecemeal*, and the goal of reassembling (associating) the disparate parts of the system that resulted again led to the emergence of resistance.

24 The foundational significance of Hamilton's couples was precisely that the symbol i did not appear in them, and was therefore absent from the attempts at triplets discussed in the previous note. A typical bridging move there was to go from couples written as (a, b) to triplets written (a, b, c).

25 NBE, 103.

have defined, and in particular about the operation of multiplication, specialised initially to that of squaring an arbitrary triplet. He works first in the purely algebraic representation and if, for clarity, we write $t = x + iy + jz$, he finds:

$$t^2 = x^2 - y^2 - z^2 + 2ixy + 2jxz + 2ijyz \qquad (1)$$

This equation follows automatically from the laws of standard algebra, coupled with the usual definition that $i^2 = -1$ and the new definition $j^2 = -1$ that was part of Hamilton's algebraic bridgehead. In this instance, then, we see that the primitive notion of modelling can be partly decomposed into two more transparent operations, bridging and *transcription*, where the latter amounts to the copying of an operation defined in the base-model – in this instance the rules of algebraic multiplication – into the system set up by the bridgehead. And this, indeed, is why I use the word 'bridgehead:' it defines a point to which attributes of the base-model can be transferred, a destination for modelling, as I put it earlier. We can note here that just as it is appropriate to think of fixing a bridgehead as an active, free, move, it is likewise appropriate to think of transcription as a sequence of passive, forced, moves, a sequence of moves – resulting here in equation 1 – that follow from what is already established concerning the base-model. And we can note, further, that the surrender of agency on Hamilton's part is equivalent to the assumption of agency by discipline. While Hamilton was indeed the person who thought through and wrote out the multiplications in question, he was not free to choose how to perform them. Anyone already disciplined in algebraic practice, then or now, can check that Hamilton (and I) have done the multiplication correctly. This then is our first example of the dance of agency in conceptual practice, in which disciplinary agency carried Hamilton (and carries us) beyond the fixing of a bridgehead.

The disciplined nature of transcription is what makes possible the emergence of resistance in conceptual practice, but before we come to that we should note that the decomposition of modelling into bridging and transcription is only partial. Equation 1 still contains an undefined quantity – the product ij – that appears in the last term of the right-hand side. This was determined neither in Hamilton's first free move nor in the forced moves that followed. The emergence of such 'gaps' is, I believe, another general feature of the modelling process: disciplinary agency is insufficent to carry through the processes of cultural extension that begin with bridging. Gaps appear throughout Hamilton's work on triplets, for example, and one typical response of his was that which I call *filling*, meaning the assignment of values to undefined terms in further free moves.[26] Resuming the initiative in the

[26] See, for example, his development of rules for the multiplication of couples: Hamilton, 'Theory of Conjugate Functions, or Algebraic Couples; With a Preliminary and Elementary Essay on Algebra as the Science of Pure Time', *Transactions of the Royal Irish Academy*, 17 (1837): 293-422, reprinted in Hamilton, *Mathematical Papers*, 80-83.

dance of agency, Hamilton could here have, say, simply assigned a value to the product ij and explored where that led him through further forced moves. In this instance, though, he proceeded differently.

The sentences that begin 'On the other hand, if this is to represent the third proportional . . .' refer to the operation of squaring a triplet in the geometrical rather than the algebraic representation. Considering a triplet as a line-segment in space, Hamilton was almost in a position to transcribe onto his new bridgehead the rules for complex multiplication summarised above in section 2, but, although not made explicit in the passage, one problem remained. While the first rule concerning the length of the product of lines remained unambiguous in three-dimensional space, the second, concerning the orientation of the product line, did not. Taken literally, it implied that the angle made by the square of any triplet with the x-axis was twice the angle made by the triplet itself – 'twice as far angularly removed from 1, 0, 0 as x, y, z' – but it in no way specified the orientation of the product line in space. Here disciplinary agency again left Hamilton in the lurch. Another gap thus arose in moving from two to three dimensions and, in this instance, Hamilton responded with a characteristic, if unacknowledged, filling move.

He further specified the rule for multiplication of line-segments in space by enforcing the new requirement that the square of a triplet remain in the plane defined by itself and the x-axis (this is the only way in which one can obtain his stated result for the square of a triplet in the geometrical representation). As usual, this move seems natural enough, but the sense of naturalness is easily shaken when taken in the context of Hamilton's prior practice. One of Hamilton's earliest attempts at triplets, for example, represented them as lines in three-dimensional space, but multiplication was defined differently in that attempt.[27] Be that as it may, this particular filling move sufficed and was designed to make possible a series of forced transcriptions from the two- to the three-dimensional versions of complex algebra that enabled Hamilton to compute the square of an arbitrary triplet. Surrendering once more to the flow of discipline, he found that the 'real part [of the corresponding line-segment] ought to be $x^2 - y^2 - z^2$ and its two imaginary parts ought to have for coefficients 2xy and 2xz.' Or, returning this result to purely algebraic notation:[28]

---

[27] Hamilton, *Lectures on Quaternions*, 139-40, cites his notes of 1830 as containing an attempt at constructing a geometrical system of triplets by denoting the end of a line-segment in spherical polar coordinates as $x = r\cos\theta$, $y = r\sin\theta\cos\phi$, $z = r\sin\theta\sin\phi$, and extending the rule of multiplication from two to three dimensions as $r'' = rr'$, $\theta'' = \theta + \theta'$, $\phi'' = \phi + \phi'$. This addition rule for the angle ϕ breaks the coplanarity requirement at issue.

[28] One route to this result is to write the triplet t in spherical polar notation. According to the rule just stated, on squaring, the length of the line-segment goes from r to r^2, the angle θ doubles, while the angle ϕ remains the same. Using standard trigonometric relations to express $\cos 2\theta$ and $\sin 2\theta$ in terms of $\cos\theta$ and $\sin\theta$ one can then return to x, y, z notation and arrive at equation 2.

$$t^2 = x^2 - y^2 - z^2 + 2ixy + 2jxz \quad (2)$$

Now, there is a simple difference between equations 1 and 2, both of which represent the square of a triplet but calculated in different ways. The two equations are identical except that the problematic term 2ijyz of equation 1 is absent from equation 2. This, of course, is just the kind of thing that Hamilton was looking for to help him in defining the product ij, and we will examine the use he made of it in a moment. First, it is time to talk about *resistance*. The two base-models that Hamilton took as his points of departure – the algebraic and geometrical representations of complex numbers – were associated in a one-to-one correspondence of elements and operations. Here, however, we see that as so far extended by Hamilton, the 3-place systems had lost this association. The definition of a square in the algebraic system (equation 1) differed from that computed via the geometrical representation (equation 2). The association of 'calculation with geometry' that Hamilton wanted to preserve had been broken; a resistance to the achievement of Hamilton's goal had appeared. And, as I have already suggested, the precondition for the emergence of this resistance was the constitutive role of disciplinary agency in conceptual practice and the consequent intertwining of free and forced moves in the modelling process. Hamilton's free moves had determined the directions that his extensions of algebra and geometry would take in the indefinitely open space of modelling, but the forced moves intertwined with them had carried those extensions along to the point at which they collided in equations 1 and 2. *This, I think, is how 'the workings of the mind lead the mind itself into problems.'* We can now move from resistance itself to a consideration of the dialectic of resistance and accommodation in conceptual practice, in other words to the mangle.

The resistance that Hamilton encountered in the disparity between equations 1 and 2 can be thought of as an instance of a generalised version of the Duhem-problem.[29] Something had gone wrong somewhere in the process of cultural extension – the pieces did not fit together as desired – but Hamilton had no principled way of knowing where. What remained for him to do was to tinker with the various extensions in question – with the various free moves he had made, and thus with the sequences of forced moves that followed from

[29] The Duhem-problem is usually formulated in terms of open-ended responses to mismatches between scientific data and theoretical predictions (Pierre Duhem, *The Aim and Structure of Physical Theory* [Princeton, 1991]). As far as I am aware, the only prior discussion of it as it bears on purely mathematical/conceptual practice is to be found in Crowe, 'Duhem,' who argues, following Lakatos, *Proofs and Refutations*, that contradictions between proofs and counterexamples need not necessarily disable the former. Crowe, 'Ten Misconceptions,' also discusses the extension of Duhem's ideas about physics to mathematics, and gestures repeatedly, though without detailed documentation or analysis, to the interactive stabilisation of axioms and theorems proved within them, and of mathematical systems and the results to which they lead. I thank Professor Crowe for drawing my attention to these essays.

them – in the hope of getting around the resistance that had arisen and achieving the desired association of algebra and geometry. He was left, as I say, to seek some accommodation to resistance. Two possible starts towards accommodation are indicated in the passage last quoted, both of which amounted to further fillings-in of Hamilton's extended algebraic system, and both of which led directly to an equivalence between equations 1 and 2. The most straightforward accommodation was to set the product ij equal to zero.[30] An alternative, less restrictive but more dramatic and eventually more far-reaching move also struck Hamilton as possible. It was to abandon the assumption of commutation between i and the new square-root of –1, j.[31] In ordinary algebra, this assumption – which is to say that ab = ba – was routine. Hamilton entertained the possibility, instead, that ij = –ji. This did not rule out the possibility that both ij and ji were zero, but even without this being the case, it did guarantee that the problematic term 2ijyz of equation 1 vanished, and thus constituted a successful accommodation to the resistance that had emerged at this stage.[32]

Hamilton thus satisfied himself that he could maintain the association between his algebraic and geometrical three-place systems by the assumption that i and j did not commute, at least as far as the operation of squaring a triplet was concerned. His next move was to consider a less restrictive version of the general operation of multiplication, working through, as above, the operation of multiplying two coplanar but otherwise arbitrary triplets. Again, he found that the results of the calculation were the same in the algebraic and geometrical representations as long as he assumed either ij = 0 or ij = –ji.[33] Hamilton then moved on to consider the fully general instance of multiplication in the new formalism, the multiplication of two arbitrary triplets.[34] As before, he began in the algebraic representation. Continuing to assume ij = –ji, he wrote:

[30] The following day, Hamilton described the idea of setting ij = 0 as 'odd and uncomfortable' (LTG, 107). He offered no reasons for this description, and it is perhaps best understood as written from the perspective of his subsequent achievement. The quaternion system preserved the geometrical rule of multiplication that the length of the product was the product of the lengths of the lines multiplied. Since in the geometrical representation both i and j have unit length, the equation ij = 0 violates this rule. Here we have a possible example of the retrospective reconstruction of accounts in the rationalisation of free moves.

[31] Pycior, *The Role of Sir William Rowan Hamilton*, 147, notes that Hamilton had been experimenting with non-commuting algebras as early as August 1842, though he then tried the relations ij = j, ji = i. Hankins, *Sir William Rowan Hamilton*, 292, detects a possible influence of a meeting between Hamilton and the German mathematician Gotthold Eisenstein in the summer of 1843.

[32] If one multiplies out the terms of equation 1 paying attention to the order of factors, the coefficient of yz in the last term on the right-hand side becomes (ij + ji); Hamilton's assumption makes this coefficient zero.

[33] NBE, 103.

[34] NBE, 103-4.

$$(a + ib + jc)(x + iy + jz) = ax - by - cz + i(ay + bx) + j(az + cx) + ij(bz - cy) \quad (3)$$

He then turned back to thinking about multiplication within the geometrical representation, where a further problem arose. Recall that in defining the operation of squaring a triplet Hamilton had found it necessary to make a filling free move, assuming that the square lay in the plane of the original triplet and the x-axis. This filling move was sufficient to lead him through a series of forced moves to the calculation of the product of two arbitrary but coplanar triplets. But it was insufficient to define the orientation in space of the product of two completely arbitrary triplets: in general, one could not pass a plane through any two triplets and the x-axis. Once more, Hamilton could have attempted a filling move here, concocting some rule for the orientation of the product line in space, say, and continuing to apply the sum rule for the angle made by the product with the x-axis. In this instance, however, he followed a different strategy.

Instead of attempting the transcription of the two rules that fully specified multiplication in the standard geometrical representation of complex algebra, he began to work only in terms of the first rule – that the length of the product line-segment should be the product of the lengths of the line-segments to be multiplied. Transcribing this rule to three dimensions, and working for convenience with squares of lengths, or 'square moduli,' rather than lengths themselves, he could surrender his agency to Pythagoras' theorem and write the square modulus of the left-hand side of equation 3 as $(a^2 + b^2 + c^2)(x^2 + y^2 + z^2)$ (another forced move).[35] Now he had to compute the square of the length of the right-hand side. Here the obstacle to the application of Pythagoras' theorem was the quantity ij again appearing in the last term. If Hamilton assumed that $ij = 0$, the theorem could be straightforwardly applied, and gave a value for the square modulus of $(ax - by - cz)^2 + (ay + bx)^2 + (az + cx)^2$. The question now was whether these two expressions for the lengths of the line segments appearing on the two sides of equation 3 were equal. Hamilton multiplied them out and rearranged the expression for the square modulus of the left-hand side, and found that it in fact differed from that on the right-hand side by a factor of $(bz–cy)^2$. Once again a resistance had arisen, now in thinking about the product of two arbitrary triplets in, alternatively, the algebraic and geometrical representations. Once more, the two representations, extended from two- to three-place systems, led to different results. And once more, Hamilton looked for some accommodation to this resistance, for some way of making the two notions of multiplication equivalent, as they were in two dimensions.

The new resistance was conditional on the assumption that $ij = 0$. The question, then, was whether some other assignment of ij might succeed in

[35] According to Pythagoras' theorem, the square modulus of a line-segment is simply the sum of the squares of the coordinates of its end points, meaning the coefficients of 1, i and j in algebraic notation.

balancing the moduli of the left- and right-hand sides of equation 3.[36] And here Hamilton made a key observation. The superfluous term in the square modulus of the left-hand side of equation 3, $(bz - cy)^2$, was the square of the coefficient of ij on the right-hand side. The two computations of the square modulus could thus be made to balance by assuming not that the product of i and j vanished, but that it was some third quantity k, a '*new imaginary*,' different again from i and j, in such a way that Pythagoras' theorem could be applied to it too.[37]

The introduction of the new imaginary k, defined as the product of i and j, thus constituted a further accommodation by Hamilton to an emergent resistance in thinking about the product of two arbitrary triplets in terms of the algebraic and geometrical representations at once, and one aspect of this particular accommodation is worth emphasising. It amounted to a drastic shift of bridgehead in both systems of representation (recall that I stressed the revisability of bridgeheads earlier). More precisely, it consisted in defining a new bridgehead leading from two-place representations of complex algebra to not three-but four-place systems – the systems that Hamilton quickly called *quaternions*. Thus, within the algebraic representation, the basic entities were extended from 2 to 4, from 1, i to 1, i, j, k, while within the geometrical representation, as Hamilton wrote the next day, 'there dawned on me the notion that we must admit, in some sense, a *fourth dimension* of space'[38] – with the fourth dimension, of course, mapped by the new k-axis.

We can consider this shift in bridgehead further in the next section; for now, we can observe that Hamilton had still not completed the initial development of quaternions. The quantity k^2 remained undefined at this stage, as did the various products of i and j with k, excepting those intrinsic to his new bridgehead ij = k. Hamilton fixed the latter products by a combination of filling assumptions and forced moves following from relations already fixed:[39]

> I saw that we had probably ik = –j, because ik = iij and $i^2 = -1$; and that in like manner we might expect to find kj = ijj = –i; from which I thought it likely that ki = j, jk = i, because it seemed likely that if ji = –ij, we should have also kj = –jk, ik = –ki. And since the order of these imaginaries is not indifferent, we cannot infer that k^2 = ijij is +1, because $i^2 \times j^2 = -1 \times -1 = +1$. It is more likely that k^2 = ijij = –iijj = –1. And in fact this last assumption is necessary, if we would conform the multiplication to the law of multiplication of moduli.

[36] Strictly speaking, this is too deterministic a formulation. The question really was whether any amount of tinkering with bridgeheads, fillings and so on could get past this point without calling up this or another resistance.
[37] NBE, 104.
[38] LTG, 108.
[39] Ibid.

Hamilton then checked whether the algebraic version of quaternion multiplication under the above assumptions, including $k^2 = -1$, led to results in accordance with the rule of multiplication concerning products of lengths in the geometrical representation ('the law of multiplication of moduli'), and found that it did. Everything in his quaternion system was thus now defined in such a way that the laws of multiplication in both the algebraic and geometrical version ran without resistance into one another. Through the move to four-place systems, Hamilton had finally found a successful accommodation to the resistances that had stood in the way of his three-place extensions. The outcome of this dialectic was the general rule for quaternion multiplication:[40]

$$(a, b, c, d)(a', b', c', d') = (a'', b'', c'', d''), \text{ where}$$

$$a'' = aa' - bb' - cc' - dd',$$

$$b'' = ab' + ba' + cd' - dc',$$

$$c'' = ac' + ca' + db' - bd',$$

$$d'' = ad' + da' + bc' - cb'.$$

With these algebraic equations, and the geometrical representation of them, Hamilton had, in a sense, achieved his goal of associating calculation with geometry. He had found vectors of extension of algebra and geometry that interactively stabilised one another, as I say, preserving in four dimensions the one-to-one association of elements and operations previously established in two dimensions. I could therefore end my narrative here. But before doing so, I want to emphasise that the qualifier 'in a sense' is significant. It marks the fact that what Hamilton had achieved was a *local* association of calculation with geometry rather than a global one. He had constructed a one-to-one correspondence between a particular algebraic system and a particular geometric system, not an all-purpose link between algebra and geometry considered as abstract, all-encompassing entities. And this remark makes clear the fact that one important aspect of Hamilton's achievement was to redefine, partially at least, the cultural space of future mathematical and scientific practice: more new associations remained to be made if quaternions were ever to be 'delocalised' and linked into the overall flow of mathematical and scientific practice, requiring work that would, importantly, have been inconceivable in advance of Hamilton's construction of quaternions.

As it happens, from 1843 onwards Hamilton devoted most of his productive energies to this task, and both quaternions and the principle of noncommutation that they enshrined were taken up progressively by many

[40] Ibid. (a, b, c, d) was Hamilton's notation for an arbitrary quaternion. In the geometrical representation, the coordinates of the end-point of a line-segment in four-dimensional space are given here; in algebraic notation this same quaternion would be written as $a + ib + jc + kd$.

sections of the scientific and mathematical communities.[41] Here I will discuss one last aspect of Hamilton's practice that can serve to highlight the locality of the association embodied in quaternions. Earlier I described Hamilton's organising aim as that of connecting calculation with geometry. And, as just discussed, quaternions did serve to bring algebraic calculation to *a* geometry – to the peculiar four-dimensional space mapped by 1, i, j and k. Unfortunately this was not the geometry for which calculation was desired. The promise of triplet – not quaternion – systems had been that they would bring algebra to bear upon the real three-dimensional world of interest to mathematicians and physicists. In threading his way through the dialectic of resistance and accommodation, Hamilton had, in effect, left that world behind. Or, to put it another way, his practice had, as so far described, served to displace resistance rather than fully to accommodate to it. Technical resistances in the development of three-place mathematical systems had been transmuted into a resistance between moving from Hamilton's four-dimensional world to the three-dimensional world of interest. It was not evident how the two worlds might be related to one another. This was one of the first problems that Hamilton addressed once he had arrived at his algebraic formulation of quaternions.

In his letter to John Graves dated 17 October 1843, Hamilton outlined a new geometrical interpretation of quaternions that served to connect them back to the world of three dimensions. This new interpretation was a straightforward but consequential redescription of the earlier four-dimensional representation. Hamilton's idea was to think of an arbitrary quaternion (a, b, c, d) as the sum of two parts: a real part, a, which was a pure real number and had no geometrical representation, and an imaginary part, the triplet, ib + jc + kd, which was to be represented geometrically as a line-segment in three-dimensional space.[42] Having made this split, Hamilton was then in a position to spell out rules for multiplication of the latter line-segments, which he summarised as follows:[43]

> Finally, we may always decompose the latter problem [the multiplication of two arbitrary triplets] into these two others; to multiply two pure imaginaries which agree in direction, and to multiply two which are at right angles with each other. In the first case, the product is a pure negative, equal to the products of the lengths or moduli with its sign changed. In the second case, the product is a pure imaginary of which the length is the product of the lengths of the factors, and which is perpendicular to both of them. The distinction between one such perpendicular and its opposite may be made by the rule of rotation [stated earlier in this letter].

---

[41] Hankins, *Sir William Rowan Hamilton*, ch. 23, Crowe, *Vector Analysis*, chs 4-7.

[42] The origin of this distinction between real and imaginary parts of quaternions lay in the differences between the equations just quoted defining quaternion multiplication: the equation defining a″ has three minus signs while those defining b″, c″ and d″ have two plus signs and only one minus.

[43] LTG, 110.

> There seems to me to be something analogous to *polarized intensity* in the pure imaginary part; and to *unpolarized energy* (indifferent to direction) in the real part of a quaternion: and thus we have some slight glimpse of a future Calculus of Polarities. This is certainly very vague, but I hope that most of what I have said above is clear and mathematical.

These strange rules for the multiplication of three-dimensional line segments – in which the product of two lines might be, depending upon their relative orientation, a number or another line or some combination of the two – served to align quaternions with mathematical and scientific practice concerned with the three-dimensional world.[44] Nevertheless, the association of algebra with geometry remained local. No contemporary physical theories, for example, spoke of entities in three-dimensional space obeying Hamilton's rules. It therefore still remained to find out in practice whether quaternions could be delocalised to the point at which they might become useful. With hindsight, one can pick out from the rules of multiplication a foreshadowing of modern vector analysis with its 'dot' and 'cross' products, and in the references to 'polarized intensity' and 'unpolarized energy' one can find a gesture towards electromagnetic theory, where quaternions and vector analysis found their first important use. But, as Hamilton wrote, unlike the mathematics of quaternions this 'slight glimpse of the future' was, in 1843, 'certainly very vague.' It was only in the 1880s, after Hamilton's death, that Josiah Willard Gibbs and Oliver Heaviside laid out the fundamentals of vector analysis, dismembering the quaternion system into more useful parts in the process.[45] This key moment in the delocalisation of quaternions was also the moment of their disintegration.

4. Concepts and the Mangle

I have come to the end of my story of Hamilton and quaternions, and the analysis that I have interwoven with the narrative is complex enough, I think, to warrant a general summary and even a little further elaboration.

My overall object in this essay has been to get to grips with the specifically conceptual aspects of scientific practice (I continue to use 'scientific' as an umbrella term which includes mathematics). My point of departure has been the traditional one, an understanding of conceptual extension as a process of modelling. Thus I have tried to show that complex algebra and its geometrical representation in the complex plane were both constitutive models in

---

[44] Note that this geometric interpretation included a handedness rule – a 'rule of rotation' – which reversed the sign of the product of perpendicular lines when the order of their multiplication was reversed, thus explaining algebraic noncommutation in much the same way as the two-dimensional geometrical representation of complex numbers had explained the 'absurd' negative and imaginary quantities.

[45] Crowe, *Vector Analysis*, ch. 5.

Hamilton's practice. But I have gone beyond the tradition in two ways. First, instead of treating modelling (metaphor, analogy) as a primitive term, I have suggested that it bears further analysis and decomposition into the three phases of bridging, transcription and filling. I have exemplified these phases and their interrelation in Hamilton's work, and I have tried to show how the openness of modelling is tentatively cut down by human discretionary choices – by human agency, traditionally conceived – in bridging and filling, and by disciplinary agency – disciplined, machine-like human agency – in transcription. I have also exemplified the fact that these two aspects of modelling – active and passive from the perspective of the human actor – are inextricably intertwined inasmuch as the object of constructing a bridgehead, for example, is, as I have stressed, to load onto it disciplined practices already established around the base model. Conceptual practice thus has the quality of a *dance of agency*, specifically in this case between the discretionary human agent and what I have been calling disciplinary agency. The constitutive part played by disciplinary agency in this dance guarantees that the free moves of human agents – bridging and filling – carry those agents along trajectories that cannot be foreseen in advance, that have to be found out in practice.

My second step beyond traditional conceptions of modelling has been to note that it does not proceed in a vacuum. Issues of cultural multiplicity surface here. My suggestion is that conceptual practice is organised around the production of associations, the making (and breaking) of connections and the creation of alignments between disparate cultural elements, where, in the present instance, the association in question was that between three-place algebras and three-dimensional geometries (initially, at least). And the key observation is that the entanglement of disciplinary agency in practice makes the achievement of such associations nontrivial in the extreme. Hamilton wanted to extend algebra and geometry into three dimensions while maintaining a one-to-one correspondence of elements and operations between them, but neither he nor anyone subsequently has been able to do so. Resistance thus emerges in conceptual practice in relation to intended associations, and precipitates the dialectics of accommodation and further resistance that I call the mangle. Now I want to discuss just what gets mangled.

Most obviously mangled in Hamilton's practice were the modelling vectors that he pursued. In the face of resistance, he tinkered with choices of bridgeheads and fillings, tuning, one can say, the directions along which complex algebra and its geometrical representation were to be extended. And, as we saw, this mangling of modelling vectors eventually (not at all necessarily) met with success. In the quaternion system, Hamilton arrived at an association of one-to-one correspondence of elements and operations between an extension of complex algebra and an extension of its geometrical representation. This achievement constituted an *interactive stabilisation* of the specific free moves and the associated forced moves that led up to it. *This* particular bridgehead, coupled with *these* particular transcriptions and fillings, defined the vector alongst which complex algebra should be extended, and similarly for the associated geometry. Exactly how existing conceptual structures should be extended was, then, the upshot of the mangle – as was the precise structure of the quaternion system that these particular extensions defined.

Here it is worth pausing to reiterate for conceptual practice two points that I have made elsewhere in respect of captures and framings of material agency. First, the precise trajectory and end-point of Hamilton's practice were in no way given in advance. Nothing prior to that practice determined its course. Hamilton had, in the real time of his mathematical work, to fix bridgeheads and fillings and to find out where they led via disciplined transcriptions. He had, further, to find out in real time just what resistances would emerge relative to intended conceptual alignments – such resistances again could not be foreseen in advance – and to make whatever accommodations he could find to them, with the success or failure of such accommodations itself only becoming apparent in practice. Conceptual practice has, therefore, to be seen as *temporally emergent*, as do its products.[46] Likewise, it is

[46] Barbara Herrnstein Smith has commented to me on the preceding sentences that 'idioms of *discovery* (things 'appearing,' agents 'finding' them) seem to dominate just where one might expect those of *construction* to emerge (for better or for worse) most emphatically.' There is a point of potential confusion here that can be clarified. I do want to insist that scientists have to 'find out,' in real-time, where practice leads, what resistances will emerge relative to which associations and alignments. But nothing in my analysis requires or supports the correspondence or Platonist realist assumption that a unique pre-existing structure ('things') are exposed or discovered in the achievement of associations. I argue at some length against correspondence realism and in favour of a non-correspondence 'pragmatic realism' (not antirealism) in *The Mangle*, ch 6. In this connection, I can remark here that when discussing Hamilton's unsuccessful attempts at constructing triplet systems, historians often invoke in a Platonist fashion later mathematical existence proofs that appear to be relevant. Thus, for example, Hankins, *Sir William Rowan Hamilton*, 438, note 2, reproduces the following quotation from the Introduction to Vol. 3 of Hamilton's collected papers (xvi): 'Thirteen years after Hamilton's death G. Frobenius proved that there exist precisely three associative division algebras over the reals, namely, the real numbers themselves, the complex numbers and the real quaternions.' One is tempted to conclude from such assertions that Hamilton's search for triplets was doomed in advance (or fated to arrive at quaternions) and that the temporal emergence of his practice and its products is therefore only apparent. Against this, one can note that proofs like Frobenius' are themselves the products of sequences of practices which remain to be examined. There is no reason to expect that analysis of these sequences would not point to the temporal emergence of the proofs themselves. Note also that these sequences were precipitated by Hamilton's practice and by subsequent work on triplets, quaternions and other many-place systems, all of which served to mark out what an 'associative division algebra over the reals' might mean. Since this concept was not available to Hamilton he cannot have been looking for new instances of it. On the defeasibility of 'proof' see Lakatos, *Proofs and Refutations* and Trevor Pinch, 'What Does a Proof Do If It Does Not Prove? A Study of the Social Conditions and Metaphysical Divisions Leading to David Bohm and John von Neumann Failing to Communicate in Quantum Physics', in *The Social Production of Scientific Knowledge. Sociology of the Sciences, I*, eds Everett Mendelsohn, Peter Weingart and Richard Whitley (Dordrecht, 1977).

appropriate to note the *posthumanist* aspect of conceptual practice as exemplified in Hamilton's work. My analysis here, as elsewhere, entails a decentring of the human subject, though this time towards disciplinary agency rather than the material agency that has been at issue in other discussions.[47] It is not, of course, the case that Hamilton as a human agent disappears from my analysis. I have not sought to reduce him to an 'effect' of disciplinary agency, and I do not think that that can sensibly be done. Hamilton's free moves were just as constitutive of his practice and its product as were his forced ones. It is rather that the centre of gravity of my account is positioned *between* Hamilton as a classical human agent, a locus of free moves, and the disciplines that carried him along. To be more precise, at the centre of my account is the dance of intertwined human and disciplinary agency that traced out the trajectory of Hamilton's practice.

So far I have been talking about the transformation of modelling vectors and formalisms in conceptual practice. But more was mangled and interactively stabilised in our example than that, and I want to consider first the intentional structure of Hamilton's work before returning to its disciplinary aspects. One must, I think, take seriously Hamilton's already quoted intention 'to connect . . . calculation with geometry, through some extension, to the space of three dimensions.' One cannot otherwise make sense of the dialectics of resistance and accommodation that steered his practice through the open-ended space of modelling and eventually terminated in the quaternion system. The point that I want to stress, however, is that we should think of specific goals and purposes as situated *in the plane* of scientific practice. They are not entities that control practice from without. Thus Hamilton's goal was only conceivable within the cultural space in which an association between complex algebra and geometry had already been constructed, and it was further transformed (to an orientation to four instead of three dimensions) in the real time of his practice, as part and parcel of the dialectic of resistance and accommodation that we have examined. Hamilton aimed at an association in three dimensions, but he finally achieved one in four, via the shift in bridgehead implicit in the introduction of the new square-root of −1 that he called k. Like the technical elements of scientific culture, then, goals themselves are always liable to mangling in practice.

From the intentional structure of human agency we can turn to to its disciplined, repetitive, machine-like aspects. I have emphasised that Hamilton was carried along in his practice by disciplinary agency, and it was crucial to my analysis that in his transcriptions he acted without discretion. Such lack of discretion is the precondition for the emergence of dialectics of resistance and accommodation. But it is worth emphasising also that Hamilton evidently did exercise discretion in choosing just which disciplines to submit himself to. Thus, throughout his practice he maintained the first part of the

[47] Pickering, *The Mangle*, chs 2, 3, 5.

geometrical rule already established for the multiplication of lines in the complex plane (that the length of the product of two line-segments was the product of their individual lengths). But it was crucial to his path to quaternions that at a certain point he simply abandoned the second part of the multiplication rule concerning the orientation of product lines in space. He did not attempt to transcribe this when thinking about the multiplication of two arbitrary triplets. Part of Hamilton's strategy of accommodation to resistance was, then, a selective and tentative modification of discipline – in this case, an *eliminative* one. Hamilton bound himself to a part but not all of established routine practice.[48]

One can also understand Hamilton's introduction of non-commuting quantities into his extension of complex algebra as a selective modification of discipline, but in this case an *additive* one. He continued to follow standard practice as far as ordinary numbers were concerned – treating their products as indifferent to the order of terms to be multiplied – but invented a quite new and non-routine rule for the multiplication of his various square-roots of –1. In such ways Hamilton both drew upon established routines to carry himself along and, as part of his accommodation to resistance, transformed those routines, eliminating or adding to them as seemed to him promising. Disciplinary agency, I therefore want to say, has again to be seen as in the plane of practice and mangled there in the very dialectics of resistance and accommodation to which it gives structure. And, further, transformed disciplines are themselves interactively stabilised in the achievement of cultural associations. That certain specific transformations of discipline rather than others should have been adopted was itself determined in the association of calculation with geometry that Hamilton eventually achieved with quaternions.

5. Science and The Mangle

I opened this essay by suggesting that one cannot claim to understand scientific practice unless one can offer an analysis of its specifically conceptual aspects, and that is what I have just sketched out. In this section I want to indicate very briefly how this analysis contributes to a more general picture of scientific practice and culture.

The present essay is part of a project which seeks to develop an understanding of science in a performative idiom, an idiom capable of recognising that the world is continually doing things and that so are we (in contrast to the traditional representational image which sees science as being, above all, in the business of representing a dead nature devoid of agency). Thus I have

[48] Similarly, in his earlier attempts to construct an algebraic system of triplets modelled on his system of couples, in the face of resistances Hamilton abandoned the established algebraic principle of unique division: Hamilton, *Mathematical Papers*, 129-31).

elsewhere paid close attention to the machines and instruments that are integral to scientific culture and practice, and concluded that we should see the machinic field of science as being very precisely adjusted in its material contours to capture and frame material agency. The exact configuration of a machine or an instrument is the upshot of a tuning process that delicately positions it within the flow of material agency, harnessing and directing the latter – domesticating it. The image that lurks in my mind seems to be that of a finely engineered valve that both regulates and directs the flow of water from a pipe (though perhaps it is some kind of a turbine). The performative idiom encourages us also to think about human agency, and the argument that I have sought to exemplify here and elsewhere is that this can be grasped along similar lines. One should think about the scale and social relations of scientific agency, and the disciplined practices of such agency, as likewise being finely tuned in relation to its performativity. And, beyond that, the engineering of the material and the human do not proceed independently of one another: in scientific culture particular configurations of material and human agency appear as interactively stabilised against one another.[49]

Once one begins to think about knowledge as well as performance the picture becomes more elaborate but its form remains the same. One can think of factual and theoretical knowledge in terms of representational chains passing through various levels of abstraction and conceptual multiplicity and terminating, in the world, on captures and framings of material agency. And, as we have seen here, conceptual structures (scientific theories and models, mathematical formalisms) can themselves be understood as positioned in fields of disciplinary agency much as machines are positioned in fields of material agency. Conceptual structures are like precisely engineered valves, too, domesticating disciplinary agency. Again, though, conceptual engineering should not be thought of as proceeding independently of the engineering of the human and material. As I have just argued, disciplines, for instance, are themselves subject to transformation in conceptual practice, and, in general, conceptual and machinic elements of culture should be seen as evolving together in empirical practice. Scientific culture, then, appears as itself a wild kind of machine built from radically heterogeneous parts, a supercyborg, harnessing material and disciplinary agency in material and human performances, some of which lead out into the world of representation, of facts and theories.[50]

I confess that I like this image of scientific culture. It helps me to fix in my mind the fact that the specific contents of scientific knowledge are always immediately tied to specific and very precisely formed fields of machines and disciplines. Above all, it helps me to focus on the fact that scientific knowl-

49 Pickering, *The Mangle*, chs 2, 3.

50 I have seen draft essays by John Law that evoke a related image of scientific and technological culture as a kind of giant plumbing system, continually under extension and repair.

edge is just one part of the picture, not analytically privileged in any way but something that evolves in an impure, posthuman, dynamics together with all of the other cultural strata of science – material, human, social (in the next section I throw metaphysical systems into this assemblage, too). This is, of course, in contrast with traditional representationalist images of science, which can hardly get the nonrepresentational strata of science into focus and which can never grasp its performative aspect.

I turn now to the question of how the supercyborg of scientific culture is extended in time. Traditional answers assert that something substantive within scientific culture (as I define it) endures through cultural extension and explains or controls it – social interests, epistemic rules, or whatever. Or perhaps something quite outside culture has the controlling role: the world itself, Nature. I have previously criticised the idea that the social can play the required explanatory role,[51] and I continue this argument below. I have also argued against against any necessarily controlling role for epistemic rules and given my own account of how 'the world itself' plays into cultural extension.[52] Here I want to stress that on my analysis *nothing* substantive explains or controls the extension of scientific culture. Existing culture is the surface of emergence of its own extension, in a process of open-ended modelling having no destination given or knowable in advance. Everything within the multiple and heterogeneous culture of science is, in principle, at stake in practice. Trajectories of cultural transformation are determined in dialectics of resistance and accommodation played out in real-time encounters with temporally emergent agency, dialectics which occasionally arrive at temporary oases of rest in the achievement of captures and framings of agency and of associations between multiple cultural extensions. I have noted, it is true, that one needs to think about the intentional structure of human agency to understand this process; vectors of cultural extension are tentatively fixed in the formulation of scientific plans and goals, and resistances have to be seen as relative to such goals. But as I have shown, plans and goals are both emergent from existing culture and at stake in scientific practice, themselves liable to mangling in dialectics of resistance and accommodation. They do not endure through, explain or control cultural extension.

So this is my overall claim about science: there is no substantive explanation to be given for the extension of scientific culture. There is however, and this is also my claim, a temporal pattern to practice that we can grasp, that we can find instantiated everywhere, and that constitutes an understanding of what it is going on. It is the pattern just described – of open-ended extension through modelling, dialectics of resistance and accommodation, and so on. And in good conscience, this pattern – the mangle – is the only explanation that I can defend of what scientific culture becomes at any moment: of the

---

[51] Pickering, *The Mangle*, section 2.5.
[52] Ibid, ch. 6.

configuration of its machines, of its facts and theories, of its conceptual structures, disciplines and social relations, and so forth. Science mangles on.

6. Mathematics, Metaphysics and the Social

The central task of this essay has been to understand how dialectics of resistance and accommodation can arise in conceptual practice. I want to end, however, by developing two subsidiary topics. It is common knowledge, amongst historians of mathematics at least, that Hamilton was as much a metaphysician as a mathematician, and that he felt that his metaphysics was, indeed, at the heart of his mathematics.[53] I therefore want to see how the relation between mathematics and metaphysics can be understood in this instance. At the same time, it happens that David Bloor has offered a clear and interesting explanation of Hamilton's metaphysics as a case study in the sociology of scientific knowledge (SSK).[54] SSK has been at the heart of developments in science studies over the past thirty years or so, and I think it will therefore be useful to try to clarify how my own analysis of practice diverges from it.[55] In what follows I focus upon issues of temporal emergence and the possibilities for a distinctively sociological explanation of science (this latter connecting directly to the posthumanism of the mangle). I begin with Bloor's account of Hamilton's metaphysics and then offer my own.

Bloor's essay on 'Hamilton and Peacock on the Essence of Mathematics' focusses upon the different metaphysical understandings of algebra articulated by Hamilton on the one side and a group of Cambridge mathematicians including Peacock on the other. We can get at this difference by returning to the foundational crisis in 19th-century algebra. As discussed in section 2 above, the geometrical representation of complex algebra was one way of defusing the crisis and giving meaning to negative and imaginary quantities. But various mathematicians did not opt for this commonsense route, preferring more metaphysical approaches. Peacock and the Cambridge mathematicians took a *formalist* line, as Bloor calls it, which suggested that mathematical symbols and the systems in which they were embedded were sufficient unto themselves, in need of no extra-mathematical foundations and subject to whatever interpretation proved appropriate to specific uses. Thus,

---

[53] Hankins, *Sir William Rowan Hamilton*, John Hendry, 'The Evolution of William Rowan Hamilton's View of Algebra as the Science of Pure Time,' *Studies in History and Philosophy of Science* 15 (1984): 63-81.

[54] Bloor, 'Hamilton and Peacock on the Essence of Algebra', in *Social History of Nineteenth Century Mathematics*, eds H. Mehrtens, H. Bos and I. Schneider (Boston, 1981).

[55] Canonical works in SSK include Barry Barnes, *T. S. Kuhn*, Bloor, *Knowledge and Social Imagery*, Harry Collins, *Changing Order: Replication and Induction in Scientific Practice* (Chicago, 1992, 2nd ed.) and Steven Shapin, 'History of Science and Its Sociological Reconstructions,' *History of Science* 20 (1982): 157-211.

from the formalist point of view, there was and could be no foundational crisis in algebra. A quotation from the mathematician George Boole sums up this position nicely:[56]

> They who are acquainted with the present state of the theory of Symbolical Algebra are aware, that the validity of the processes of analysis does not depend upon the interpretation of the symbols which are employed, but solely upon the laws of their combination. Every system of interpretation which does not affect the truth of the relations supposed, is equally admissible, and it is thus that the same process may, under one scheme of interpretation, represent the solution of a question on the properties of numbers, under another, that of a geometrical problem, and under a third, that of a problem of dynamics or optics.

Hamilton disagreed. He thought that mathematical symbols and operations must have some solid foundations that the mind latches onto – consciously or not – in doing algebra. And, as Bloor puts it: 'Hamilton's metaphysical interests placed him securely in the Idealist tradition. He adopted the Kantian view that mathematics is synthetic *a priori* knowledge. Mathematics derives from those features of the mind which are innate and which determine *a priori* the general form that our experience must take. Thus geometry unfolds for us the pure form of our intuition of space. Hamilton then said that if geometry was the science of pure space, then algebra was the science of pure time.'[57]

And, indeed, Hamilton developed his entire theory of complex algebra explicitly in such terms. In his 'Theory of Conjugate Functions, or Algebraic Couples; With a Preliminary and Elementary Essay on Algebra as the Science of Pure Time,' first read to the Royal Irish Academy in 1833, he showed how positive real algebraic variables – denoted a, b, c, etc – could be regarded as 'steps' in time (rather than magnitudes of material entities), and how negative signs in front of them could be taken as denoting reversals of temporality, changing before into after. He also elaborated the system of couples mentioned earlier. Written (a, b), these couples transformed like the usual complex variables under the standard mathematical operations but, importantly, the problematic symbol 'i' was just absent from them. Hamilton's claim was thus to have positively located and described the foundations of complex algebra in our intuitions of time and its passing.[58]

---

[56] Boole, *Mathematical Analysis of Logic* (1847), quoted in Ernest Nagel, *Teleology Revisited and Other Essays in the Philosophy and History of Science* (New York, 1979), 166.

[57] Bloor, 'Hamilton and Peacock,' 204.

[58] Hendry, 'The Evolution of William Rowan Hamilton's View of Algebra,' offers a subtle analysis of the early development of Hamilton's Kantianism and suggests that Hamilton might possibly have developed his theory of couples independently of it, hitching the metaphysics to the algebra 'as a vehicle through which to get the essay published' (64). This observation creates no problems for my argument in what follows. I do not insist that practice always has a metaphysical flavour, but I do want to insist that metaphysics, when relevant, is subject to change and transformation in practice.

This much is well known, but Bloor takes the argument one step further: 'I am interested in why men who were leaders in their field, and who agreed about so much at the level of technical detail, nevertheless failed to agree for many years about the fundamental nature of their science. I shall propose and defend a sociological theory about Hamilton's metaphysics and the divergence of opinion about symbolical algebra to which he was a party.'[59] Bloor's idea is thus, first, that the technical substance of algebra did not determine its metaphysical interpretation, and therefore, second, that we need to invoke something other than technical substance – namely, the social – to explain why particular individuals and groups subscribed to the metaphysical positions that they did. This is a standard opening gambit in SSK, and Bloor follows it up by discussing the different social positions and visions of Hamilton and the Cambridge formalists and explaining how particular metaphysical views serve to buttress them.[60] According to Bloor, Hamilton was aligned with Coleridge and his circle and, more broadly, with 'the interests served by Idealism'[61] – conservative, holistic, reactionary interests opposing the growing materialism, commercialism and individualism of the early 19th century and the consequent breakdown of the traditional social order. As Bloor explains it, Hamilton's idealism assimilated mathematics to the Kantian category of Understanding. Understanding in turn was understood to be subordinate to the higher faculty of Reason. And, on the plane of human affairs, Reason was itself the province not of mathematics but of religion and the church. Thus the 'practical import' of Hamilton's idealism 'was to place mathematics as a profession in a relation of general subordination to the Church. Algebra, as Hamilton viewed it, would always be a reminder of, and a support for, a particular conception of the social order. It was symbolic of an "organic" social order of the kind which found its expression in Coleridge's work on Church and State.'[62]

So, Hamilton's social vision and aspirations structured his metaphysics. As far as the Cambridge group of mathematicians was concerned, the same pattern was repeated but starting from a different point. In mathematics and beyond they were both 'reformers and radicals' and 'professionals' keen to assert their autonomy from traditional sources of authority like the church.[63] Their formalism and its opposition to the foundationalism of people like

---

[59] Bloor, 'Hamilton and Peacock,' 203.

[60] The only respect in which Bloor's essay is untypical of SSK is that he stops short, explaining metaphysics without pressing on into the technical substance of science. He does remark, however, that 'should it transpire that this metaphysics is indeed relevant to technical mathematics, then my ideas may help to illuminate these matters as well' (ibid., 206). I am more concerned with the overall form of Bloor's argument than with its restriction to metaphysics.

[61] Ibid., 220.

[62] Ibid., 217.

[63] Ibid., 222, 228.

Hamilton, then, served this end, defining mathematics as the special province of mathematicians. It was an anti-metaphysics, one might say, which served to keep metaphysicians and the church out. Thus Bloor's analysis of the differences between the two parties over the foundations of mathematics. As he summarises it:[64]

> Stated in its broadest terms, to be a formalist was to say: 'we can take charge of ourselves.' To reject formalism was to reject this message. These doctrines were, therefore, ways of rejecting or endorsing the established institutions of social control and spiritual guidance, and the established hierarchy of learned professions and intellectual callings. Attitudes towards symbols were themselves symbolic, and the messages they carried were about the autonomy and dependence of the groups which adopted them.

I have no quarrel with Bloor's arguments as rehearsed so far. I have no knowledge of the social locations and aspirations of the parties concerned that would give me cause to doubt the existence of the social-metaphysical-mathematical correlations he outlines. But still, something peculiar happens towards the end of Bloor's essay. He concludes by stating that 'I do not pretend that this account is without problems or complicating factors' and then lists them.[65] For the rest of this section we will be concerned with just one complicating factor.

'It is necessary,' Bloor remarks, 'to notice and account for the fact that Hamilton's opposition to Cambridge formalism seemed to decline with time. In a letter to Peacock dated Oct. 13, 1846, Hamilton declared that his view about the importance of symbolical science "may have approximated gradually to yours." Interestingly,' Bloor remarks, 'Hamilton also noted some four years later "how much the course of time has worn away my political eagerness."'[66] The structure of these sentences is, I think, characteristic of what SSK, the sociology of scientific knowledge, looks like when brought to bear upon empirical studies. Note first that the shift in Hamilton's metaphysics is viewed as a 'problem.' It appears that way to Bloor because he wants to understand the social as not just a correlate of the metaphysical but as a kind of cause.[67] The social is the solid, reliable foundation that holds specific metaphysical positions in place in an otherwise open-ended space. Any drifting of Hamilton's metaphysics threatens this understanding, and Bloor therefore tries to recoup this drift, by qualifying it as perhaps apparent – 'seemed to decline' – and then by associating it with a decline in Hamilton's 'political eagerness.' Perhaps Hamilton's social situation and views changed first and gave rise to Hamilton's concessions to formalism, seems to be Bloor's

[64] Ibid., 228.
[65] Ibid., 228.
[66] Ibid., 229.
[67] Bloor, *Knowledge and Social Imagery*, lists causal social explanation as the first distinguishing mark of the 'strong programme' in SSK.

message (though the dates hardly look promising). If so, Bloor's causal arrow running from the social to the metaphysical would be secure.[68]

In what follows, I want to offer a different interpretation of Hamilton's metaphysical wandering, but before that I want to comment further on Bloor's general position. Three points bear emphasis. First, although I earlier described SSK as tending to regard the social as a non-emergent cause of cultural change in science, it is clear that Bloor *does* recognise here that the social can itself change with time. This is precisely how he hopes to cope with the problem of changing metaphysics. But second, he offers no examination or analysis of how the social changes. The social, I want to say, is treated as an at most *quasi-emergent* category, both in this essay and in the SSK canon in general.[69] The gaze of SSK only ever catches a fixed image of the social in the act of structuring the development of the technical and metaphysical strata of science. SSK always seems to miss the movie in which the social is itself transformed.[70]

---

[68] Thus Bloor's text immediately following the previous quotation continues: 'A corresponding and opposite movement took place in Whewell's life. Here, *in obliging conformity with my thesis*, it is known that as Whewell moved to the right . . . he increasingly moved away from the symbolical approach in his mathematical writings' ('Hamilton and Peacock,' 229-30, emphasis mine).

[69] See the works cited in note 55.

[70] In its early development, SSK was articulated against philosophical positions that rancorously opposed the suggestion that there was *anything* significantly social about scientific knowledge. A concentration on situations where the social could plausibly be regarded as both fixed and as explanatory of metaphysical and technical developments therefore fulfilled a strategic argumentative function for SSK. SSK's endless deferral of any enquiry into how the social might itself evolve seems strange, though, even given that background. It is, I suspect, part and parcel of SSK's almost principled refusal to interrogate key sociological concepts like 'interest.' Thus, in 1977 we find Barry Barnes writing in *Interests and the Growth of Knowledge* (London, 1977), 78, that 'new forms of activity arise not because men are determined by new ideas, but because they actively deploy their knowledge in a new context, as a resource to further their interests,' but then, on the last page of the book, he shuffles interests out into unexplored regions of social theory with the remark that, 'I have deliberately refrained from advancing any precise definitions of "interest" and "social structure;" this would have had the effect of linking the claims being advanced to particular schools of thought within sociological theory. Instead, I have been content, as it were, to latch the sociology of knowledge into the ongoing general trends of social thought' (ibid., 86). Nothing has changed in the intervening years. Interests, and *de facto* their dynamics, are still left out in the cold by SSK. In a recent essay review of Latour's *Science in Action*, Shapin writes: 'One must . . . welcome any pressure that urges analysts further to refine, define, justify and reflect upon their explanatory resources. If there is misunderstanding, by no means all the blame needs to be laid at Latour's door [sic!]. "Interest-explanation" does indeed merit further justification' ('Following Scientists Around,' *Social Studies of Science* 18 (1988): 549). And replying to his critics in the 'Afterword' to the second edition of *Knowledge and Social Imagery*, Bloor writes that, 'Undeniably the terminology of interest explanations is intuitive, and much about them awaits clarification' (171). In chapter 3 of that same book, Bloor advances the Durkheimian argument that resistance to the strong programme arises from a sacred quality attributed to science in modern society. Perhaps in SSK the social has become the sacred.

Bloor's essay, then, exemplifies an important difference between SSK and the mangle: in contrast to SSK, I would argue that the social should in general be seen as in the plane of practice, both feeding into technical practice and being emergently mangled there, rather than as a fixed origin of unidirectional causal arrows. Third, it is characteristic of SSK that Bloor does not even consider the possibility that there might be any explanation for Hamilton's metaphysical shift *other than* a change in the social. In contrast, I now want to offer an explanation that refers this shift not outwards to the social but inwards, towards Hamilton's technical practice.

Bloor says that Hamilton's opposition to formalism 'seemed' to decline, but the evidence is that there was no 'seeming' about it. As Hamilton put it in another letter written in 1846 to his friend Robert Graves:[71]

> I feel an increased sympathy with, and fancy that I better understand, the Philological School [Bloor's formalists]. It enables me to see better the high functions of language, to trace more distinctly and more generally the influence of signs over thoughts, and to understand an answer which I hazarded some years ago to a question of yours, What did I suppose to be the *Science of Pure Kind*? namely, that I supposed it must be the *Science of Symbols*.

1846, in fact, seems to be an important date in Hamilton's metaphysical biography. It was just around then that he began to indicate in various ways that his position had changed. One might suspect, therefore, that Hamilton's worries about metaphysical idealism had their origins in his technical practice around quaternions in the early 1840s.[72] And this suspicion is supported by the fact that Hamilton's technical writings on quaternions – specifically the preface to his first book on the subject, the massive *Lectures on Quaternions* of 1853 – contain several explicit discussions of his past and present metaphysical stances. We can peruse a few and try to make sense of what happened.

Hamilton's preface to the *Lectures* takes the form of a historical introduction to his thought and to related work of other mathematicians, and one striking feature of it is the tone of regret and retraction that Hamilton adopts whenever the Science of Pure Time comes up. The preface begins with a summary of his early work on couples, which he introduces with the remark that: 'In this manner I was led, many years ago, to regard Algebra as the SCIENCE OF PURE TIME . . . If I now reproduce a few of the opinions put forward in

[71] Quoted in Nagel, *Teleology Revisited*, 189.

[72] Hankins, *Sir William Rowan Hamilton*, 310, briefly connects Hamilton's metaphysical shift with his technical practice along the lines elaborated below. I should mention that Hamilton's Kantianism had a second string besides his thinking about time – namely, a concern with triadic structures grasped in relation to the Trinity (ibid., 285-91). Besides possible utility, then, Hamilton's searches for triplets and his concern with three-dimensional geometry have themselves a metaphysical aspect. My focus here, though, is with his overall move away from Kantianism towards formalism.

that early Essay, it will be simply because they may assist the reader to place himself in that *point of view*, as regards the first elements of *algebra*, from which a passage was gradually made by me to that comparatively *geometrical* conception which it is the aim of this volume to unfold. And with respect to anything unusual in the *interpretations* thus proposed, for some simple and elementary notations, it is my wish to be understood as not at all insisting on them as *necessary*, but merely proposing them as consistent amongst themselves, and preparatory to the study of quaternions, in at least one aspect of the latter.'[73] So much for *a priori* knowledge.

Later, Hamilton verges upon apology for mentioning his old metaphysics: 'Perhaps I ought to apologise for having thus ventured here to reproduce (although only historically . . .) a view so little supported by scientific authority. I am very willing to believe that (though not unused to calculation) I may have habitually attended too little to the *symbolical* character of Algebra, as a Language, or organized system of *signs*: and too much (in proportion) to what I have been accustomed to consider its *scientific* character, as a Doctrine analogous to Geometry, through the Kantian parallelism between the *intuitions* of Time and Space.'[74] Later still, Hamilton speaks positively about the virtues of formalism and their integration into his own mathematical practice, saying that he 'had attempted, in the composition of that particular series [of papers on quaternions understood as quotients of lines in three-dimensional space, published between 1846 and 1849], to allow a more prominent influence to the general *laws of symbolical language* than in some former papers of mine; and that to this extent I had on this occasion sought to imitate the *Symbolical Algebra* of Dr Peacock.'[75]

Far from being situated on the opposite side of a metaphysical gulf from Peacock, then, by 1846 Hamilton was *imitating* Peacock's formalist approach in his technical practice (without, I should add, entirely abandoning his earlier Kantianism). And to understand why, we need, I think, to look more closely at that practice. In the very long footnote that begins with the apology for mentioning the Science of Pure Time, Hamilton actually goes on to assert that he could have developed many of the aspects of the quaternion system to be covered in the rest of the book within his original metaphysical framework, and that this line of development 'would offer no result which was not perfectly and easily *intelligible*, in strict consistency with that *original* thought (or intuition) of time, from which the whole theory should (on this supposition) be evolved . . . Still,' he continues,[76]

> I admit fully that the actual *calculations* suggested by this [the Science of Pure Time], or any other view, must be performed according to some fixed *laws of*

[73] Hamilton, *Lectures on Quaternions*, 117-18.
[74] Ibid., 125.
[75] Ibid., 153.
[76] Ibid., 125-26.

> *combination of symbols*, such as Professor De Morgan has sought to reduce, for ordinary algebra, to the smallest possible compass . . . and that in following out such *laws* in their symbolical consequences, uninterpretable (or at least uninterpreted) *results* may be expected to arise. . . [For example] in the passage which I have made (in the Seventh Lecture), from *quaternions* considered as *real* (or as geometrically *interpreted*), to *biquaternions* considered as *imaginary* (or as geometrically *uninterpreted*), but as symbolically *suggested* by the generalization of the quaternion formulae, it will be perceived . . . that I have followed a *method of transition*, from *theorems proved* for the *particular* to *expressions assumed* for the *general*, which bears a very close *analogy* to the methods of Ohm and Peacock: although I have *since* thought of a way of *geometrically interpreting the biquaternions* also.

Now, I am not going to exceed my competence by trying to explain what biquaternions are and how they specifically fit into the story, but I think one can get an inkling from this quotation of how and why Hamilton's metaphysics changed. While Hamilton had found it possible calmly to work out his version of complex algebra on the basis of his Kantian notions about time, in his subsequent mathematical practice leading through quaternions he was, to put it crudely, flying with the seat of his pants. He was struggling through dialectics of resistance and accommodation, reacting as best he could to the exigencies of technical practice, without much regard to or help from any *a priori* intuitions of the inner meanings of the symbols he was manipulating. The variety of the bridging and filling moves that he made on the way to quaternions that I reviewed above, for example, hardly betray any 'strict consistency' with an '*original* thought (or intuition).' Further, what guided Hamilton through the open-ended space of modelling was, I argued, disciplinary agency – the replaying of established *formal* manipulations in new contexts marked out by bridging and filling. And, at the level of products rather than processes, a similar situation obtained. Hamilton continually arrived at technical results and then had to scratch around for interpretations of them – starting with the search for a three-dimensional geometric interpretation of his initial four-dimensional formulation of quaternions, and ending up in the quotation just given with biquaternions ('I have since thought of a way'). Moreover, Hamilton proved to be able to think of several ways of interpreting his findings. In the preface to the *Lectures* he discusses three different three-dimensional geometrical interpretations, one of which (not that mentioned above at the end of section 3) forms the basis for his exposition of quaternions in the body of the book.[77] Formal results followed by an indefinite number of interpretations: this is a description of formalist metaphysics.

[77] Ibid., 145-54. Of one of these systems, Hamilton wrote: 'It seemed (and still seems) to me natural to connect this *extra-spatial unit* [the non-geometrical part of the quaternion] with the conception of TIME.' But then he reverted to the formalist mode: 'Whatever may be thought of these abstract and semi-metaphysical *views*, the *formulae* . . . are in any event a sufficient *basis* for the erection of a CALCULUS of quaternions' (ibid., 152).

So, there is a *prima facie* case for understanding the transformation in Hamilton's metaphysics in the mid-1840s as an accommodation to resistances arising in technical-metaphysical practice. A tension emerged between Hamilton's Kantian *a priorism* and his technical practice, to which he responded by attenuating the former and adding to it an important dash of formalism. My suggestion is, therefore, that we should see metaphysics as yet another heterogeneous element of the culture that scientists operate in and on. Like the technical culture of science, like the conceptual, like the social, and like discipline, metaphysics is itself at stake in practice, and just as liable to temporally emergent mangling there in interaction with all of those other elements. That is the positive conclusion of this section as far as my analysis of practice is concerned.

Comparatively, I have tried to show how my analysis differs from SSK in its handling of a specific example. Where SSK necessarily looks outwards from metaphysics (and technical culture in general) to quasi-emergent aspects of the social for explanations of change (and stability), I have looked inwards, to technical practice itself. There is an emergent dynamics there that goes unrecognised in SSK. I have, of course, said nothing on my own account about the transformation in Hamilton's 'political eagerness' that Bloor mentions. Having earlier argued for the mangling of the social, I find it quite conceivable that Hamilton's political views might also have been emergently mangled and interactively stabilised alongside his metaphysics in the evolution of the quaternion system. On the other hand they might not. I have no more information on this topic than Bloor – but at least the mangle can indicate a way past the peculiar quasi-emergent vision of the social that SSK offers us.

16

Mathematics as Objective Knowledge and as Human Practice

EDUARD GLAS

Mathematics is the product of a communal practice, but at the same time this product becomes partially autonomous from the practice that produced it. Although created by ourselves, mathematical objects are not entirely transparent to us: they possess objective properties and give rise to problems that are certainly not our own inventions. These statements are easily recognizable as aspects of Popper's theory of objective knowledge. Popper saw mathematics (as well as science, art and other sociocultural institutions) as an evolutionary product of the intellectual efforts of humans who, by objectivizing their creations and trying to solve the often unintended and unexpected problems arising from those creations, produce new mathematical objects, problems and critical arguments.

Mathematical propositions are proved by logically compelling arguments, independently of empirical evidence and therefore free of the interpretational ambiguities that make empirical scientific knowledge essentially uncertain. Real alternatives, in the sense of mutually incompatible theories generating conflicting claims about the truth or falsity of particular statements, seem not to exist in mathematics. Those who subscribe to this image of mathematics often conclude that social processes could not possibly play a constitutive role in the development of mathematical knowledge. On the other hand those who, like social constructivists and adherents to the 'strong programme', insist on the social nature of the mathematical enterprise, mostly begin by challenging the said image.

I will argue that in order to acknowledge the social dimension of mathematics there is no need to question the objectivity and partial autonomy of mathematical knowledge in the Popperian sense. It suffices to shift our focus from the ways in which new truths are derived to the ways in which new problems are conceived and tackled. Indeed, there is more to mathematics than the mere accumulation of true statements. Mathematicians are not interested just in truths (much less in truisms), but in truths that provide answers to questions that are worthwhile and promising in the contemporary scene of inquiry.

Which problems are considered promising and worthwhile is, of course, not independent of the communal practice involved. Mathematical development is to a certain extent shaped by the shared conceptions of problems, aims and values of a research community. Differentiation between communities of practitioners with varying conceptions of what are the relevant questions, what the appropriate ways of tackling them, what the right criteria for appraising success, etc., provides ample room for intellectual variation and selection, and thus may affect the developmental pattern of the discipline. I will present a historical example that shows this evolutionary mechanism at work.

My presentation will consist of two parts. In the first part, I will discuss and defend Popper's theory of the evolution of objective mathematical knowledge, as an important alternative to the foundationist schools of formalism, intuitionism and platonism. For Popper, the theory of knowledge ultimately boils down to the theory of problem solving, and it is from this perspective that the sociocultural side of mathematics can fruitfully be approached, as I will try to show in the second part.

1. Fallibilism

Popper is not usually regarded as a philosopher of mathematics. As mathematical propositions fail to forbid any observable state of affairs, his demarcation criterion clearly divides mathematics from empirical science, and Popper was primarily concerned with empirical science. When speaking of a Popperian philosophy of mathematics, we mostly immediately think of Lakatos, who is usually considered to have applied and extended Popper's philosophy of *science* to mathematics. Like Lakatos, Popper saw considerable similarity between the methods of mathematics and of science – he held most of mathematics to be hypothetico-deductive (Popper 1984, p. 70) – and he thought highly of his former pupil's quasi-empiricist approach to the logic of mathematical development (Popper 1981, p. 136-7, 143, 165). His own views of the matter, however, are not to be identified with those of Lakatos, nor does their significance consist only in their having prepared the ground for the latter's methodological endeavours.

Popper never developed his views of mathematics systematically. However, scattered throughout his works, and often in function of other discussions, there are many passages which together amount to a truly Popperian philosophy of mathematics. This side of Popper's philosophy has remained rather underexposed, especially as compared with the excitement aroused by Lakatos's work, many of whose central ideas were developments of Popperian views, not only of science, but more specifically of mathematics as well (cf. Glas 2001).

Already, in *Logik der Forschung*, Popper had argued that we should never save a threatened theoretical system by ad hoc adjustments, 'conventionalist stratagems', that reduce its testability (Popper 1972, p. 82-3) – a view which was to be exploited by Lakatos to such dramatic effect in the dialogues of *Proofs and Refutations*, under the heads of monster barring, exception barring, and monster adjustment. In *Conjectures and Refutations*, Popper had shown how the critical method can be applied to pure mathematics. Rather than questioning directly the status of mathematical truths, he tackled mathematical absolutism from a different angle. Mathematical truths may possess the greatest possible (though never absolute) certainty, but mathematics is not just accumulation of truths. Theories essentially are attempts at solving certain problems, and they are to be critically assessed, evaluated, and tested, by their ability to adequately solve the problems that they address, especially in comparison with possible rivals (Popper 1969, p. 197-9, 230).

This form of critical fallibilism obviously differed from Lakatos's quasi-empiricism, among other things by avoiding the latter's considerable problems with identifying the 'basic statements' that can act as potential falsifiers of mathematical theories. Even so, Lakatos's referring to what he called Popper's 'mistake of reserving a privileged infallible status for mathematics' (Lakatos 1976, p. 139 footnote) seems unjust. Claiming immunity to one kind of refutation – empirical – is not claiming immunity to all forms of criticism, much less infallibility. As a matter of fact, Popper did not consider anything, including logic itself, entirely certain and incorrigible (Popper 1984, p. 70-2).

2. Objectivity

Central to Popper's philosophy of mathematics was a group of ideas clustering around the doctrine of the relative autonomy of knowledge 'in the objective sense' – in contradistinction to the subjective sense of the beliefs of a knowing subject. Characteristic of science and mathematics is that they are formulated in a descriptive and argumentative language, and that the problems, theories, and errors contained in them stand in particular relations, which are independent of the beliefs that humans may have with respect to them. Once objectivized from their human creators, mathematical theories have an infinity of entailments, some entirely unintended and unexpected, that transcend the subjective consciousness of any human – and even of all humans, as is shown by the existence of unsolvable problems (Popper 1981, p. 161). In this sense, no human subject can ever completely 'know' the objective content of a mathematical theory, that is, including all its unforeseeable and unfathomable implications.

It is of course trivially true that knowledge in the said objective sense can subsist without anybody being aware of it, for instance in the case of totally forgotten theories that are later recaptured from some written source. It also has significant effects on human consciousness – even observation depends on judgements made against a background of objective knowledge – and through it on the physical world (for instance in the form of technologies).

Human consciousness thus typically acts as a mediator between the abstract and the concrete, or the world of culture and the world of nature. To acknowledge that linguistically expressed knowledge can subsist without humans, that it possesses independent properties and relationships, and that it can produce mental and also – indirectly – physical effects, is tantamount to saying that it in a way exists. Of course, it does not exist in the way in which we say that physical or mental objects or processes exist: its existence is of a 'third' kind. As is well known, Popper coined the expression 'third world' (or 'world 3', as he later preferred) to refer to this abstract realm of objectivized products of human thought and language.

Popper's insisting upon the crucial distinction between the objective (third-world) and the subjective (second-world) dimension of knowledge enabled him to overcome the traditional dichotomies between those philosophies of mathematics that hold mathematical objects to be human constructions, intuitions, or inventions, and those that postulate their objective existence. His 'epistemology without a knowing subject' accounts for how mathematics can at once be autonomous *and* man-made, that is, how mathematical objects, relations and problems can be said in a way to exist independently of human consciousness *although* they are products of human (especially linguistic) practices. Mathematics is a human activity, and the product of this activity, mathematical knowledge, is a human creation. Once created, however, this product assumes a partially autonomous and timeless status (it 'alienates' itself from its creators, as Lakatos would have it), that is, it comes to possess its own objective, partly unintended and unexpected properties, irrespective of when, if ever, humans become aware of them.

Popper regarded mathematical objects – the system of natural numbers in particular – as products of human language and human thought: acquiring a language essentially means being able to grasp objective thought *contents*. The development of mathematics shows that with new linguistic means new kinds of facts and in particular new kinds of problems can be described. Unlike what apriorists like Kant and Descartes held, being human constructions does not make mathematical objects completely transparent, *clair et distinct*, to us. For instance, as soon as the natural numbers had been created or invented, the distinctions between odd and even, and between compound and prime numbers, and the associated problem of the Goldbach conjecture came to exist objectively: Is any even number greater than 2 the sum of two primes? Is this problem solvable or unsolvable? And if unsolvable, can its unsolvability be proved? (Popper 1984, p. 34). These problems in a sense have existed ever since humankind possessed a number system, although during many centuries nobody had been aware of them. Thus we can make genuine *discoveries* of independent problems and new hard facts about our own creations, and of objective (not merely intersubjective) truths about these matters.

Nothing mystical is involved here. On the contrary, Popper brought the platonist heaven of ideal mathematical entities down to earth, characterizing it as objectivized *human* knowledge. The theory of the third world at once accounts for the working mathematician's strong feeling that (s)he is dealing with something real, and explains how human consciousness can have access to abstract objects. As we have seen, these objects are not causally inert: for instance, by reading texts we become aware of some of their objective contents and the problems, arguments, etc., that are contained in them, so that the platonist riddle of how we can gain knowledge of objects existing outside space and time does not arise. Of course, speaking of causality here is using this notion in a somewhat peculiar, not in a mechanistic sense. That reading texts causes in us a certain awareness of what is contained in those texts is just a plain fact, for whose acceptance no intricate causal theory of language understanding is needed.

Cultural artefacts like mathematics possess their own partially autonomous properties and relationships, which are independent of our awareness of them: they have the character of hard facts that are to be *discovered.* In this respect they are very much like physical objects and relations, which are not unconditionally 'observable' either, but are only apprehended in a language which already incorporates many theories in the very structure of its usages. Like mathematical facts, empirical facts are thoroughly theory-impregnated and speculative, so that a strict separation between what traditionally has been called the analytic and the synthetic elements of scientific theories is illusory. The effectiveness of pure mathematics in natural science is miraculous only to a positivist, who cannot imagine how formulas arrived at entirely independently of empirical data can be adequate for the formulation of theories supposedly inferred from empirical data. But once it is recognized that the basic concepts and operations of arithmetic and geometry have been designed originally for the practical purpose of counting and measuring, it is almost trivial that all mathematics based on them remains applicable exactly to the extent that natural phenomena resemble operations in geometry and arithmetic sufficiently to be conceptualized in (man-made) terms of countable and measurable things, and thus to be represented in mathematical language. In mathematics and physics alike, theories are often put forward as mere speculations, mere possibilities, the difference being that scientific theories are to be tested directly against empirical material, and mathematical theories only indirectly, if and in so far as they are applied in physics or otherwise (Popper 1969, p. 210, 331).

3. Interaction

It is especially the (dialectic) idea of *interaction* and partial *overlap* between the three worlds that makes Popper's theory transcend the foundationist programmes. Clearly, objective knowledge (at world-3 level) – the objective contents of theories – can exist only if those theories have been materially realized in texts (at world-1 level), which cannot be written nor be read without involving human consciousness (at world-2 level). Put somewhat bluntly, platonists acknowledge only a third world as the realm to which all mathematical truths pertain, strictly separated from the physical world; intuitionists locate mathematics in a second world of mental constructions and operations, whereas formalists reduce mathematics to rule-governed manipulation with 'signs signifying nothing', that is, mere material (first-world) 'marks'. In all these cases, reality is split up into at most two independent realms (physical and ideal or physical and mental), as if these were the only possible alternatives. Popper's tripartite world view surpasses physicalist or mentalist reductionism as well as physical/mental dualism, emphasizing that there are *three* partially autonomous realms, intimately coupled through feed-back. The theory of the interaction between all three worlds shows how these seemingly incompatible mathematical ontologies can be reconciled and their mutual oppositions superseded (Popper 1984, p. 36-37; cf Niiniluoto 1992).

The notion of a partially autonomous realm of objective knowledge has been criticized, most elaborately by O'Hear in his Popper monograph (O'Hear 1980). O'Hear does not deny that objectivized mathematical theories have partly unforeseeable and inevitable implications, but he does not consider this sufficient reason for posing what he calls 'an autonomous non-human realm of pure ideas'. Popper, of course, always spoke of a *partially* autonomous realm, not of 'pure ideas', but especially of fallible theories, problems, tentative solutions and critical arguments. O'Hear, however, argues that Popper's theory is misleading because it implies that we are not in control of world 3 but are, on the contrary, completely controlled *by* it (*ibid*., p. 183, 207). For relationships in world 3 are of a *logical* character and this seems to imply that they are completely beyond our control. On O'Hear's construal, Popper allowed only a human-constructive input at the very beginning of the history of mathematics – the phase of primitive concepts connected with counting and measuring – after which logic took over and developments were no longer under human control. World 3 would be entirely autonomous rather than only partially autonomous, and mathematicians would be passive analyzers rather than active synthesizers of mathematical knowledge – almost the opposite of Popper's earlier emphasis on the active role of the subject in observation and theory formation. I think that these conclusions rest on a misunderstanding of the logical character of relationships in world 3.

To stress the objective and partly autonomous dimension of knowledge is not to lose sight of the fact that it is created, discussed, evaluated, tested and modified by human beings. Popper regarded world 3 above all as a product of intelligent human practice, and especially of the human ability to express and criticize arguments in language. The objectivity of mathematics rests, as does that of all science, upon the criticizability of its arguments, so on language as the indispensable medium of critical discussion (Popper 1981, p. 136-137). Indeed, it is from language that we get the idea of 'logical consequence' in the first place, on which the third world so strongly depends. But mathematics is not *just* language, and neither is it *just* logic: there are such things as extra-logical mathematical objects. And although critical discussion depends on the use of discursive language, mathematics is not bound to one particular *system* of logic. O'Hear (1980, p. 191-198) rightly argued that there is room for choices to fit our pre-systematic intuitions and even physical realities (he for instance discusses deviating logics to fit quantum mechanics). But the possibility of alternative logics does not invalidate the idea of logical consequence as such, it does not make one or the other of alternative logical systems illogical. The choice of a *specific* logical system for mathematics or science has itself to be decided by 'logical' argumentation (in the *general* sense of the term).

Although the third world arises together with argumentative language, it does not consist exclusively of linguistic forms but contains also non-linguistic objects. As is well known, the concept of number, for instance, can be axiomatically described in a variety of ways, which all define it only up to isomorphism. That we have different logical explications of number does not mean that we are talking about different objects (nor that the numbers with which our ancestors worked were entirely different from ours). We must distinguish between numbers as third-world objects and the fallible and changing theories that we form about these objects. That the third-world objects themselves are relatively autonomous means that our intuitive grasp of them is always only partial, and that our theories about them are essentially incomplete, unable to capture fully their infinite richness.

4. Popperian Dialectic

The idea that the third world of objective mathematical knowledge is partly autonomous does not at all imply that the role of mathematicians is reduced to passive observation of a pre-given realm of mathematical objects and structures – no more than that the autonomy of the first world would reduce the role of physicists to passive observation of physical states of affairs. On the contrary, the growth of mathematical knowledge is almost entirely due to the constant feed-back or 'dialectic' between human creative action upon the third world and the action of the third world upon human thought. Popper characterized world 3 as the (evolutionary) product of the rational efforts of humans who, by trying to eliminate contradictions in the extant body of knowledge, produce new theories, arguments, and problems, essentially along the lines of what he called 'the critical interpretation of the (non-Hegelian) dialectic schema: $P_1 \rightarrow TT \rightarrow EE \rightarrow P_2$' (Popper 1981, p. 164). P_1 is the initial problem situation, that is, a problem picked out against a third-world background. TT is the first tentative theoretical solution, which is followed by error elimination (EE), its severe critical examination and evaluation in comparison with any rival solutions. P_2 is the new problem situation arising from the critical discussion, in which the 'experiences' (that is, the failures) of the foregoing attempts are used to pinpoint both their weak and their strong points, so that we may learn how to improve our guesses.

Every rational theory, whether mathematical or scientific or metaphysical, is rational on Popper's view exactly 'in so far as it tries to solve certain problems. A theory is comprehensible and reasonable only in its relation to a given problem situation, and it can be discussed only by discussing this relation' (Popper 1969, p. 199). In mathematics as in science, it is always problems and tentative problem solutions that are at stake: 'only if it is an answer to a problem – a difficult, a fertile problem, a problem of some depth – does a truth, or a conjecture about the truth, become relevant to science. This is so in pure mathematics, and it is so in the natural sciences' (*ibid.*, p. 230). Popper clearly did not view mathematics as a formal language game, but as a rational problem solving activity based, like all rational pursuits, on speculation and criticism.

Although they have no falsifiers in the logical sense – they do not forbid any singular spatiotemporal statement – mathematical theories (as well as logical, philosophical, metaphysical and other non-empirical theories) can nevertheless be critically assessed for their ability to solve the problems in response to which they were designed, and accordingly improved along the lines of the *situational* logic or dialectic indicated above. In particular, mathematical and other 'irrefutable' theories often provide a basis or framework for the development of scientific theories that *can* be refuted (Popper 1969, chapter 8) – a view which later was to inspire Lakatos's notion of research programmes with an 'irrefutable' hard core (Lakatos 1978, p. 95).

Most characteristic of Popper's approach to mathematics was his focussing entirely on the dynamics of conceptual change through the dialectic process outlined, replacing the preoccupation of the traditional approach with definitions and explications of meanings. Interesting formalizations are not attempts at clarifying meanings but at solving problems – especially eliminating contradictions – and this has often been achieved by *abandoning* the attempt to clarify, or make exact, or explicate the intended or intuitive meaning of the concepts in question – as illustrated in particular by the development and rigorization of the calculus (Popper 1983, p. 266). From his objectivist point of view, epistemology becomes the theory of problem solving, that is, of the construction, critical discussion, evaluation, and critical testing, of competing conjectural theories. In this, everything is welcome as a source of inspiration, including intuition, convention and tradition, especially if it suggests new problems. Most creative ideas are based on intuition, and those that are not are the result of criticism of intuitive ideas (Popper 1984, p. 69). There is no sharp distinction between intuitive and discursive thought. With the development of discursive language, our intuitive grasp has become utterly different from what it was before. This has become particularly apparent from the twentieth-century foundation crisis and ensuing discoveries about incompleteness and undecidability. Even our logical intuitions turned out to be liable to correction by discursive mathematical reasoning (*ibid.* p. 70).

5. Socially Conditioned Change

So, mathematics is primarily conceived as a problem solving practice, and – as Popper explicitly stated – anything is welcome as a source of inspiration, especially if it suggests new problems. I will now briefly discuss a case of *socially* conditioned mathematical change, that is, a case in which social processes were the main sources of conceptual innovation, and argue that it is perfectly well possible to acknowledge the social nature of the mathematical enterprise without denying its objectivity and partial autonomy (for a detailed account, see Glas 2003).

Among the important driving forces of mathematical development are concrete, often scientific or technological problems. The calculus, for instance, was developed primarily as an indispensable tool for the science of mechanics. Mathematicians were well aware of its lack of rigor and other fundamental shortcomings, but its impressive problem solving power was reason enough not to abandon it. Instead, eighteenth-century mathematicians tried to perfect the calculus by detaching it from its geometric roots and reformulating it as a linguistic system based on deductions from proposition to proposition, without any appeals to figure-based reasoning. This was achieved by interpreting variables as non-designated quantities and by introducing the notion of function, which replaced the study of curves (cf Ferraro 2001). In the last quarter of the century, Condillac's view that language was constitutive of thought, and the *langue des calculs* its highest manifestation, was shared by many intellectuals, among them the most prominent French mathematicians of the time, Lagrange and Laplace (cf Glas 1986, p. 251–256).

A new chapter in the history of mathematics began when the new, machine-driven industrial technologies gave rise to an entirely new type of problems, and with it to a new mathematical approach, which entered into competition with the established analytical doctrine. It was standard procedure in analytical mechanics to deduce from the principles of mechanics the particular rules of equilibrium and motion in such devices as the lever, the crank, and the pulley. These 'machines', however, were idealized to the point of ignoring all material aspects of real technical devices. The approach was therefore of very limited use to the actual practice of mechanical engineering. Engineers still worked mainly with empirical rules of thumb based on trial and error. This problem – supplying an exact scientific foundation to the practice of engineering – lay at the root of the new course that mathematics embarked on.

The founders of this new approach to mathematics were Carnot and Monge, who both developed new forms of geometry concurrently with their involvement with engineering problems. Before the revolution they were not highly regarded as mathematicians, and without the military needs of the revolution they could scarcely have stood up to the competition of the leading analytical and anti-geometric style of thinking. It was their personal engagement in revolutionary politics that eventually put Carnot and Monge in a position to carry through a radical educational reform – embodied in the Ecole Polytechnique – which was essential to the formation of a new community of mathematically versed engineers (*ingénieurs savants*), who shared their particular views of the problems, aims and methods of mathematics.

Carnot initiated the science of machines as a domain in itself, and developed a new geometry concurrently with it. Finding the operational principles of machines required a new conception of geometry, less static and figure-bound than the classical version, not concerned with the fixed properties of immutable forms but with the much more general problem of possible movements in spatial configurations. The new geometry was focussed on discovering the general principles of transformation of spatial systems rather than on deducing the properties of particular figures from a set of pre-established principles (see Gillispie 1971).

Carnot's work (Carnot 1783) did not read like the eighteenth-century analytical mechanics that culminated in Lagrange's *Mécanique analytique* (Lagrange 1811 [1788]). It in fact appears barely to have been read at all and its author was not known as a scientist before the revolution. His work presupposed the competence of persons versed in abstract scientific thought, yet was written in a geometric idiom that was not suited to arouse the interest of mathematicians, who were primarily concerned with the further perfection of analysis by purging it of all remnants of figure-based reasoning. It apparently was addressed at scientifically versed engineers like himself, an intended audience which at the time existed, if at all, only *in statu nascendi*.

As an officer of the army's engineering corps (*génie*), Carnot had been educated at the military *Ecole du Génie* (School of Engineering) at Mézières, where his fellow revolutionary Monge had been his teacher. The latter's *Géométrie descriptive* (Monge 1811 [1799]) originated in the same practical engineering context and it likewise gathered geometric subject matter under a point of view at once more general than classical geometry and operational in tackling engineering problems.

In Monge's new approach to geometry, the objects of research were not defined by the particular *forms* of geometric figures, but by the *methods* used for generating and interrelating spatial configurations. Apart from pure, projective geometry, descriptive geometry furnished the basis of major developments in analytical geometry, differential geometry and pure analysis (cf Glas 1986, pp. 256–261). The constant association of analytical expressions with situations studied in geometry – on which Monge placed so high a pedagogic value – was of the greatest consequence to the image of mathematics as a whole and set the stage for what we now call 'modern' geometry. It was the particular combination of synthetic and analytic qualities, bringing analysis and geometry to bear on each other in entirely new ways, that made Monge's approach truly novel and accounts for the fecundity of its leading ideas in the exploration of various new territories.

Monge himself linked the aim and object of his geometry directly to its indispensability as a language-tool for modern industrial practice, based on division of labour and therefore having to rely on cooperation and communication between all the heads and hands involved in technical-industrial projects. Besides providing forceful means of tackling problems of design, construction and deployment of machines with the mathematical precision required by the new industrial technologies, descriptive geometry would serve as the indispensable common language of communication between all participants in the productive order of society, who otherwise would remain divided by boundaries of class, profession, and function.

Monge and Carnot both developed their mathematics concurrently with their involvement with engineering problems. Paradoxically, despite the practical and applied nature of the problems that they envisaged, the mathematics that they developed on this basis stands out by its generality and purity. It was not of a lesser standard than the authoritative analytical approach, but was oriented towards different cognitive aims: integrating formal and functional features of spatial systems rather than deducing the consequences of pre-established principles. Like Carnot's science of geometric motion, Monge's descriptive geometry was the intellectual response to the new problems, connected with the rise of machine-driven industrial technologies, that faced their professional community.

6. Alternative Practices

Whatever may be thought of the intellectual value of Monge's and Carnot's contributions to mathematics, this is certainly not a case of 'superior minds' bending the course of an entire discipline by force of reason alone. Their intellectual achievements were inextricably tied up with their professional, social, and political engagement, which the vicissitudes of the revolution allowed them to put into effect by the creation of the Ecole Polytechnique. It was through this educational reform that their work became exemplary for a whole new generation of mathematicians, who under their inspiration opened up fresh and fertile fields of inquiry (whereas the analytical 'language' view made mathematics to appear very nearly completed).

The great changes in mathematics that the birth of the new community engendered – Boyer, for instance, speaks of a geometric and an analytical 'revolution' (Boyer 1968, p. 510) – are only understandable in virtue of the institutional, social and political development of the profession of engineering. This is not to say that the causal arrow points only in one direction, from the social to the intellectual. The intellectual and the social developments were mutually constitutive, but in this particular (perhaps exceptional) case the changed socio-political and institutional role and organization of the professional community of engineers should at least be given explanatory priority as *conditio sine qua non*. For without these developments, the conceptual innovations that Monge and Carnot carried through would have missed their target; they would not have found an appreciative audience and could scarcely have had any impact on the course of mathematical development, as evidenced by their almost total failure to make any impression on the rest of the scientific world prior to the revolution.

The case should certainly not be reduced to a simple conflict of interests between separate specialties, geometry and analysis. Although the followers of the analytical main stream were in a sense 'against' geometry (to the point of making the discipline nearly extinct), Carnot and Monge were not at all 'against' analysis, quite the countary. Like their analytical colleagues, they considered it ideal for the representation and calculation of variations, and therefore indispensable for any engineer. Carnot made abundant use of analysis, and Monge even contributed considerably to its progress. We in fact owe the modern 'analytical geometry' in large measure to Monge's purely analytical characterization of lines, surfaces and solids in space.

Monge and Carnot placed themselves outside the ruling analytical tradition, not because they were geometers rather than analysts, but because they found themselves confronted with altogether different sorts of problems. They were not so much concerned with the further 'linguistic' perfection of analysis as with the concrete problems that faced their own professional (engineering) community, problems that could not be handled adequately by the contemporary analytical mechanics. The geometry that they developed in response to these problems differed fundamentally from the classical version in that its objects were defined in terms of operations and transformations, not in terms of particular types of figures. The main reason for eighteenth-century mathematicians to abandon geometry had been its relying on figure-based reasoning instead of logical deduction. Rather than joining in with this general rejection, Carnot and Monge developed a new conception of geometry altogether, detaching it from consideration of particular figures and redefining it as the study of spatially extended structures and their transformations. Although of course figures figured prominently in Carnot's and Monge's works, their mode of reasoning was not figure-bound. Figures were just heuristic means of investigation, useful to direct and support the geometric reasoning, which in itself proceeded at a level of abstraction and generalization that went far beyond what could be represented figuratively. Their conceptual innovation made the classical distinctions between analytic and synthetic methods obsolete; indeed, it was precisely the intimate unity of synthetic and analytic reasoning that made their approach truly novel and fruitful.

Under the *Empire*, the mathematicians of the analytic tradition regained much of the territory they had lost to the geometric innovators in the revolutionary days. Laplace in particular, who had not been 'seen' by the revolutionaries, was highly regarded by the emperor Napoleon, who clearly was more sensitive to the value of the 'old' tradition in point of respectability and prestige (cf Bradley 1975). The leading role of Laplace in the 'imperial' reformation of the Polytechnic is reflected in the changing relative positions of geometry and analysis in its programme. The time tables of the school show how geometry dropped from 50 hours in 1795 to 27.5 in 1812, whereas analysis and analytical mechanics in this period rose from 8 to 46, also to the cost of chemistry (Dhombres 1989, p. 572). But the turning of the tide, and even the eventual expelling of Carnot and Monge from the scientific institutions of France under the Restoration, could not make undone what had happened to mathematics. It had integrated the spirit of the revolutionary method and had become itself an important element of intellectual and social change.

Until deep into the nineteenth century, a remarkably sharp division subsisted between two groups of mathematicians, the one taking its inspiration from Carnot and Monge and primarily motivated by constructive problems in a technological context, the other more in line with Lagrange and Laplace and chiefly concerned with analytical problems in a general scientific setting (cf Grattan-Guinness 1993, pp. 408–411, who lists seventeen mathematicians in each group).

7. Concluding Discussion

The case underscores the explanatory priority of communal practices in accounting for a type of conceptual change in mathematics that is fundamental to its advance. It was their belonging to a different professional group that accounts for the different viewpoints, problems, aims, values, methods, and approaches of Carnot and Monge, as compared with the received analytical tradition. There never was disagreement about the particular contents of mathematical theories, the correctness of theorems and proofs, and the like. But the two schools differed fundamentally on such issues as what were the questions most worth asking, the methods most appropriate for handling them adequately, and the right criteria for appraising progress. There was no crisis in the extant doctrine, no accumulation of insuperable problems that demanded an entirely new approach. At most it can be said that the reigning tradition had largely exhausted itself: it was regarded as completed, and interesting new discoveries were no longer considered possible. The revolution was not sparked off by deep epistemological worries that led to a replacement of theories; instead, it has to be characterized as a replacement of research communities, the emergence of a new community of mathematically versed engineers, setting themselves radically different sorts of problems, and demanding different methods to solve them.

As a social process, this particular revolution in mathematics shared some of the features of a Kuhnian revolution, but in other respects it was quite different. The practices of the old and the new 'school' were incommensurable by being *at cross-purposes*, rather than in disagreement about the truth or falsity of each other's results. There was no problem of theory choice – the central problem in Kuhn's account of scientific revolutions – but a fundamental shift of evaluative standards, which accounts for the impossibility of resolving the differences by logical argumentation alone. The change was not induced by serious epistemological problems, but by a radical change in the social conditions under which mathematicians worked, the most significant sign of which was the moving of the (military) engineers – with their characteristic view of the objects, problems, aims and values of mathematics – to the centre of state power.

The case shows that in order to characterize mathematical change as a social process, there is no need to question the objectivity and rationality of mathematics in the sense that Popper gave to these notions. A shift of focus, away from the ways in which new truths are derived, and towards the ways in which new problems are conceived and handled, is sufficient. Indeed, unlike the Hegelian dialectic, the Popperian dialectic does not start off with theses but with problems. Problems are the initial and the central motives in the development of mathematics, which is understood primarily as a problem solving practice. It is the dynamics of problem situations, rather than the statics of definitions and theorems, that is characteristic of the growth of mathematical knowledge.

Mathematics is a social practice, shaped by the ways in which its practitioners view the problems that face their community. It is through language that mathematicians can lay out their thoughts objectively, in symbolic form, and then develop, discuss, test and improve them. Humankind has used descriptive and argumentative language to create a body of objective knowledge, stored in libraries and handed down from generation to generation, which enables us to profit from the trials and errors of our ancestors. Characteristic of science and mathematics is that they are formulated in objective language, and that the problems, theories, and errors contained in them stand in logical relations that – *pace* Kuhn – are independent of individual or collective beliefs and other mental states that humans may have with respect to the contents involved.

References

Boyer, C. B., 1968, *A History of Mathematics* (New York: Wiley)

Bradley, M., 1975, 'Scientific Education versus Military Training: The Influence of Napoleon Bonaparte on the Ecole Polytechnique', *Annals of Science* **32**, 415–449.

Carnot, L. N. M., 1783, *Essai sur les machines en général* (Dijon: no publisher)

Dhombres, J. & Dhombres, N., 1989, *Naissance d'un nouveau pouvoir: sciences et savants en France, 1793–1824* (Paris: Payot)

Ferraro, G., 2001, 'Analytical Symbols and Geometrical Figures in Eighteenth-Century Calculus', *Studies in History and Philosophy of science* **32**, 535–555

Gillispie, C. C., 1971, *Lazare Carnot Savant* (Princeton: Princeton University Press)

Glas, E., 1986, 'On the Dynamics of Mathematical Change in the Case of Monge and the French Revolution', *Studies in History and Philosophy of Science* **17**, 249–268

Glas, E., 2001, 'The Popperian Programme and Mathematics', *Studies in History and Philosophy of Science* **32**, 119–137, 355–376

Glas, E., 2003, 'Socially Conditioned Mathematical Change: The Case of the French Revolution', *Studies in History and Philosophy of Science* **34**, in press.

Grattan-Guinness, I., 1993, 'The ingénieur-savant, 1800–1830: A Neglected Figure in the History of French Mathematics and Science', *Science in Context* **6**, 405–433

Lagrange, J. L., 1811, *Mécanique analytique*, nouvelle édition, orig. 1788 (Paris: Courcier)

Lakatos, I., 1976, *Proofs and Refutations: The Logic of Mathematical Discovery*, ed. J. Worrall and G. Currie (Cambridge: Cambridge University Press)

Lakatos, I., 1978, *the Methodology of Scientific Research Programmes* (Philosophical Papers Vol. 1), ed. J. Worrall and G. Currie (Cambridge: Cambridge University Press)

Monge, G., 1811, *Géométrie descriptive*, nouvelle édition, orig. 1799 (Paris: Klostermann)

Niiniluoto, I., 1992, 'Reality, Truth, and Confirmation in Mathematics – Reflections on the Quasi-Empiricist Programme', pp. 60–77 in Echeverria, J., Ibarra, A. and Mormann, T. (eds.), *The Space of Mathematics* (Berlin, New York: De Gruyter)

O'Hear, A., 1980, *Karl Popper* (London: Routledge & Kegan Paul)

Popper, K. R., 1969, *Conjectures and Refutations: The Growth of Scientific Knowledge*, third edition (London: Routledge)

Popper, K. R., 1972, *The Logic of Scientific Discovery*, sixth edition (London: Hutchinson)

Popper, K. R., 1981, *Objective Knowledge: An Evolutionary Approach*, revised edition (Oxford: Clarendon)

Popper, K. R., 1983, *Realism and the Aim of Science*, ed. W. W. Bartley (Totowa: Rowen and Littlefield)

Popper, K. R., 1984, *Auf der Suche nach einer besseren Welt* (München: Piper)

17

The Locus of Mathematical Reality: An Anthropological Footnote

LESLIE A. WHITE

"He's [the Red King's] dreaming now," said Tweedledee: "and what do you think he's dreaming about?"

Alice said, "Nobody can guess that."

"Why, about you!" Tweedledee exclaimed, clapping his hands triumphantly. "And if he left off dreaming about you, where do you suppose you'd be?"

"Where I am now, of course," said Alice.

"Not you!" Tweedledee retorted contemptuously. "You'd be nowhere. Why, you're only a sort of thing in his dream!"

"If that there King was to wake," added Tweedledum, "you'd go out bang!-just like a candle."

"I shouldn't!" Alice exclaimed indignantly. "Besides, if I'm only a sort of thing in his dream, what are you, I should like to know?"

"Ditto," said Tweedledum.

"Ditto, ditto!" cried Tweedledee.

He shouted this so loud that Alice couldn't help saying "Hush! You'll be waking him, I'm afraid, if you make so much noise."

"Well, it's no use your talking about waking him," said Tweedledum, "when you're only one of the things in his dream. You know very well you're not real."

"I am real!" said Alice, and began to cry.

"You won't make yourself a bit realler by crying," Tweedledee remarked: "there's nothing to cry about."

"If I wasn't real," Alice said-half laughing through her tears, it all seemed so ridiculous- "I shouldn't be able to cry."

"I hope you don't suppose those are real tears?" Tweedledum interrupted in a tone of great contempt.

-Through the Looking Glass

Do mathematical truths reside in the external world, there to be discovered by man, or are they man-made inventions? Does mathematical reality have an existence and a validity independent of the human species or is it merely a function of the human nervous system?

Opinion has been and still is divided on this question. Mrs. Mary Somerville (1780-1872), an Englishwoman who knew or corresponded with such men as Sir John Herschel, Laplace, Gay Lussac, W. Whewell, John Stuart Mill, Baron von Humboldt, Faraday, Cuvier, and De Candolle, who was herself a scholar of distinction[1],' expressed a view widely held when she said,[2]

> "Nothing has afforded me so convincing a proof of the unity of the Deity as these purely mental conceptions of numerical and mathematical science which have been by slow degrees vouchsafed to man, and are still granted in these latter times by the Differential Calculus, now superseded by the Higher Algebra, all of which must have existed in that sublimely omniscient Mind from eternity."

Lest it be thought that Mrs. Somerville was more theological than scientific in her outlook, let it be noted that she was denounced, by name and in public from the pulpit by Dean Cockburn of York Cathedral for her support of science.[3]

In America, Edward Everett (1794-1865), a distinguished scholar (the first American to win a doctorate at Gottingen), reflected the enlightened view of his day when he declared,[4]

> "In the pure mathematics we contemplate absolute truths which existed in the divine mind before the morning stars sang together, and which will continue to exist there when the last of their radiant host shall have fallen from heaven."

In our own day, a prominent British mathematician, G. H. Hardy, has expressed the same view with, however, more technicality than rhetorical flourish.[5]

> "I believe that mathematical reality lies outside us, and that our function is to discover or observe it, and that the theorems which we prove, and which we describe grandiloquently as our 'creations' are simply our notes of our observations.[6]"

---

[1] She wrote the following works, some of which went into several editions: *The Mechanism of the Heavens, 1831* (which was, it seems, a popularization of the *Mécanique Céleste* of Laplace); *The* Connection of *the Physical Sciences, 1858; Molecular and Microscopic Science, 1869; Physical Geography, 1870.*

[2] *Personal Recollections of Mary Somerville,* edited by her daughter, Martha.

[3] *ibid., p. 375.* See, also, A. D. White, *The History of the Warfare of Science with Theology &c, Vol. I, p. 225,* ftn. (New York, *1930* printing).

[4] Quoted by E. T. Bell in *The Queen of the Sciences, p. 20* (Baltimore, *1931*).

[5] G. H. Hardy, *A Mathematician's Apology, pp. 63-64* (Cambridge, England; *1941*).

[6] The mathematician is not, of course, the only one who is inclined to believe that his creations are discoveries of things in the external world. The theoretical physicist, too, entertains this belief. "To him who is a discoverer in this field," Einstein observes, "the products of his imagination appear so necessary and natural that he regards them, and would like to have them regarded by others, not as creations of thought but as given realities." ("On the Method of Theoretical Physics," in *The World as I See it, p. 30;* New York, *1934*).

Taking the opposite view we find the distinguished physicist, P. W. Bridgman, asserting that "it is the merest truism, evident at once to unsophisticated observation, that mathematics is a human invention."[7] Edward Kasner and James Newman state that "we have overcome the notion *that mathematical truths have an existence independent and apart from* our own minds. It is even strange to us that such a notion could ever have existed."[8]

From a psychological and anthropological point of view, this latter conception is the only one that is scientifically sound and valid. There is no more reason to believe that mathematical realities have an existence independent of the human mind than to believe that mythological realities can have their being apart from man. The square root of minus one is real. So were Wotan and Osiris. So are the gods and spirits that primitive peoples believe in today. The question at issue, however, is not, Are these things real?, but 'Where is the locus of their reality?' It is a mistake to identify reality with the external world only. Nothing is more real than an hallucination.

Our concern here, however, is not to establish one view of mathematical reality as sound, the other illusory. What we propose to do is to present the phenomenon of mathematical behavior in such a way as to make clear, on the one hand, why the belief in the independent existence of mathematical truths has seemed so plausible and convincing for so many centuries, and, on the other, to show that all of mathematics is nothing more than a particular kind of primate behavior, Many persons would unhesitatingly subscribe to the proposition that "mathematical reality must lie either within us, or outside us." Are these not the only possibilities? As Descartes once reasoned in discussing the existence of God, "it is impossible we can have the idea or representation of anything whatever, unless there be somewhere, *either in us or out of us,* an original which comprises, in reality[9] (emphasis ours). Yet, irresistible though this reasoning may appear to be, it is, in our present problem, fallacious or at least treacherously misleading. The following propositions, though apparently precisely opposed to each other, are equally valid; one is as true as the other: 1. "Mathematical truths have an existence and a validity independent of the human mind," and 2. "Mathematical truths have no existence or validity apart from the human mind." Actually, these propositions, phrased as they are, are misleading because the term "the human mind" is used in two different senses. In the first statement, "the human mind" refers to the individual organism; in the second, to the human species. Thus both propositions can be, and actually are, true. Mathematical truths exist in the cultural tradition into which the individual is born, and so enter his mind from the outside. But apart from cultural tradition, mathematical concepts

---

[7] P. W. Bridgman, *The Logic of Modern Physics, p. 60* (New York, 1927).

[8] Edward Kasner and James Newman, *Mathematics and the Imagination, p.* 359 (New York, 1940).

[9] *Principles of Philosophy,* Pt. I, Sec. XVIII, p. 308, edited by J. Veitch (New York, 1901).

have neither existence nor meaning, and of course, cultural tradition has no existence apart from the human species. Mathematical realities thus have an existence independent of the individual mind, but are wholly dependent upon the mind of the species. Or, to put the matter in anthropological terminology: mathematics in its entirety, its "truths" and its "realities," is a part of human *culture,* nothing more. Every individual is born into a culture which already existed and which is independent of him. Culture traits have an existence outside of the individual mind and independent of it. The individual obtains his culture by learning the customs, beliefs, techniques of his group. But culture itself has, and can have, no existence apart from the human species. Mathematics, therefore-like language, institutions, tools, the arts, etc.-is the cumulative product of ages of endeavor of the human species.

The great French savant Emile Durkheim (1858-1917) was one of the first to make this clear. He discussed it in the early pages of *The Elementary Forms of the Religious Life.*[10] And in *The Rules of Sociological Method*[11] especially he set forth the nature of *culture*[12] and its relationship to the human mind. Others, too, have of course discussed the relationship to the human mind. Others, too, have of course discussed the relationship between man and culture,'[13] but Durkheim's formulations are especially appropriate for our present discussion and we shall call upon him to speak for us from time to time.

Culture is the anthropologist's technical term for the mode of life of any people, no matter how primitive or advanced. It is the generic term of which *civilization is* a specific term. The mode of life, or culture, of the human species is distinguished from that of all other species by the use of symbols. Man is the only living being that can freely and arbitrarily impose value or meaning upon any thing, which is what we mean by "using symbols." The most important and characteristic form of symbol behavior is articulate speech. All cultures, all of civilization, have come into being, have grown and developed, as a consequence of the symbolic faculty, unique in the human species.[14]

---

[10] *Les Formes Elémentaires de la Vie Religieuse* (Paris, 1912) translated by J. W. Swain (London, 1915). Nathan Altshiller-Court refers to Durkheim's treatment of this point in "Geometry and Experience," (Scientific Monthly, Vol. LX, No. 1, pp. 63-66, Jan., 1945).

[11] *'Les Règles de la Méthode Sociologique* (Paris, 1895; translated by Sarah A. Solovay and John H. Mueller, edited by George E. G. Catlin; Chicago, 1938).

[12] Durkheim did not use the term *culture.* Instead he spoke of the "collective consciousness," "collective representations," etc. Because of his unfortunate phraseology Durkheim has been misunderstood and even branded mystical. But it is obvious to one who understands both Durkheim and such anthropologists as R. H. Lowie, A. L. Kroeber and Clark Wissler that they are all talking about the same thing: culture.

[13] See, e.g., E. B. Tylor, *Anthropology* (London, 1881); R. H. Lowie, *Culture and Ethnology,* New York, 1917; A. L. Kroeber, "The Superorganic," (American Anthropologist, Vol. 19, pp. 163-213; 1917); Clark Wissler, *Man and Culture,* (New York, 1923).

[14] See, White, Leslie A., "The Symbol: the Origin and Basis of Human Behavior," (Philosophy of Science, Vol. 7, pp. 451-463; 1940; reprinted in ETC., a Review of General Semantics, Vol. I, pp. 229-237; 1944).

Every culture of the present day, no matter how simple or primitive, is a product of great antiquity. The language, tools, customs, beliefs, forms of art, etc., of any people are things which have been handed down from generation to generation, from age to age, changing and growing as they went, but always keeping unbroken the connection with the past. Every people lives not merely in a habitat of mountains or plains, of lakes, woods, and starry heavens, but in a setting of beliefs, customs, dwellings, tools, and rituals as well. Every individual is born into a man-made world of culture as well as the world of nature. But it is the culture rather than the natural habitat that determines man's thought, feelings, and behavior. To be sure, the natural environment may favor one type of activity or render a certain mode of life impossible. But whatever man does, as individual or as society, is determined by the culture into which he, or they, are born.[15] Culture is a great organization of stimuli that flows down through the ages, shaping and directing the behavior of each generation of human organisms as it goes. Human behavior is response to these cultural stimuli which seize upon each organism at birth-indeed, from the moment of conception, and even before this- and hold it in their embrace until death- and beyond, through mortuary customs and beliefs in a land of the dead.

The language a people speaks is the response to the linguistic stimuli which impinge upon the several organisms in infancy and childhood. One group of organisms is moulded by Chinese-language stimuli; another, by English. The organism has no choice, and once cast into a mould is unable to change. To learn to speak a foreign language without accent after one has matured, or even, in most cases, to imitate another dialect of his own language is exceedingly difficult if not impossible for most people. So it is in other realms of behavior. A people practices polygyny, has matrilineal clans, cremates the dead, abstains from eating pork or peanuts, counts by tens, puts butter in their tea, tattoos their chests, wears neckties, believes in demons, vaccinates their children, scalps their vanquished foes or tries them as war criminals, lends their wives to guests, uses slide rules, plays pinochle, or extracts square roots if the culture into which they were born possesses these traits. It is obvious, of course, that people do not choose their culture; they inherit it. It is almost as obvious that a people behaves as it does because it possesses a certain type of culture- or more accurately, is possessed by it.

To return now to our proper subject. Mathematics is, of course, a part of culture. Every people inherits from its predecessors, or contemporary neighbors, along with ways of cooking, marrying, worshipping, etc., ways of counting, calculating, and whatever else mathematics does. Mathematics is, in fact, a form of behavior: the responses of a particular kind of primate organism to a set of stimuli. Whether a people counts by fives, tens, twelves or

[15] Individuals vary, of course, in their constitutions and consequently may vary in their responses to cultural stimuli.

twenties; whether it has no words for cardinal numbers beyond 5, or possesses the most modern and highly developed mathematical conceptions, their mathematical behavior is determined by the mathematical culture which possesses them.

We can see now how the belief that mathematical truths and realities lie outside the human mind arose and flourished. They do lie outside the mind of each individual organism. They enter the individual mind as Durkheim says from the outside. They impinge upon his organism, again to quote Durkheim, just as cosmic forces do. Any mathematician can see, by observing himself as well as others, that this is so. Mathematics is not something that is secreted, like bile; it is something drunk, like wine. Hottentot boys grow up and behave, mathematically as well as otherwise, in obedience to and in conformity with the mathematical and other traits in their culture. English or American youths do the same in their respective cultures. There is not one iota of anatomical or psychological evidence to indicate that there are any significant innate, biological or racial differences so far as mathematical or any other kind of human behavior is concerned. Had Newton been reared in Hottentot culture he would have calculated like a Hottentot. Men like G. H. Hardy, who know, through their own experience as well as from the observation of others, that mathematical realities enter the mind from the outside, understandably-but erroneously-conclude that they have their origin and locus in the external world, independent of man. Erroneous, because the alternative to "outside the human mind," the individual mind, that is, is not "the external world, independent of man," but culture, the body of traditional thought and behavior of the human species.

Culture frequently plays tricks upon us and distorts our thinking. We tend to find in culture direct expressions of "human nature" on the one hand and of the external world on the other. Thus each people is disposed to believe that its own customs and beliefs are direct and faithful expressions of man's nature. It is "human nature," they think, to practice monogamy, to be jealous of one's wife, to bury the dead, drink milk, to appear in public only when clad, to call your mother's brother's children "cousin," to enjoy exclusive right to the fruit of your toil, etc., if they happen to have these particular customs. But ethnography tells us that there is the widest divergence of custom among the peoples of the world: there are peoples who loathe milk, practice polyandry, lend wives as a mark of hospitality, regard inhumation with horror, appear in public without clothing and without shame, call their mother's brother's children "son" and "daughter," and who freely place all or the greater portion of the produce of their toil at the disposal of their fellows. There is no custom or belief that can be said to express "human nature" more than any other.

Similarly it has been thought that certain conceptions of the external world were so simple and fundamental that they immediately and faithfully expressed its structure and nature. One is inclined to think that yellow, blue, and green are features of the external world which any normal person would distinguish until he learns that the Creek and Natchez Indians did not distinguish yellow from green; they had but one term for both. Similarly, the Choctaw, Tunica, the Keresan Pueblo Indians and many other peoples make no terminological distinction between blue and green.[16]

The great Newton was deceived by his culture, too. He took it for granted that the concept of *absolute space* directly and immediately corresponded to something in the external world; space, he thought, is something that has an existence independent of the human mind. "I do not frame hypotheses," he said. But the concept space is a creation of the intellect as are other concepts. To be sure, Newton himself did not create the hypothesis of absolute space. It came to him from the outside, as Durkheim properly puts it. But although it impinges upon the organism *comme les forces cosmiques*, it has a different source: it is not the cosmos but man's culture.

For centuries it was thought that the theorems of Euclid were merely conceptual photographs, so to speak, of the external world; that they had a validity quite independent of the human mind; that there was something necessary and inevitable about them. The invention of non-Euclidean geometries by Lobatchewsky, Riemann and others has dispelled this view entirely. It is now clear that concepts such as space, straight line, plane, etc., are no more necessary and inevitable as a consequence of the structure of the external world than are the concepts green and yellow- or the relationship term with which you designate your mother's brother, for that matter.

To quote Einstein again:[17]

"We come now to the question: what is a priori certain or necessary, respectively in geometry (doctrine of space) or its foundations? Formerly we thought everything; nowadays we think-nothing. Already the distance-concept is logically arbitrary; there need be no things that correspond to it, even approximately."

Kasner and Newman say that "non-Euclidean geometry is proof that mathematics . . . is man's own handiwork, subject only to the limitations imposed by the laws of thought."[18]

Far from having an existence and a validity apart from the human species, all mathematical concepts are "free inventions of the human intellect," to use a phrase with which Einstein characterizes the concepts and fundamental principles of physics.[19] But because mathematical and scientific concepts have always entered each individual mind from the outside, everyone until recently has concluded that they came from the external world instead of from man-

---

[16] Cf. "Keresan Indian Color Terms," by Leslie A. White, Papers of the Michigan. *Academy of Science, Arts, and Letters, Vol. XXVIII, pp. 559-563; 1942 (1943).*

[17] Article "Space-Time." Encyclopaedia Britannica, 14th edition.

[18] Op. *cit., p.* 359

[19] "On the Method of Theoretical Physics," in The World as I See It, p. 33 (New York, 1934).

made culture. But the concept of culture, as a scientific concept, is but a recent invention itself.

The cultural nature of our scientific concepts and beliefs is clearly recognized by the Nobel prize winning physicist, Erwin Schrödinger, in the following passage:[20]

"Whence arises the widespread belief that the behavior of molecules is determined by absolute causality, whence the conviction that the contrary is unthinkable? Simply from the custom, inherited through thousands of years, of thinking causally, which makes the idea of undetermined events, of absolute, primary causalness, seem complete nonsense, a logical absurdity," (Schrodinger's emphases).

Similarly, Henri Poincaré asserts that the axioms of geometry are mere "conventions," i.e., customs: they "are neither synthetic a priori judgments nor experimental facts. They are conventions[21]

We turn now to another aspect of mathematics that is illuminated by the concept of culture. Heinrich Hertz, the discoverer of wireless waves, once said:[22]

"One cannot escape the feeling that these mathematical formulas have an independent existence and an intelligence of their own, that they are wiser than we are, wiser even than their discoverers [sic], that we get more out of them than was originally put into them."

Here again we encounter the notion that mathematical formulas have an existence "of their own," (i.e., independent of the human species), and that they are "discovered," rather than man-made. The concept of culture clarifies the entire situation. Mathematical formulas, like other aspects of culture, do have in a sense an "independent existence and intelligence of their own." The English language has, in a sense, "an independent existence of its own." Not independent of the human species, of course, but independent of any individual or group of individuals, race or nation. It has, in a sense, an "intelligence of its own." That is, it behaves, grows and changes in accordance with principles which are inherent in the language itself, not in the human mind. As man becomes self-conscious of language, and as the science of philosophy matures, the principles of linguistic behavior are discovered and its laws formulated.

So it is with mathematical and scientific concepts. In a very real sense they have a life of their own. This life is the life of culture, of cultural tradition. As Durkheim expresses it:[23] "Collective ways of acting and thinking have a reality outside the individuals who, at every moment of time, conform to it. These ways of thinking and acting exist in their own right." It would be quite

---

[20] *Science and the Human Temperament*, p. 115 (London, 1935).

[21] "On the Nature of Axioms," in *Science and Hypothesis*, published in *The Foundations of Science* (The Science Press, New York, 1913).

[22] Quoted by E. T. Bell, *Men of Mathematics*, p. 16 (New York, 1937).

[23] *The Rules* 0/ *Sociological Method,* Preface to 2nd edition, p. lvi.

possible to describe completely and adequately the evolution of mathematics, physics, money, architecture, axes, plows, language, or any other aspect of culture without ever alluding to the human species or any portion of it. As a matter of fact, the most effective way to study culture scientifically is to proceed *as if* the human race did not exist. To be sure it is often convenient to refer to the nation that first coined money or to the man who invented the calculus or the cotton gin, But it is not necessary, nor, strictly speaking, relevant. The phonetic shifts in Indo-European as summarized by Grimm's law have to do solely with linguistic phenomena, with sounds and their permutations, combinations and interactions. They can be dealt with adequately without any reference to the anatomical, physiological, or psychological characteristics of the primate organisms who produced them. And so it is with mathematics and physics. Concepts have a life of their own. Again to quote Durkheim, "when once born, [they] obey laws all their own. They attract each other, repel each other, unite, divide themselves and multiply. . . ."[24]

Ideas, like other culture traits, interact with each other, forming new syntheses and combinations. Two or three ideas coming together may form a new concept or synthesis. The laws of motion associated with Newton were syntheses of concepts associated with Galileo, Kepler and others. Certain ideas of electrical phenomena grow from the "Faraday stage," so to speak, to those of Clerk Maxwell, H. Hertz, Marconi, and modern radar. "The application of Newton's mechanics to continuously distributed masses *led inevitably* to the discovery and application of partial differential equations, which in their turn first provided the language for the laws of the field-theory,"[25] (emphasis ours). The theory of relativity was, as Einstein observes, "no revolutionary act, but the natural continuation of a line that can be traced through centuries."[26] More immediately, "the theory of Clerk Maxwell and Lorentz led inevitably to the special theory of relativity."[27] Thus we see not only that any given thought system is an outgrowth of previous experience, but that certain ideas lead inevitably to new concepts and new systems. Any tool, machine, belief, philosophy, custom or institution is but the outgrowth of previous culture traits. An understanding of the nature of culture makes clear, therefore, why Hertz felt that "mathematical formulas have an independent existence and an intelligence of their own."

His feeling that "we get more out of them than was originally put into them," arises from the fact that in the interaction of culture traits new syntheses are formed which were not anticipated by "their discoverers," or which

---

[24] *The Elementary Forms of the Religious Life, p.* 424. See also *The Rules of Sociological Method,* Preface to 2nd edition, p. ii, in which he says "we need to investigate . . . the manner in which social representations [i.e., culture traits] adhere to and repel one another, how they fuse or separate from one another."

[25] Einstein, "The Mechanics of Newton and their Influence on the Development of Theoretical Physics," in *The World as I See It, p.* 58.

[26] "On the Theory of Relativity," in *The World as I See It, p.* 69.

[27] Einstein, "The Mechanics of Newton *&C,* p. 57.

contained implications that were not seen or appreciated until further growth made them more explicit. Sometimes novel features of a newly formed synthesis are not seen even by the person in whose nervous system the synthesis took place. Thus Jacques Hadamard tells us of numerous instances in which he failed utterly to see things that "ought to have struck . . . [him] blind."[28] He cites numerous instances in which he failed to see "obvious and immediate consequences of the ideas contained"[29] in the work upon which he was engaged, leaving them to be "discovered" by others later.

The contradiction between the view held by Hertz, Hardy and others that mathematical truths are discovered rather than man-made is thus resolved by the concept of culture. They are both; they are discovered but they are also man-made. They are the product of the mind of the human species. But they are encountered or discovered by each individual in the mathematical culture in which he grows up. The process of mathematical growth is, as we have pointed out, one of interaction of mathematical elements upon each other. This process requires, of course, a basis in the brains of men, just as a telephone conversation requires wires, receivers, transmitters, etc. But we do not need to take the brains of men into account in an explanation of mathematical growth and invention any more than we have to take the telephone wires into consideration when we wish to explain the conversation it carries. Proof of this lies in the fact of numerous inventions (or "discoveries") in mathematics made simultaneously by two or more person working independently.[30]

---

[28] Jacques Hadamard, *The Psychology of Invention in the Mathematical Field, p. 50* (Princeton, *1945*).

[29] *ibid., p. 51.*

[30] The following data are taken from a long and varied list published in *Social Change,* by Wm. F. Ogburn (New York, *1923), pp. 90-102,* in which simultaneous inventions and discoveries in the fields of chemistry, physics, biology, mechanical invention, etc., as well as in mathematics, are listed.

Law of inverse squares: Newton, *1666;* Halley, *1684.*

Introduction of decimal point: Pitiscus, *1608-12;* Kepler, *1616;* Napier, *1616-17.*

Logarithms: Burgi, *1620;* Napier-Briggs, *1614.*

Calculus: Newton, *1671;* Leibnitz, *1676.*

Principle of least squares: Gauss, *1809;* Legendre, *1806.*

A treatment of vectors without the use of co-ordinate systems: Hamilton, *1843; Grassman, 1843;* and others, *1843.*

Contraction hypothesis: H. A. Lorentz, *1895;* Fitzgerald, *1895.*

The double theta functions: Gopel, 1847; Rosenhain, 1847.

Geometry with axiom contradictory to Euclid's parallel axiom: Lobatchevsky, 1836-40; Bolyai, 1826-33; Gauss, 1829.

The rectification of the semi-cuba! parabola: Van Heuraet, 1659; Neil, 1657; Fermat, 1657-59.

The geometric law of duality: Oncelet, 1838; Gergone, 1838.

As examples of simultaneity in other fields we might cite: Discovery of oxygen: Scheele, 1774; Priestley, 1774.

Liquefaction of oxygen: Cailletet, 1877; Pictet, 1877.

Periodic law: De Chancourtois, 1864; Newlands, 1864; Lothar Meyer, 1864.

If these discoveries really were caused, or determined, by individual minds, we would have to explain them as coincidences. On the basis of the laws of chance these numerous and repeated coincidences would be nothing short of miraculous. But the culturological explanation makes the whole situation clear at once. The whole population of a certain region is embraced by a type of culture. Each individual is born into a pre-existing organization of beliefs, tools, customs and institutions. These culture traits shape and mould each person's life, give it content and direction. Mathematics is, of course, one of the streams in the total culture. It acts upon individuals in varying degree, and they respond according to their constitutions. Mathematics is the organic behavior response to the mathematical culture.

But we have already noted that within the body of mathematical culture there is action and reaction among the various elements. Concept reacts upon concept; ideas mix, fuse, form new syntheses. This process goes on throughout the whole extent of culture although more rapidly and intensively in some regions (usually the center) than in others (the periphery). When this process of interaction and development reaches a certain point, new syntheses[31] are formed of themselves. These syntheses are, to be sure, real events, and have location in time and place. The places are of course the brains of men. Since the cultural process has been going on rather uniformly over a wide area and population, the new synthesis takes place simultaneously in a number of brains at once. Because we are habitually anthropocentric in our thinking we tend to say that these men made these discoveries. And in a sense, a biological sense, they did. But if we wish to explain the discovery as an event in the growth of mathematics we must rule the individual out completely. From this standpoint, the individual did not make the discovery at all. It was something that happened to him. He was merely the place where the lightning struck. A simultaneous "discovery" by three men working "independently" simply means that cultural-mathematical lightning can and does strike in more than one place at a time. In the process of cultural growth, through invention or discovery, the individual is merely the neural medium in which the "culture"[32] of ideas grows. Man's brain is merely a catalytic agent, so to speak, in the cultural process. This process cannot exist independently of neural tissue, but the function of man's nervous system is merely to make possible the interaction and re-synthesis of cultural elements.

To be sure individuals differ just as catalytic agents, lightning conductors or other media do. One person, one set of brains, may be a better medium for

---

Law of periodicity of atomic elements: Lothar Meyer, 1869; Mendeleff, 1869.

Law of conservation of energy: Mayer, 1843; Joule, 1847; Helmholz, 1847; Colding, 1847; Thomson, 1847.

A host of others could be cited. Ogburn's list, cited above, does not pretend to be complete.

[31] Hadamard entitles one chapter of his book "Discovery as a Synthesis."

[32] We use "culture" here in its bacteriological sense: a culture of bacilli growing in a gelatinous medium.

the growth of mathematical culture than another. One man's nervous system may be a better catalyst for the cultural process than that of another. The mathematical cultural process is therefore more likely to select one set of brains than another as its medium of expression. But it is easy to exaggerate the role of superior brains in cultural advance. It is not merely superiority of brains that counts. There must be a juxtaposition of brains with the interactive, synthesizing cultural process. If the cultural elements are lacking, superior brains will be of no avail. There were brains as good as Newton's in England 10,000 years before the birth of Christ, at the time of the Norman conquest, or any other period of English history. Everything that we know about fossil man, the prehistory of England, and the neuro-anatomy of homo sapiens will support this statement. There were brains as good as Newton's in aboriginal America or in Darkest Africa. But the calculus was not discovered or invented in these other times and places because the requisite cultural elements were lacking. Contrariwise, when the cultural elements are present, the discovery or invention becomes so inevitable that it takes place independently in two or three nervous systems at once. Had Newton been reared as a sheep herder, the mathematical culture of England would have found other brains in which to achieve its new synthesis. One man's brains may be better than another's, just as his hearing may be more acute or his feet larger. But just as a "brilliant" general is one whose armies are victorious, so a genius, mathematical or otherwise, is a person in whose nervous system an important cultural synthesis takes place; he is the neural locus of an epochal event in culture history.[33]

The nature of the culture process and its relation to the minds of men is well illustrated by the history of the theory of evolution in biology. As is well known, this theory did not originate with Darwin. We find it in one form or another, in the neural reactions of many others before Darwin was born: Buffon, Lamarck, Erasmus Darwin, and others. As a matter of fact, virtually all of the ideas which together we call Darwinism are to be found in the writings of J. C. Prichard, an English physician and anthropologist (1786-1848). These various concepts were interacting upon each other and upon current theological beliefs, competing, struggling, being modified, combined, re-synthesized, etc., for decades. The time finally came, i.e., the stage of development was reached, where the theological system broke down and the risng tide of scientific interpretation inundated the lands.

Here again the new synthesis of concepts found expression simultaneously in the nervous systems of two men working independently of each other: A. R. Wallace and Charles Darwin. The event had to take place when it did. If

[33] The distinguished anthropologist, A. L. Kroeber, defines geniuses as "the indicators of the realization of coherent patterns of cultural value," *Configurations of Culture Growth, p. 839* (Berkeley, *1944*).

Darwin had died in infancy, the cultural process would have found another neural medium of expression.

This illustration is especially interesting because we have a vivid account, in Darwin's own words, of the way in which the "discovery" (i.e., the synthesis of ideas) took place:

> "In October 1838," Darwin wrote in his autobiographic sketch, "that is, fifteen months after I had begun my systematic enquiry, *I happened to read for amusement* 'Maithus on Population,' and being well prepared to appreciate the struggle for existence which everywhere goes on from long-continued observation of the habits of animals and plants, it at once struck me that under these circumstances favourable variations would tend to be preserved, and unfavourable ones to be destroyed. The result of this would be the formation of a new species. *Here then I had at last got a theory by which to work . . .*" (emphasis ours),

This is an exceedingly interesting revelation. At the time he read Malthus, Darwin's mind was filled with various ideas, (i.e., he had been moulded, shaped, animated and equipped by the cultural milieu into which he happened to have been born and reared-a significant aspect of which was independent means; had he been obliged to earn his living in a "counting house" we might have had "Hudsonism" today instead of Darwinism). These ideas reacted upon each other, competing, eliminating, strengthening, combining. Into this situation was introduced, *by chance,* a peculiar combination of cultural elements (ideas) which bears the name of Malthus, Instantly a reaction took place, a new synthesis was formed-"here at last he had a theory by which to work." Darwin's nervous system was merely the place where these cultural elements came together and formed a new synthesis. It was something that *happened* to Darwin rather than something he *did.*

This account of invention in the field of biology calls to mind the well-known incident of mathematical invention described so vividly by Henri Poincaré. "One evening, after working very hard on a problem but without success," he writes:[34] "contrary to my custom, I drank black coffee and could not sleep. Ideas rose in crowds; I felt them collide until pairs interlocked, so to speak, making a stable combination. By the next morning I had established the existence of a class of Fuchsian functions . . . I had only to write out the results, which took but a few hours."

Poincaré further illustrates the process of culture change and growth in its subjective (i.e., neural) aspect by means of an imaginative analogy.[35] He imagines mathematical ideas as being something like "the hooked atoms of Epicurus. During complete repose of the mind, these atoms are motionless, they are, so to speak, hooked to the wall." No combinations are formed. But in mental activity, even unconscious activity, certain of the atoms "are

---

[34] "Mathematical Creation," in *Science and Method,* published in *The Foundations of Science,* p. 397 (The Science Press; New York and Garrison, 1913). I

[35] ibid., p. 393.

detached from the wall and put in motion. They flash in every direction through space . . . like the molecules of a gas Then their mutual impacts may produce new combinations." This is merely a description of the subjective aspect of the cultural process which the anthropologist would describe objectively (i.e., without reference to nervous systems). He would say that in cultural systems, traits of various kinds act and react upon each other, eliminating some, reinforcing others, forming new combinations and syntheses. The significant thing about the loci of inventions and discoveries from the anthropologist's standpoint is not quality of brains, but relative position within the culture area: inventions and discoveries are much more likely to take place at culture centers, at places where there is a great deal of cultural interaction, than on the periphery, in remote or isolated regions.

If mathematical ideas enter the mind of the individual mathematician from the outside, from the stream of culture into which he was born and reared, the question arises, where did culture in general, and mathematical culture in particular, come from in the first place? How did it arise and acquire its content?

It goes without saying of course that mathematics did not originate with Euclid and Pythagoras- or even with the thinkers of ancient Egypt and Mesopotamia. Mathematics is a development of thought that had its beginning with the origin of man and culture a million years or so ago. To be sure, little progress was made during hundreds of thousands of years. Still, we find in mathematics today systems and concepts that were developed by primitive and preliterate peoples of the Stone Ages, survivals of which are to be found among savage tribes today. The system of counting by tens arose from using the fingers of both hands. The vigesimal system of the Maya astronomers grew out of the use of toes as well as fingers. *To calculate is* to count with *calculi,* pebbles. A *straight line* was a *stretched linen cord,* and so on.

To be sure, the first mathematical ideas to exist were brought into being by the nervous systems of individual human beings. They were, however, exceedingly simple and rudimentary. Had it not been for the human ability to give these ideas overt expression in symbolic form and to communicate them to one another so that new combinations would be formed, and these new syntheses passed on from one generation to another in a continuous process of interaction and accumulation, the human species would have made no mathematical progress beyond its initial stage. This statement is supported by our studies of anthropoid apes. They are exceedingly intelligent and versatile. They have a fine appreciation of geometric forms, solve problems by imagination and insight, and possess not a little originality.[36] But they cannot express their neuro-sensory-muscular concepts in overt symbolic form. They cannot communicate their ideas to one another except by gestures, i.e., by *signs* rather than *symbols*. Hence ideas cannot react upon one another in their

[36] See, W. Köhler's *The Mentality 0l Apes* (New York, 1931).

minds to produce new syntheses. Nor can these ideas be transmitted from one generation to another in a cumulative manner. Consequently, one generation of apes begins where the preceding generation began. There is neither accumulation nor progress.[37]

Thanks to articulate speech, the human species fares better. Ideas are cast into symbolic form and given overt expression. Communication is thus made easy and versatile. Ideas now impinge upon nervous systems from the outside. These ideas react upon each other within these nervous systems. Some are eliminated; others strengthened. New combinations are formed, new syntheses achieved. These advances are in turn communicated to someone else, transmitted to the next generation. In a relatively short time, the accumulation of mathematical ideas has gone beyond the creative range of the individual human nervous system *unaided by cultural tradition*. From this time on, mathematical progress is made by the interaction of ideas already in existence rather than by the creation of new concepts by the human nervous system alone. Ages before writing was invented, individuals in all cultures were dependent upon the mathematical ideas present in their respective cultures. Thus, the mathematical behavior of an Apache Indian is the response that he makes to stimuli provided by the mathematical ideas in his culture. The same was true for Neanderthal man and the inhabitants of ancient Egypt, Mesopotamia and Greece. It is true for individuals of modern nations today.

Thus we see that mathematical ideas were produced originally by the human nervous system when man first became a human being a million years ago. These concepts were exceedingly rudimentary, and the human nervous system, *unaided by culture,* could never have gone beyond them regardless of how many generations lived and died. It was the formation of a cultural tradition which made progress possible. The communication of ideas from person to person, the transmission of concepts from one generation to another, placed in the minds of men (i.e., stimulated their nervous systems) ideas which through interaction formed new syntheses which were passed on in turn to others.

We return now, in conclusion, to some of the observations of G. H. Hardy, to show that his conception of mathematical reality and mathematical behavior is consistent with the culture theory that we have presented here and is, in fact, explained by it.

"I believe that mathematical reality lies outside us,"[38] he says. If by "us" he means "us mathematicians individually," he is quite right. They do lie outside each one of us; they are a part of the culture into which we are born.

---

[37] See Leslie A. White, 'On the Use of Tools by Primates" (*Journ. of Comparative Psychology*, Vol. 34, pp. 369-374, Dec. 1942). This essay attempts to show that the human species has a highly developed and progressive material culture while apes do not, although they can use tools with skill and versatility and even invent them, because man, and not apes, can use symbols.

[38] *A Mathematician's Apology, p. 63.*

Hardy feels that "in some sense, mathematical truth is part of *objective* reality,"[39] (my emphasis, L.A.W.). But he also distinguishes "mathematical reality" from "physical reality," and insists that "pure geometries are *not* pictures . . . [of] the spatio-temporal reality of the physical world."[40] What then is the nature of mathematical reality? Hardy declares that "there is no sort of agreement . . . among either mathematicians or philosophers"[41] on this point. Our interpretation provides the solution. Mathematics does have objective reality. And this reality, as Hardy insists, is *not* the reality of the physical world. But there is no mystery about it. Its reality is cultural: the sort of reality possessed by a code of etiquette, traffic regulations, the rules of baseball, the English language or rules of grammar.

Thus we see that there is no mystery about mathematical reality. We need not search for mathematical "truths" in the divine mind or in the structure of the universe. Mathematics is a kind of primate behavior as languages, musical systems and penal codes are. Mathematical concepts are man-made just as ethical values, traffic rules, and bird cages are manmade. But this does not invalidate the belief that mathematical propositions lie outside us and have an objective reality. They do lie outside us. They existed before we were born. As we grow up we find them in the world about us. But this objectivity exists only for the individual. The locus of mathematical reality is cultural tradition, i.e., the continuum of symbolic behavior. This theory illuminates also the phenomena of novelty and progress in mathematics. Ideas interact with each other in the nervous systems of men and thus form new syntheses. If the owners of these nervous systems are aware of what has taken place they call it invention as Hadamard does, or "creation," to use Poincaré's term. If they do not understand what has happened, they call it a "discovery" and believe they have found something in the external world. Mathematical concepts are independent of the individual mind but lie wholly within the mind of the species, i.e., culture. Mathematical invention and discovery are merely two aspects of an event that takes place simultaneously in the cultural tradition and in one or more nervous systems. Of these two factors, culture is the more significant; the determinants of mathematical evolution lie here. The human nervous system is merely the catalyst which makes the cultural process possible.

---

[39] "Mathematical Proof," p. 4 (*Mind*, Vol. *38, pp. 1-25, 1929*).

[40] *A Mathematician's Apology, pp.* 62-63, 65.

[41] *ibid., p. 63.*

18

Inner Vision, Outer Truth

REUBEN HERSH

There is an old conundrum, many times resurrected: why do mathematics and physics fit together so surprisingly well? There is a famous article by Eugene Wigner, or at least an article with a famous title: "The Unreasonable Effectiveness of Mathematics in Natural Sciences." After all, pure mathematics, as we all know, is created by fanatics sitting at their desks or scribbling on their blackboards. These wild men go where they please, led only by some notion of 'beauty', 'elegance', or 'depth', which nobody can really explain. Wigner wrote, 'It is difficult to avoid the impression that a miracle confronts us here, quite comparable in its striking nature to the miracle that the human mind can string a thousand arguments together without getting itself into contradictions, or to the two miracles of the existence of laws of nature and of the human mind's capacity to divine them.'

In Lobachevsky's non-Euclidean geometry, or Cayley's matrix theory, and Galois' and Jordan's group theory, and the algebraic topology of the mid-twentieth century, pure mathematics seemed to have left behind any physical interpretation or utility. And yet, physicists later found these 'useless' mathematical abstractions to be just the tools they needed.

Freeman Dyson writes, in his Foreword to Monastyrsky's *Riemann, Topology, and Physics*, of 'one of the central themes of science, the mysterious power of mathematical concepts to prepare the ground for physical discoveries which could not have been foreseen or even imagined by the mathematicians who gave the concepts birth.'

On page 135 of that book, there is a quote from C. Yang, co-author of the Yang-Mills equation of nuclear physics, speaking in 1979 at a symposium dedicated to the famous geometer, S.-S. Chern.

"Around 1968 I realised that gauge fields, non-Abelian as well as Abelian ones, can be formulated in terms of nonintegrable phase factors, i.e., path-dependent group elements. I asked my colleague Jim Simons about the mathematical meaning of these nonintegrable phase factors, and he told me they are related to connections with fibre bundles. But I did not then appreciate that the fibre bundle was a deep mathematical concept. In 1975 I invited Jim Simons to give to the theoretical physicists at Stony Brook a series of lectures on differential forms and fibre bundles. I am grateful to him that he accepted the invitation and I was among the beneficiaries. Through these lectures T. T. Wu and I finally understood the concept of nontrivial bundles and the Chern-Weil theorem, and realized how beautiful and general the theorem is. We were thrilled to appreciate that the nontrivial bundle was exactly the concept with which to remove, in monopole theory, the string difficulty which had been bothersome for over forty years [that is, singular threads emanating from a Dirac monopole].

"When I met Chern, I told him that I finally understood the beauty of the theory of fibre bundles and the elegant Chern-Weil theorem. I was struck that gauge fields, in particular, connections on fibre bundles, were studied by mathematicians without any appeal to physical realities. I added that it is mysterious and incomprehensible how you mathematicians would think this up out of nothing. To this Chern immediately objected. 'No, no, this concept is not invented-it is natural and real."

Why does this happen?

Is there some arcane psychological principle by which the most original and creative mathematicians find interesting or attractive just those directions in which Nature herself wants to go? Such an answer might be merely explaining one mystery by means of a deeper mystery.

Or perhaps the "miracle" is an illusion. Perhaps for every bit of abstract purity that finds physical application, there are a dozen others that find no such application, but instead eventually die, disappear and are forgotten. This second explanation could even be checked out, by a doctoral candidate in the history of mathematics. I have not checked it myself. My gut feeling is that it is false. It seems somehow that most of the mainstream research in pure mathematics does eventually connect up with physical applications.

Here is a third explanation, a more philosophical one that relies on the very nature of mathematics and physics. Mathematics evolved from two sources, the study of numbers and the study of shape, or more briefly, from arithmetic and visual geometry. These two sources arose by abstraction or observation from the physical world. Since its origin is physical reality, mathematics can never escape from its inner identity with physical reality. Every so often, this inner identity pops out spectacularly when, for example, the geometry of fiber bundles is identified as the mathematics of the gauge field theory of elementary particle physics. This third explanation has a satisfying feeling of philosophical depth. It recalls Leibnitz's "windowless monads", the body and soul, which at the dawn of time God set forever in tune with each other. But this explanation, too, is not quite convincing. For it implies that all mathematical growth is predetermined, inevitable. Alas, we know that is not so. Not all mathematics enters the world with that stamp of inevitability. There is also "bad" mathematics, that is, pointless, ugly, or trivial. This sad fact forces us to admit that in the evolution of mathematics there is an element of human choice, or taste if you prefer. Thereby we return to the mystery we started with. What enables certain humans to choose better than they have any way of knowing?

A good rule in mathematical heuristics is to look at the extreme cases – when a small parameter becomes zero, or a large parameter becomes infinite. Here, we are studying the way that discoveries in "pure"mathematics sometimes turn out to have important, unexpected uses in science (especially physics). I would like to use the same heuristic – "look at the extreme cases". But in our present discussion, what does that mean, "extreme case"? Of course, we could give this expression many different meanings. I propose to mean "extremely simple". To start with, let's take counting, that is to say, the natural numbers.

These numbers were a discovery in mathematics. It was a discovery that much later became important in physics and other sciences. For instance, one counts the clicks of a Geiger counter. One counts the number of white cells under a microscope. Yet the original discovery or invention of counting was not intended for use in science; indeed, there was no "science" at that early date of human culture.

So let us take this possibly childish example, and ask the same question we might ask about a fancier, more modern example. What explains this luck or accident, that a discovery in "pure mathematics" turns out to be good for physics?

Whether we count and find the planets seven, or whether we study the n-body problem, where n is some positive integer, we certainly do need and use counting – the natural numbers – in physics and every other science.

This remark seems trivial. Such is to be expected in the extreme cases. We do not usually think of arithmetic as a special method or theory, like tensors, or groups, or calculus. Arithmetic is the all-pervasive rock bottom essence of mathematics. Of course it is essential in science; it is essential in everything. There is no way to deny the obvious fact that arithmetic was invented without any special regard for science, including physics; and that it turned out (unexpectedly) to be needed by every physicist.

We are therefore led again to our central question, "How could this happen?" How could a mathematical invention turn out, unintentionally, after the fact, to be part of physics? In this instance, however, of the counting numbers, our question seems rather lame. It is not really surprising or unexpected that the natural numbers are essential in physics or in any other science or non-science. Indeed, it seems self-evident that they are essential everywhere. Even though in their development or invention, one could not have foreseen all their important uses.

So to speak, when one can count sheep or cattle or clam shells, one can also count (eventually) clicks of a Geiger counter or white cells under a microscope. Counting is counting. So in our first simple example, there really is no question, 'How could this happen?' Its very simplicity makes it seem obvious how 'counting in general' would become, automatically and effortlessly, 'counting in science'.

Now let's take the next step. The next simplest thing after counting is circles. Certainly it will be agreed that the circle is sometimes useful. The Greeks praised it as 'the heavenly curve'. According to Otto Neugebauer, "Philosophical minds considered the departure from strictly uniform circular motion the most serious objection against the Ptolemaic system and invented extremely complicated combinations of circular motions in order to rescue the axiom of the primeval simplicity of a spherical universe" *(The Exact Sciences in Antiquity)*. I. B. Cohen wrote, "The natural motion of a body composed of aether is circular, so that the observed circular motion of the heavenly bodies is their natural motion, according to their nature, just as motion upward or downward in a straight line is the natural motion for a terrestrial object" *(The Birth of a New Physics)*.

And here is a more detailed account of the circle in Greek astronomy: "Aristotle's system, which was based upon earlier works by Eudoxus of Cnidos and Callippus, consisted of 55 concentric celestial spheres which rotated around the earth's axis running through the center of the universe. In the mathematical system of Callippus, on which Aristotle directly founded his cosmology of concentric spheres, the planet Saturn, for example, was assigned a total of four spheres, to account for its motion 'one for the daily motion, one for the proper motion along the zodiac or ecliptic, and two for its observed retrograde motions along the zodiac" (E. Grant, *Physical Science in the Middle Ages)*.

In recent centuries, other plane curves have become familiar. But the circle still holds a special place. It is the 'simplest', the starting point in the study of more general curves. Circular motion has special interest in dynamics. The usual way to specify a neighborhood of a given point is by a circle with that point as center.

So we see that the knowledge of circles which we inherited from the Greeks (with a few additions) is useful in many activities today, including physics and the sciences. I suppose this is one reason why 10th grade students are required to study Euclidean geometry.

Again, we return to the same question. How can we explain this 'miracle'?

Few people today would claim that circles exist in nature. Any seemingly circular motion turns out on closer inspection to be only approximately circular.

Not only that. The notion of a circle is not absolute. If we define distance otherwise, we get other curves. To the Euclidean circle we must add non-Euclidean "circles". If the Euclidean circle retains a central position, it does so because we choose – for reasons of simplicity, economy, convenience, tradition – to give it that position.

We see, then, two different ways in which a mathematical notion can enter into science. We can put it there, as Ptolemy put circles into the planetary motion. Or we can find it there, as we find discreteness in some aspect or other of every natural phenomenon.

Let's take one last example, a step up the ladder from the circle. I mean the conic sections, especially the ellipse. These curves were studied by Apollonius of Perga (262-200 B.C.) as the "sections" (or "cross sections" as we would say) of a right circular cone. If you cut the cone with a cutting plane parallel to an element of the cone, you get a parabola. If you tilt the cutting plane toward the direction of the axis, you get a hyperbola. If you tilt it the other way, against the direction of the axis, you get an ellipse.

This is "pure mathematics", in the sense that it has no contact with science or technology. Today we might find it somewhat impure, since it is based on a visual model, not on a set of axioms.

The interesting thing is that nearly 2,000 years later, Kepler announced that the planetary orbits are ellipses. (There also may be hyperbolic orbits, if you look at the comets.)

Is this a miracle? How did it happen that the very curves Kepler needed to describe the solar system were the ones invented by Apollonius some 1800 or 1900 years earlier?

Again, we have to make the same remarks we did about circles. Ellipses are only approximations to the real orbits. Engineers using earth satellites nowadays need a much more accurate description of the orbit than an ellipse. True, Newton proved that 'the orbit' is exactly an ellipse. And today we reprove it in our calculus classes. In order to do that, we assume that the earth is a point mass (or equivalently, a homogeneous sphere). But you know and I know (and Newton knew) that it is not.

Kepler brought in Apollonius's ellipse because it was a good approximation to his astronomical data. Newton brought in Apollonius's ellipse because it was the orbit predicted by his gravitational theory (assuming the planets are point masses, and that the interactive attraction of the planets is 'negligible'). Newton used Kepler's (and Apollonius's) ellipses in order to justify his gravitational theory. But what if Apollonius had never lived? Or what if his eight books had been burned by some fanatic a thousand years before? Would Newton have been able to complete his work?

We can imagine three different scenarios: (1) Kepler and Newton might have been defeated, unable to progress; (2) they might have gone ahead by creating conic sections anew, on their own; (3) they might have found some different way to study the dynamics of the planets, doing it without ellipses.

Scenario three is almost inconceivable. Anyone who has looked at the Newtonian theory will see that the elliptic trajectory is unavoidable. Without Apollonius, one might not know that this curve could be obtained by cutting a cone. But that fact is quite unnecessary for the planetary theory. And surely somebody would have noticed the connection with cones (probably Newton himself).

Scenario one, that Newton would have been stuck if not for Apollonius, is *quite* inconceivable. He, like other mathematical physicists since his time, would have used what was available and created what he needed to create. While Apollonius' forestalling Kepler and Newton is remarkable and impressive, from the viewpoint of Newton's mechanics it is inessential. In the sequence of events that led to the Newtonian theory, what mattered were the accumulation of observations by Brahe, the analysis of data by Kepler, and the development by Barrow and others of the "infinitesimal calculus". The theory of the conic sections, to the extent that he needed it, could have been created by Newton himself. In other words, scenario two is the only believable one.

If a mathematical notion finds repeated use, in many branches of science, then such repeated use may testify to the universality, the ubiquitousness, of a certain physical property – as discreteness, in our first example. On the other hand, the use of such a mathematics may only be witness to our preference for a certain picture or model of the world, or to a mental tradition which we find comfortable and familiar. And also, perhaps, to the amiability or generosity of nature, which allows us to describe her in the manner we choose, without being "too far" from the truth.

What then of the real examples – matrices, groups, tensors, fiber bundles, connections? Maybe they mirror or describe physical reality "by lucky accident", so to speak, since the physical application could not have been foreseen by the inventors.

On the other hand, maybe they are used as a matter of mere convenience – we understand them because we invented them, and they work "well enough".

Maybe we are not even able to choose between these two alternatives. To do so would require knowledge of how nature "really" is, but all we can ever have are data and measurements and hypotheses in which we put more or lees credence.

In fact, it may be deceptive to pose the two alternatives – true to nature, like the integers, or an imposed model, like the circle. Any useful theory must be both. Understandable – i.e., part of our known mathematics, either initially or ultimately – and also "reasonably" true to the facts, the data. Both aspects – man-made and also faithful to reality –must be present.

These self-critical remarks do not make any simplification in our problem.

The problem is, to state it for the last time, how is it that mathematical inventions made with no regard for scientific application turn out so often to be useful in science?

We have two alternative explanations, suggested by our two primitive examples, counting and circles. Example one, counting, leads to explanation one: That certain fundamental features of nature are found in many different parts of physics or science; that a mathematical structure which faithfully captures such a fundamental feature of nature will necessarily turn out to be applicable in science.

According to explanation two, (of which the circle was our simple example), there are several different ways to describe or "model" mathematically any particular physical phenomenon. The choice of a mathematical model may be based more on tradition, taste, habit, or convenience, than on any necessity imposed by the physical world. The continuing use of such a model (circles, for example) is not compelled by the prevalence of circles in nature but only by a preference for circles on the part of human beings—, scientists, in particular.

What conclusion can we make from all this? I offer one. It seems to me that there is not likely to be any universal explanation of all the surprising fits between mathematics and physics. It seems clear that there are at least two possible explanations; in each instance, we must decide which explanation is most convincing. Such an answer, I am afraid, will not satisfy our insistent hankering for a single simple explanation. Perhaps we will have to do without one.

Bibliography

1. I. Bernard Cohen, *The Birth of a New Physics.* (Doubleday & Company, 1960).
2. Edward Grant, *Physical Science in the Middle Ages.* (Cambridge University Press, 1977).
3. Michael Monastyrsky, *Riemann, Topology, and Physics.* (Birkhauser Boston, 1987).
4. O. Neugebauer, *The Exact Sciences in Antiquity.* (Dover Publications, Inc., 1969).
5. Mark Steiner, *The Applicability of Mathematics as a Philosophical Problem.* Harvard University Press, 1988.
6. Eugene Paul Wigner, *Symmetries and Reflections.* (Bloomington, Indiana University Press, 1967).